High Energy Spin Physics
Volume 2: Workshops

W. Meyer E. Steffens W. Thiel (Eds.)

High Energy Spin Physics

Volume 2: Workshops

Proceedings of the 9th International Symposium
Held at Bonn, FRG, 6-15 September 1990

With 236 Figures

Springer-Verlag
Berlin Heidelberg New York
London Paris Tokyo
Hong Kong Barcelona
Budapest

Priv. Doz. Dr. Werner Meyer
Physikalisches Institut, Universität Bonn,
Nußallee 12, W-5300 Bonn 1, Fed. Rep. of Germany

Priv. Doz. Dr. Erhard Steffens
Max-Planck-Institut für Kernphysik,
Saupfercheckweg 1, W-6900 Heidelberg, Fed. Rep. of Germany

Dr. Werner Thiel
Physikalisches Institut, Universität Bielefeld,
Universitätsstraße, W-4800 Bielefeld, Fed. Rep. of Germany

ISBN-13:978-3-642-76663-3 e-ISBN-13:978-3-642-76661-9
DOI: 10.1007/978-3-642-76661-9

Softcover reprint of the hardcover 1st edition 1991

Typesetting: Camera ready by contributors

55/3140-543210 – Printed on acid-free paper

Preface

The 9th International Symposium on High Energy Spin Physics was held in Bonn, 6-15 September 1990, with the Physikalisches Institut der Universität Bonn as the host.

The symposium was preceded by a series of four workshops on

- polarized electron sources and electron spin polarimeters
- Siberian snakes and polarization in circular machines
- polarized gas targets
- polarized solid targets.

160 participants from 11 countries, among them many young physicists, came together and discussed mainly technological spin problems. The high level of participation indicates that workshops combined with the symposium are attractive not only for people who plan and prepare polarized beams and targets but also for experimentalists and theorists dealing with high energy spin physics.

At these workshops many very interesting and important recent results were presented and reviewed. Thus we hope these proceedings will be valuable to many researchers in these fields.

The Organizing Committee would like to thank all participants, in particular the speakers and the session chairmen, for their contributions to the workshops and for helping to create a lively and stimulating atmosphere. Special thanks go to the organizers – W. Haeberli, S. Mango, E. Reichert, E. Steffens, W. Thiel, U. Wienands – for their cooperation in preparing and running these workshops.

We gratefully acknowledge the enthusiastic help of the members of our institute in preparing and running the conference and the workshops, especially Mrs. D. Faßbender, Mrs. E. Wendorf, Mrs. J. Wetzel, and Dr. U. Idschok.

The Symposium was sponsored by the Bundesministerium für Forschung und Technologie, the Deutscher Akademischer Austauschdienst, the Ministerium für Wissenschaft und Forschung des Landes Nordrhein-Westfalen, the Deutsche Forschungemeinschaft, and the University of Michigan. Finally, we thank Springer-Verlag for their assistance and cooperation in publishing these proceedings.

Bonn
February 1991

W. Meyer

Organizing Committee

K.-H.	Althoff (Chair)	Bonn	P.	Kroll	Wuppertal
W.	Meyer (Co-Chair)	Bonn	W.	Pfeil	Bonn
G.	Anton	Bonn	T.	Reichelt	Bonn
M.	Huber	Bonn	E.	Reichert	Mainz
D.	Husmann	Bonn	H.	Rollnik	Bonn
U.	Idschok	Bonn	W.	Schwille	Bonn
K.	Kilian	Jülich	K.G.	Steffen	DESY
G.	Knop	Bonn	E.	Steffens	Heidelberg

International Advisory Committee

A.D.	Krisch (Chair)	Michigan	D.	Kleppner	MIT
J.	Arvieux	Saturne	H.J.	Lipkin	Weizmann
J.	Buon	Orsay	A.	Masaike	Kyoto
O.	Chamberlain	Berkeley	Yu.K.	Pilipenko	Dubna
E.D.	Courant	Brookhaven	C.Y.	Prescott	SLAC
G.R.	Court	Liverpool	A.N.	Skrinsky	Novosibirsk
G.	Fidecaro	CERN	L.D.	Soloviev	Serpukhov
K.J.	Heller	Minnesota	K.G.	Steffen	DESY
V.W.	Hughes	Yale	E.W.	Vogt	TRIUMF

Sponsors

- Bundesministerium für Forschung und Technologie
- Ministerium für Wissenschaft und Forschung des Landes Nordrhein-Westfalen
- Deutsche Forschungsgemeinschaft (DFG)
- Deutscher Akademischer Austauschdienst (DAAD)
- The University of Michigan

Contents

Part I Polarized Electron Sources and Electron Spin Polarimeters

Part II Siberian Snakes and Polarization in Circular Machines

Part III **Polarized Gas Targets**

Part I

Polarized Electron Sources and Electron Spin Polarimeters

The SLC Polarized Electron Source*

J.E. Clendenin

Stanford Linear Accelerator Center, Stanford University,
Stanford, CA 94309, USA

A polarized electron source consisting of a 3-electrode photocathode gun and a flashlamp-pumped dye laser has been designed and built for the SLC and is currently undergoing commissioning. The source is described, and the operating configuration is discussed. The present status of the source and future plans are briefly indicated.

1. Introduction

The use of GaAs as a photocathode in an electron gun for producing a polarized beam for injection into a high energy electron accelerator was pioneered at SLAC.[1] The original GaAs source, which is still being used at SLAC to study methods of achieving higher polarization, was designed to produce peak currents exceeding 100 mA in a 1.6 μsec pulse at repetition rates of up to 180 Hz. The gun had a two electrode structure with the cathode biased at 60 to 70 kV. The basic configuration of this electron gun has been repeated in many other accelerator centers now producing or planning to produce polarized electron beams.

The SLC requires a single S-band bunch[2] of 5×10^{10} e^- per pulse at the interaction region to collide with a similar positron bunch at rates of up to 120 Hz. The injector utilizes subharmonic bunching to compress a 2 ns pulse into a fraction of an S-band cycle. The subharmonic buncher operates at the sixteenth subharmonic of the 2856 MHz accelerating RF of the linac. Given the efficiency of the buncher and the remaining sections of the SLC, the gun itself is required to produce up to 2×10^{11} e^- per bunch which is equivalent to a peak current of about 16 A. The transverse emittance of the initially accelerated beam is dominated by an emittance growth in the buncher of about a factor of 10.

The SLC positron beam originates with an electron beam from the same gun generating the colliding electron beam. At 120 Hz, the two gun pulses must be accelerated by the same linac RF pulse. Since the overall positron production efficiency is about unity, these two gun pulses also must be approximately equal in intensity.

2. Description of the Source

To meet SLC requirements, a new GaAs photocathode gun was designed to operate at up to 200 kV to minimize longitudinal space charge effects at high intensities.[3] A

⋆ Work supported by Department of Energy contract DE–AC03–76SF00515.

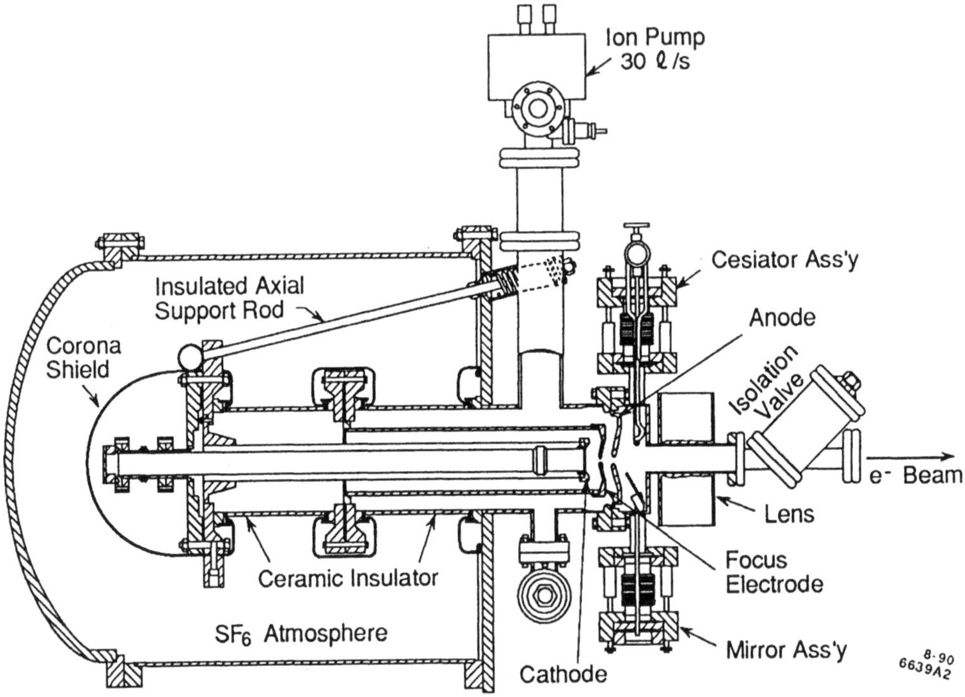

Figure 1: The SLC Polarized Electron Gun. The GaAs cathode sits in the middle of a larger stainless steel cathode structure and can be separately retracted for heat treatment. The cesiator and mirrors are on linear manipulators. For cesiating the cathode, they are positioned along the beamline. The cesium beam passes through a hole in the mirror. During the cesiation process, the cathode is biased to about 50 VDC. The photoemission is monitored with an electrometer. Typically an ordinary incadescent lamp is used to illuminate the cathode. The light from the side-mounted lamp is bounced from the mirror to the anode, back to the mirror, and finally to the cathode.

focusing electrode was placed between the cathode and anode to allow the transverse emittance to be adjusted. The focusing electrode also limits the electric field at the cathode to < 3 MV/m at the full 200 kV bias, although fields as high as 12 MV/m are present on other portions of the electrode surfaces that are exposed to Cs. The gun is shown in Fig. 1. A schematic of the SLC injector is shown in Fig. 2. The source specifications are given in Table 1.

The light source is a Candela flashlamp-pumped dye laser that can be operated at up to 120 Hz with a maximum power output of 70 kW in a 600 ns pulse. For the SLC, 2 ns pulses are selected from the laser beam using a Pockels cell and two crossed polarizers. The Pockels cell is driven by a planar triode amplifier. The fast pulse is achieved by use of a ferrite-loaded pulse compression line and pulse forming network. Additional electro-optical systems are utilized to provide intensity stabilization and to allow remote control and monitoring of intensity and timing. Timing control with low jitter is provided by the SLC timing system. The entire laser operation is designed to be monitored and controlled remotely.

The predicted maximum electron polarization from a GaAs photocathode is 50%. By utilizing GaAs crystal layers less than 0.4 mm thick, this polarization can actually

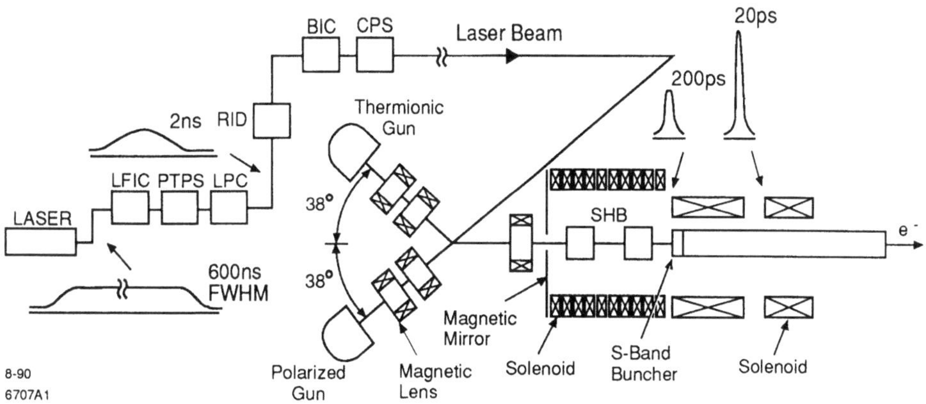

Figure 2: Schematic of the SLC injector. There are two gun ports, to one of which the SLC thermonic gun is presently connected and operating. The 600 ns laser pulse traverses the following electro-optic devices: LFIC (laser feedback intensity control), PTPS (pulse-to-pulse stabilizer), and LPC (laser pulse chopper), at which point the laser pulse is 2 ns as shown. Next is the RID (ratio of intensity driver), BIC (bunch intensity control), and CPS (circular polarization sign), after which the laser beam is directed to the gun cathode. The temporal structure of the polarized gun electron pulse follows the laser pulse and thus is also ~2 ns FWHM. The phase space of the photocathode electron beam is designed to be similar to that of the thermionic gun. Two stages of electron bunch compression are shown: at ~200 ps after the SHB (subharmonic buncher), and at ~20 ps in the accelerator section where the beam is fully relativistic. There is no significant depolarization during the bunching process.

Table 1. SLC requirements for the polarized source.

Item	Require-ments	Present Status	Units
Laser repetition rate	120	120	Hz
Maximum gun current	2×10^{11}	(a)	e^-/bunch
Pulse-to-pulse intensity jitter (σ)	< 0.5	~ 3	%
Pulse-to-pulse timing jitter (σ)	< 15	< 30	ps
Pulse duration (FWHM)	2	< 3	ns
Pre- and post-bunch charge in accelerated beam	$< 5 \times 10^8$	(b)	e^-/bunch
Pre- and post-bunch charge after Damping Ring	$< 1 \times 10^8$	(c)	e^-/bunch
Bunch separation (nominal at 120 Hz)	61.625	(d)	ns

(a) Present laser intensity sufficient if Q.E. > 0.5%.

(b) Present laser optics sufficient if charge is integrated over 600 ns.

(c) Present laser optics sufficient if charge integrated over 30 ns.

(d) Temporal pulse separation with present laser optical system completely controllable through the SLC timing system.

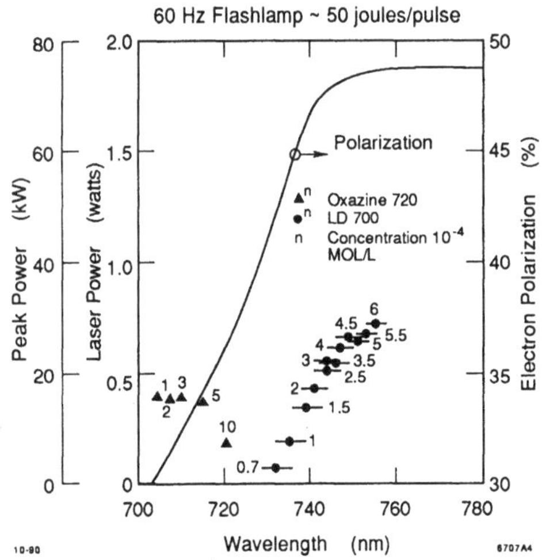

Figure 3: Laser output power and polarization of thin GaAs as a function of wavelength for two dyes of several molecular concentrations. The polarization curve shown here is taken from the "within 5 h after activation" curve shown in Fig. 2 of Ref. 4.

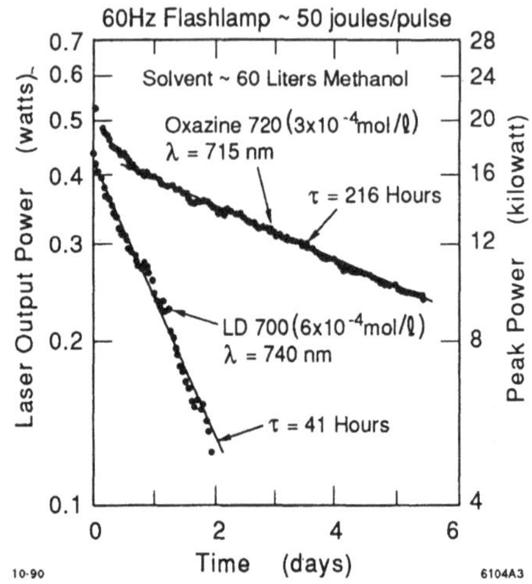

Figure 4: Measured laser outputpower as a function of time for two different dyes at optimum concentrations.

be achieved even at room temperature.[4] However, a laser operating at or slightly above 760 nm is required. To date, the most promising dye for the laser is Oxazine 720. Utilizing this dye, a high-power light beam can be produced while the dye retains a long lifetime. However, the laser output power peaks at about 715 nm, a wavelength for which the polarization from a thin GaAs cathode is less than 35%. Figure 3 shows as a function of wavelength the electron polarization for thin GaAs on the righthand ordinate, and the laser intensity for two dyes at various concentrations on the lefthand ordinate. Figure 4 shows the laser intensity vs. time for both dyes at optimal concentrations.[5] By adding about 10% Al to the GaAs, the roll-off in the polarization curve can be shifted downward[6] to about 720 nm. The net result is that a polarization of about 40% is expected for the SLC source operating with the present dye laser and an AlGaAs cathode. (Depolarization effects during the acceleration and transport of the beam to the interaction region are expected to be relatively small.[7])

3. Present Status and Future Plans

The SLC polarized gun was installed alongside the SLC thermionic gun at the linac injector in the spring of 1986. The dye laser and the HV power supplies for the gun were not installed at the injector until the fall of 1989. Initial testing of the source in this configuration indicated excessive field emission at high voltage that, in turn, resulted in contamination of the cathode. The gun has been removed from the linac injector and is presently in the laboratory undergoing modification and additional testing.

Research with new cathodes to increase the polarization of the electron beam continues at SLAC. In addition, a new, tunable, solid-state laser, which is expected to be purchased during the coming year, will permit use of thin GaAs cathodes and correspondingly higher polarization. SLAC is also building a second polarized gun for the SLC. A system to rapidly exchange the two guns will be incorporated into the design in order to increase the efficiency of the polarized beam system.

Acknowledgements

The polarized electron source described above is the work of many individuals, including E. L. Garwin, S. Kaiser, T. Maruyama, K. Moffeit, G. Mundy, F. Perrier (now at Saclay), D. C. Schultz, C. K. Sinclair (now at CEBAF), M. Woods, and D. Wright of SLAC; and R. Prepost of University of Wisconsin.

References

1. C. Y. Prescott, W. B. Atwood, and R. L. A. Cottrell, et al., *Phys. Lett.* **77B** (1978) 347.
2. Wakefields produced by charge in the accelerated beam outside a single S-band cycle of 350 ps disrupt the SLC beam.
3. C. K. Sinclair and R. H. Miller, *IEEE Trans. Nucl. Phys.* **NS–28** (1981) 2649.
4. T. Maruyama, R. Prepost, and E. L. Garwin, et al., *Appl. Phys. Lett.* **55** (1989) 1686.
5. The dye measurements were presented by F. Perrier at the Polarized Electron Sources and Polarimeters Workshop, 8th Int. Symp. on High Energy Spin Physics, Minneapolis, 1988.
6. T. Maruyama, E. L. Garwin, and R. Prepost, et al., "Electron Spin-Polarization in Photoemission from Thin AlGaAs," SLAC–PUB–5367 (1990).
7. K. C. Moffeit, *High-Energy Spin Physics,* ed. K. J. Heller, *AIP Conf. Proc. 187* (New York, 1989) 901.

The Orsay Polarized Electron Source

S. Essabaa, C.G. Aminoff, J. Arianer, and I. Brissaud*

Institut de Physique Nucléaire, F-91406 Orsay Cedex, France
*Present address: Department of Technical Physics,
Helsinki University of Technology, SF-02150 Espoo, Finland

A polarized electron source is developed at Orsay to equip existing european accelerators within the next two years. We have chosen to adapt the flowing helium afterglow source working at Rice University : it is able to provide a very high polarization (approximately 90 %) delivering currents higher than 1 μA [1]. We are investigating new ways to improve the intensity performance to get a reliable device as required for an accelerator.

SOURCE DESCRIPTION

A. The principle of the Rice source has been described in different papers [1]. Here we summarize the main features. Figure 1 shows the scheme of the Orsay source which is very similar to the Rice one. A high purity helium jet passes through a Laval nozzle in a microwave discharge cavity to produce helium metastable atoms 2^3S. The helium flow rate is about 1500 sccm for a flow tube pressure of 0.150 torr. The metastable $He(2^3S)$ atoms are then oriented by optical pumping with an infrared light laser (λ = 1.083 μm). This light is circularly polarized by a quaterwave plate. The quantization axis is defined by a weak magnetic field (5 G) parallel to the laser light direction. In the following step a Penning reaction strips the polarized electrons from the metastable atoms :

$$He(2^3S)\uparrow\uparrow + CO_2 \rightarrow He(1^1S)\uparrow\downarrow + CO_2^+\uparrow + e^-\uparrow$$

In this reaction, the weakly bound electron is ejected from the 2s orbit while an electron from the CO_2 molecule is captured into the hole of the 1s orbit of the metastable atom . CO_2 is used as reactant gas because of its high Penning cross-section and of its electron cooling efficiency. The polarized electrons are then extracted through a biased aperture and transported by optics (three cylindrical lenses) to a Faraday cup or a Mott polarimeter [2] through a high vacuum chamber.

B. The polarized electron intensity is proportional to the triplet 2^3S metastable atom density and to the flow velocity. This density may be increased by raising the helium jet flow and by reducing the length of the pyrex tube located between the microwave cavity and the irradiation

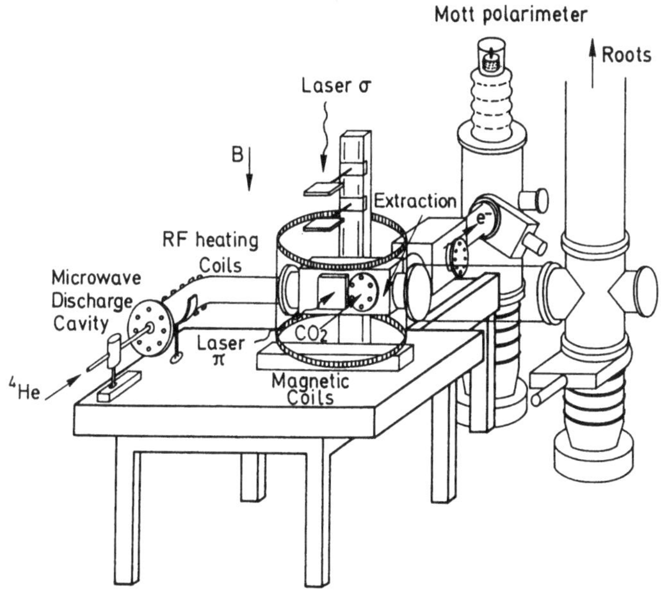

Fig. 1 : Schematic diagram of the Orsay source

chamber. At the same time, the pumping power of the roots blower would be multiplied by a factor 2 at least to increase the flow velocity. In order to improve the extraction efficiency, a high vacuum is expected in the optics region to reduce the risk of gas discharge. The most serious parasitic effect is the trapping of the X-rays emitted in the decay of the 2 P states. These radiations are not polarized causing a polarization degradation. This effect may be partially avoided by using a more limited interaction region to reduce the trapping probability and a more powerful laser. We are investigating in these ways.

CHARACTERISTICS OF THE ORSAY SOURCE

In the Rice device the orientation of He metastable atoms is obtained by optical pumping through D_1 transition $2^3S \rightarrow 2^3P_1$ (figure 2). The Doppler

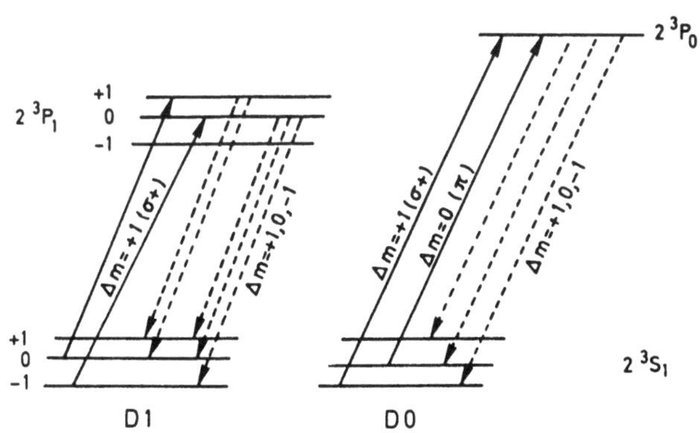

Fig. 2 : Helium optical pumping process with D_0 and D_1 transitions.

linewidth of the 2^3P levels is approximately 1.7 GHz at 300°K. A particular care must be taken to avoid excitation of the D_2 transition $2^3S \rightarrow 2^3P_2$ which generates an opposite polarization. Keep in mind that the separation between the 2^3P_1 and 2^3P_2 levels is only 2.4 GHz. In order to avoid a slight tuning of the laser wavelength to the blue side of the D_0 transition, we have chosen to use the optical pumping of D_0 transition $2^3S \rightarrow 2^3P_0$ as suggested by Leduc et al. [3] and successfully demonstrated by Schearer et al. [4] recently. The polarization of 2^3S metastable atoms saturates at 50 % if circularly polarized D_0 light (σ+) is used alone. If one adds linearly polarized D_0 light (π) perpendicular to the field direction, near 100 % polarization of metastable atoms is reached in a more confortable manner (figure 2).

Presently our laser is operational. It is similar to the one developed by Leduc's group [5]. The lasing material is a LNA crystal excited by krypton discharge lamps in a Microcontrole 904 YAG laser head. The laser cavity end mirrors are plane ; one is a high reflector (99.9 %), the other one is used as the output coupler (99.4 %). The cavity length is 35 cm. Laser line narrowing and wavelength tuning are achieved by inserting uncoated etalon plates (0.3 and 1 mm thick) in the cavity. In order to minimize the walk-off losses in the etalons, we tune the laser wavelength by changing the temperature of the 1 mm etalon instead of tilting it. Using this technique, we have increased the output power up to 4 watts on the He resonance at 1.083 μm. The spectral width is of the order of 3 GHz. These characteristics are well suited for efficient optical pumping on the D_0 line with multimode laser.

CONCLUSIONS

Our goals are :

a) to find results close to the Rice ones in a reproducible way before the end of the year 1990,
b) to try to improve these performances,
c) to build a computer controlled source usable within two years.

The adaptation to an accelerator remains to be done.

ACKNOWLEDGMENTS

We wish to express our acknowledgments to G.K. Walters, F.B. Dunning, G. Rutherford and G. Lynn for many helpful discussions and comments on "*their source*". Many thanks are due to M. Leduc for her illimited support to this project. We thank also G. Lampel and D. Campbell for the use of the polarimeter drawings. L. Schearer provided us the preliminary results on the Do transition experiments before publication ; we wish to thank him. It is a pleasure to acknowledge for many fruitful discussions P. Camus, J. Hamel, R. Pelletier and J. Pinard. We are indebted to R. Breuil, A. Ferro and L. Kotfila for their technical support.

(1) L.G. Gray, K.W. Giberson, C. Cheng, R.S. Keiffer, F.B. Dunning and G.K. Walters, Rev. Sci. Instrum., 54(1983)271 ; G.H. Rutherford, J.M. Ratliff, J.G. Lynn, F.B. Dunning and G.K. Walters, Rev. Sci. Instrum., in press (1990) and references herein.

(2) D.M. Campbell, C. Hermann, G. Lampel and R. Owen, J. Phys. E18 (1985)664 ; T.J. Moravec, F.B. Dunning and G.K. Walters, Rev. Sci. Instr. 50(1979)5 and references herein.

(3) M. Leduc, private communication.

(4) L.D. Schearer and P. Tin, submitted to publication (1990) and private communication.

(5) C.G. Aminoff, C. Larat, M. Leduc and F. Laloë, Rev. Phys. Appl. 24(1989)827 ; C.G. Aminoff, C. Larat, M. Leduc, B. Viana and D. Vivien, Journ. of Lum. (1990) to be published.

Source of Polarized Electrons for MAMI B

H.G. Andresen[1], *K. Aulenbacher*[2], *M. Ertel*[1], *E. Reichert*[2], *and K.H. Steffens*[1]

[1]Institut für Kernphysik, Johannes Gutenberg Universität, Postfach 3980, W-6500 Mainz, Fed. Rep. of Germany
[2]Institut für Physik, Johannes Gutenberg Universität, Postfach 3980, W-6500 Mainz, Fed. Rep. of Germany

A source of polarized electrons has been set up in order to inject polarized electrons into the 855 MeV c.w. electron accelerator MAMI B at the Mainz nuclear physics institute. It is based on photoemission of the 3/5 semiconductor GaAsP and will provide a d.c. current of 100μA with a beam emittance of 1π·mm·mrad, and a polarization of about 40%.

1 Demands on a polarized source for MAMI B

The source shall provide injection of an intense beam of polarized electrons into the MAMI accelerator. Nuclear physics experiments with a c.w. beam of longitudinally polarized electrons at 855MeV offer new possibilities for carrying out precision measurements of interesting physical observables e.g. the electric form factor of the neutron. In order to carry out such an experiment in a reasonable time, the source has to fulfill the following requirements:

- Beam intensity(d.c.): Because of the h.f. chopping done at the beginning of the injection LINAC the d.c. current has to be ten times higher than the mean current at the target position. The design value of the target current is 10μA, so it follows that $i_{d.c.} = 100\mu A$.
- Beam emittance: MAMI requires an emittance of 1π·mm·mrad at 100keV to assure a reliable operation of the accelerator.
- Polarization: Design value is 40% of longitudinal polarization at target position, a quick reversal of P is necessary.
- Under these conditions the experiment will last 100 hours. The source system should continuously operate during this time.

2 Operation principle and design considerations

2.1 New features

Our source is based on photoemission from the 3/5 semiconductor $GaAs_{.68}P_{.32}$. This kind of source has worked well in several nuclear physics experiments so far. One example was an experiment carried out by our group on parity violation in quasielastic scattering from

^9Be [1],[2]. This was done at our former LINAC–accelerator, with a duty cycle of 10^{-4}. In order to switch to the new c.w. accelerator several important improvements have to be achieved (see table): The mean current - 25μA during the old experiments - has to be raised to 100 μA (d.c.). The second constringent factor is the emittance of 1π·mm·mrad. The beam emittance of the old source was 25 times higher. By dividing mean current and emittance, one finds that the beam brightness of the new source has to be 100 times higher than that of the old one. If one would wish to use the full current capability of MAMI the brightness would have to be raised again by a factor of ten. However this will not be required in the first phase of our experiments. Another change is in the beam energy of the gun: It has to be increased from 44keV to 100keV.

parameter	LINAC-source(achieved)	MAMI-source(design)
crystal	$GaAs_{0.62}P_{0.38}$	$GaAs_{0.62}P_{0.38}$
mean current	25μA	100μA
emittance	25π·mm·mrad	1 π·mm·mrad
beam brightness	1μA/π·mm·mrad	100μA/π·mm·mrad
polarization	43%	>40%
quantum efficiency	0.6%	0.6%
operation time	>100 hours	100 hours
beam energy	44keV	100keV
time structure	50Hz,4μs	d.c.
polarization reversal	pockels cell	pockels cell

2.2 Optimization of P^2·I

The figure of merit for a polarized source is the square of the polarization multiplied by the available current. It depends on the wavelength of the radiation that is used to get photoemission from the cathode, because the polarization of the photoemitted beam decreases with the photon energy whereas the quantum efficiency of the cathode (number of electrons emitted per incident photon) increases. For the LINAC–source one finds an optimum photon energy at 1.94 eV (640 nm). The polarization is then about 43% and the quantum efficiency 0.6%, corresponding to 3mA per Watt of laser power. As the same cathode material is used for the MAMI source a dye laser working at 640 nm is choosen as a light source for the new setup. The laser delivers up to one watt of d.c. power at 640nm. With this power, it will be able to cope with decreasing quantum efficiency during operation, and with losses in the optical elements.

2.3 Requirements for long operation time

The usual trouble with photocathodes is the limited operation time due to degradation of the Negative Electron Affinity (NEA–) layer at the surface of the crystal. This layer is necessary to give the polarized electrons the possibilty of being emitted. The degradation of this less than one atomic layer thick coating results in an immediate decrease of quantum efficiency. Therefore any contamination of the crystal surface with other substances has to be avoided. This contamination may arise from at least two causes:

a) adsorption of atoms out of the residual gas of the vacuum system.

b) adsorption of atoms emitted from the vacuum chamber, when the 100keV beam hits the walls (electron stimulated desorption).

Both cases can be avoided, if a) extreme ultra high vacuum conditions are realized and b) the electron beam transmission is very high. Consequently the whole apparatus has to be build in bakeable UHV-technology. The second requirement can be fulfilled, if the electron–optics is carefully designed and has great apertures to allow a high transmission.

3 Experimental setup

Figure 1 shows two important parts of our apparatus: The source and the laser system. All the components are in their operational place. The Laser system consists of an Argon-ion-Laser with a d.c. power of up to 15 watts and a dye laser. The Ar-laser beam is focussed on a dye stream which is emitted from a nozzle in the dye laser resonator. Due to the high energy density at the focal point of the argon laser the conversion efficiency of the dye is very high, more than 20% of the argon power are converted to 640nm radiation. 1 Watt of d.c. power at the correct wavelength can be obtained with moderate

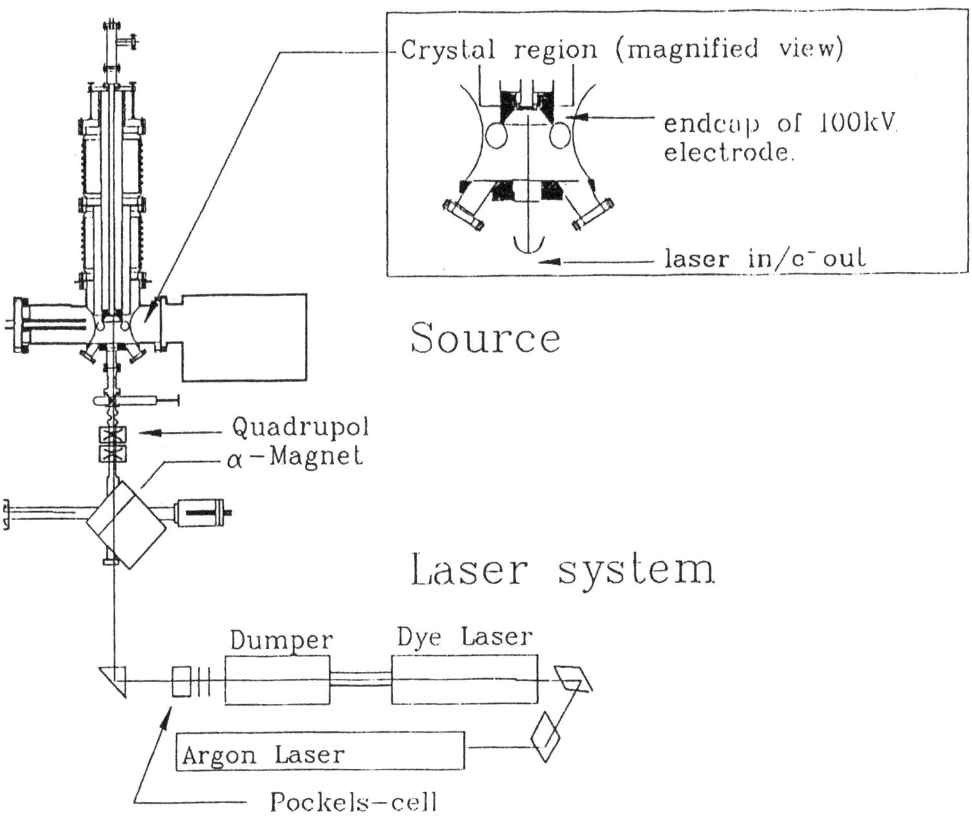

Figure 1: Experimental setup of laser system and source

pumping power, what guarantees a reliable continous operation. The dye laser resonator contains an electro optical element (cavity dumper) by which the beam may be chopped. This may be needed for accelerator tuning or time of flight experiments. The circular polarization of the laser light is generated by a pockels cell. This offers the advantage of a fast reversal of electron beam polarization by reversing the circular polarization of the Laser beam. After leaving the polarization optics the laser beam is deflected into the source. The crystal region of the source is shown in a magnified view. The electrons are emitted in opposite direction to the laser beam. The electron optics was optimized by computersimulation, we used an improved version of the SLAC/EGUN code[3]. It is a simple design, consisting only of two electrodes. The crystal is mounted at the bottom of the -100keV electrode. The specific (Pierce–like) geometry of the endcap of the electrode provides the desired beam parameters at the anode. The beginning of the beamline may also be seen in the figure. The main electron optical elements are quadrupoles and 270° bends, so called alpha magnets. The beam transport line to the accelerator is very long in our case because we need access to the source during operation and this was not available in the vicinity of the MAMI injection system. So the beam has to be transported over a distance of 20 meters before being injected into MAMI.

4 Experimental results

The source has been in operation together with the first 11 meters of the beamline. The source vacuum during operation is less than $1{\cdot}10^{-10}$ mbar. The beam emittance was measured using the same equipment that was utilized for determining the MAMI emittance limits. The emittance of our beam will be low enough if the laser beamspot on the crystal surface is about 1mm in diameter. This can be easily achieved by using a focussing lens. The beam parameters for a focussed laser beam may be seen in the following table:

horizontal	vertical
$\beta = .408m$	$\beta = .342m$
$\alpha = -.54$	$\alpha = -.41$
$\gamma = 3.15/m$	$\gamma = 3.41/m$
$\epsilon = 0.59\pi \cdot mm \cdot mrad$	$0.39\pi \cdot mm \cdot mrad$

The results are in agreement with the computersimulation. We succeded in transporting the beam through the 11 meter line, a transmission of more than 95% was achieved. The highest operated d.c. current was 250μA.

5 Outlook

All components that are necessary to connect the source to the accelerator are in fabrication. This includes the spin manipulation system which will be integrated in the remaining 9 meters of beamline. We will be able to build up these 9 meters until the beginning of next year. The first injection of polarized electrons in MAMI will be possible in spring 91.

References

[1] W.Heil et al. *Nucl.Phys.***B327** (1989)

[2] W. Hartmann et. al.: A source of polarized electrons based on photoemission of GaAsP. *Nuclear Instruments and methods***A286** p.1-8 (1990)

[3] W. B. Herrmannsfeldt *SLAC-report 166* Stanford (1979)

Status of the Kharkov Linac Polarized Electron Source

V.L. Agranovich, A.V. Glamazdin, V.G. Gorbenko, A.P. Prudius, and P.V. Sorokin

Kharkov Institute of Physics and Technology, SU-310108 Kharkov, USSR

In this note we briefly describe the GaAs polarized electron source, which is the full-scale injector prototype for the electron linear accelerator of the Kharkov Institute of Physics and Technology (KIPT).

The polarized source parameter are determined, on the one hand by the input parameters of the injector of the KIPT linac and, on the other hand, they depend on the photocathodes and the technical characteristics of the laser. The design goals for the source are summarized in Table I. Fig. 1 presents an overall view of the polarized electron source. The activation (preparation) chamber consists of the photocathode storage wheel with 10 nests for crystal holders. One of the nests is isolated from the chamber body and is used for the photocathode activation and for the photocurrent measurement. The cathode is activated at this position by oxygen or fluorine and cezium. The chosen photocathode is transported to a cleaning oven or to the cathode unit of the photogun chamber by means of two manipulators. The chambers are evacuated by ion pumps with liquid nitrogen cooling. The limiting pressure is 10^{-11} Torr, the pumping speed is 150 liters per second. The manipulators are evacuated with an additional pump. In the photogun chamber there is a three-electrode Piers gun. Its cathode unit has a nest to accomodate any crystal holder from the storage wheel. The cathode is supplied by negative voltage up to 100 kV. The anode is under the potential of the ground.

Table I. Design Goals for the Polarized Electron Source

Injection energy, keV	70
Beam crossover, mm · mmrad*	10
Intensity** , μA	1 ... 50
Polarization, %	40
Cathode efficiency, %	2
Pulse frequency, Hz	50, 100, 300
Pulse duration, μsec	1 ... 2
Helicity reversal	optical
Cathode lifetime, hrs	24
Time for the cathode replacement, hrs	1

*The cathode of 8 mm diameter

**Depending on the laser type

The preliminary pumping is carried out with a magnetic ion diode pump of 400 l/s and with a turbo-molecular pump of 500 l/s with a liquid nitrogen trap. The pump of the gun chamber can be replaced by a cryogenic helium pump with a pumping speed of 1000 l/s.

The electron optics system is similar to the one of the source PEGGY-1 /1/. It includes the 90° - bending magnet, three units with magnetic lenses, correctors, the chamber with a screen and the Faraday cup and also the spin rotator - Wien filter. All the lenses are coupled. The bending magnet chamber and the screen and Faraday cup chamber are evacuated with the ion triode pump

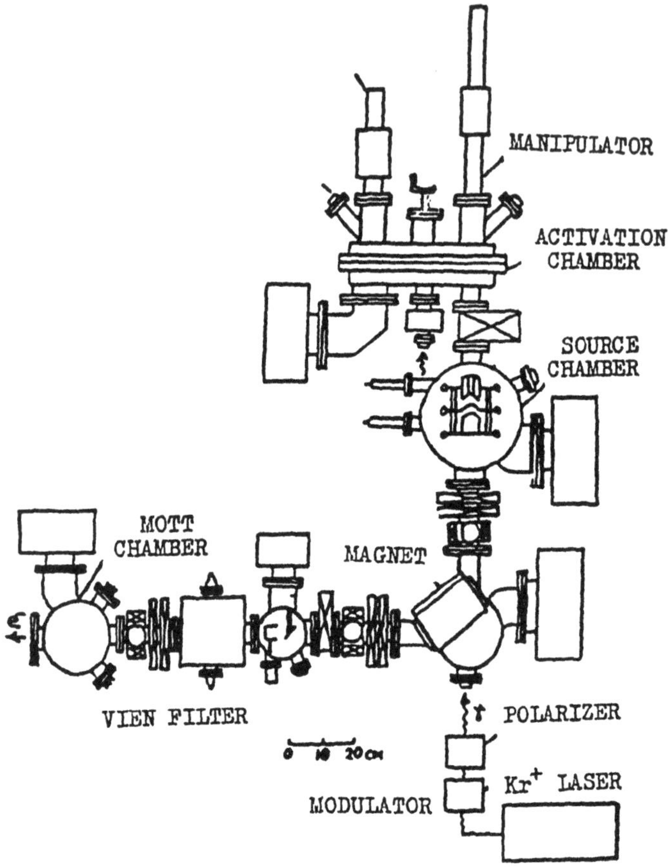

Fig. 1. GaAs Source of Polarized Electrons

of 150 l/s and with the ion diode pump of 25 l/s, accordingly. This ensures differential pumping with a pressure of 10^{-8} Torr in the Mott chamber.

The electron beam polarization is measured under the Mott scattering on thin gold targets at 100 keV. The gold targets with thickness of about 250, 500, and 750 $\mu g/cm^2$ and a screen are situated in a rotating wheel. The semiconductor detectors of scattered electrons are arranged at $\pm 120^o$ to the incident beam and are on the plane which is perpendicular to the median plane of the Wien filter. The 250 l/s diode ion magnetic pump is used for Mott scattering chamber pumping. An additional valve connects this chamber and the preliminary vacuum system.

The source of circularly photons uses a Kr^+-ion CW laser with a wavelength $\lambda = 647$ nm and a beam power of 1 W. The major components of the optics system are: linear polarizer - Glan-Thompson prism, an electro-optical shutter and a prism polarizer at the output. The pulse time of the electric supply of the shutter is 2 ... 3 μs, the frequency is 50 ... 300 pulses per second. There are circular polarizers of several types: a 1/4 wavelength plate of mica or suphire for $\lambda = 647$ nm, the KDP-crystal, which has a longitudinal electro-optical effect. There are a prism polarizer in the source for the circular polarization measurements and a laser intensity spectral light meter.

The spectral photoemission characteristics of GaAs, GaAsP and GaAlAs cathodes will be investigated with two liquid lasers at a frequency of 1 ... 50 pluses per second and the pulse length 1 ... 2 μs at different powers. The light power during the pulse time is stabilised similarly as in work /2/.

The polarized electron source is carried to completion and is under investigation now. The following results for the source are obtained:

- the pressure of $2 \cdot 10^{-10}$ Torr has been reached after 150 hours of heating at 350° C;
- the oven has been tested at temperatures up to 700° C, the pressure was lower than 10^{-8} Torr, if the temperature increase was slow;
- cesium and oxygen sources have been checked up by a mass-spectrometer;
- the electron optics system has been investigated using the electron beam from the field-emission cathode;
- the semi-conductor detectors in the Mott scattering chamber have also been calibrated with the field-emission beam:
- the low energy source of the Leningrad Polytechnical Institute has been used at electron energies of 1.3 ... 6 eV at different temperatures for the GaAs, GaAlAs, GaAsP photocathode investigations;
- the quality factor P^2I (I - quantum efficiency of the cathode) for GaAsP (100) photocathode at room temperature is found to be $4 \cdot 10^{-4}$ at electron polarization of 40% /3/.

We hope to get the polarized beam by the end of this year.

References

/1/ J.W. Robson, Programm Electrostatic Optics, SLAC-TN-76-34, 1976, 56p.

/2/ W. Hartmann, D. Conrath, W. Gasteyer et al. A source of polarized electron based on photoemission of GaAsP. Nucl. Inst. and Meth. in Phys. Ros. 1990, vA 286, p. 1-8.

/3/ Ju. P. Jashin, A. I. Klimin, Ju. A. Mamajev et al. Spin-polarized photoemission from GaAsP. PhTT (Sovjet), 1987, v29, N5, p.1491.

Photocathode Studies for an Ultracold Electron Beam Device

S. Zwickler, D. Habs, P. Krause, R. Neumann, D. Schwalm, and A. Wolf

Physikalisches Institut der Universität Heidelberg and
Max-Planck-Institut für Kernphysik, Postfach 103980,
W-6900 Heidelberg, Fed. Rep. of Germany

An additional electron cooling device will be constructed for the Heidelberg Test Storage Ring TSR. The electron temperatures should be two orders of magnitude lower than those achieved with a conventional electron cooler. First experimental results concerning the GaAs cathode are presented, especially measurements of the energy spread in longitudinal direction at low and high currents and calculations of the acceleration section.

1. INTRODUCTION

By using a cold GaAs photocathode instead of a thermocathode, an ultracold electron beam with a rest-frame energy spread of about two orders of magnitude smaller than in present electron cooling devices may be obtained [1]. At heavy ion storage rings like the TSR or the ESR such an electron beam device will be of great use both for electron cooling of ion beams to even lower temperatures and for the study of electron-ion interactions, where narrow resonances can be scanned with high resolution.

The work on the ultracold electron beam has been directed towards the development of the photocathode mounted in a special ultrahigh-vacuum setup, and towards the design of an adequate beam transport system which preserves the low energy spread of the electrons.

2. PRINCIPLES OF OPERATION

The circular GaAs crystal of 2 mm diameter (Zn doped up to $10^{19}/cm^3$) is activated by covering it with cesium and oxygen/NF_3 layers leading to a negative electron affinity. Electrons lifted from the valence to the conduction band by laser light can leave this cold cathode and have a narrow energy distribution, if the photon energy is slightly higher than the bandgap[2]. In the experiments we use a 7mW HeNe-laser, a laser-diode (wavelength 810nm, max.power 300mW), and a TiSapphire-laser of up to 400mW power with a wavelength that can be tuned between 790 and 900nm. The electrons are accelerated by two electrodes and guided in a magnetic field of 300 Gauss to form a parallel beam. Several crystals have been installed in the UHV test setup and gave quantum yields between 0.1 and 1% for infrared laser light and 3% for HeNe-laser light. In this way it was possible to produce the projected current of 1 mA already with 130 mW incident light. Unfortunately the lifetime of the photocathode showed a strong dependence on the light intensity, independent of the original quantum yield or the integrated extracted current.

Figure 1 displays the fast decrease of the quantum yield at a laser power of 100 mW which corresponds to a current of 0.8 mA. In this case a contamination of the crystals surface due to a high electron current hitting the surface of the vacuum chamber could be excluded because the electrons did not leave the cathode except during the short period when the quantum yield was measured. Hence,

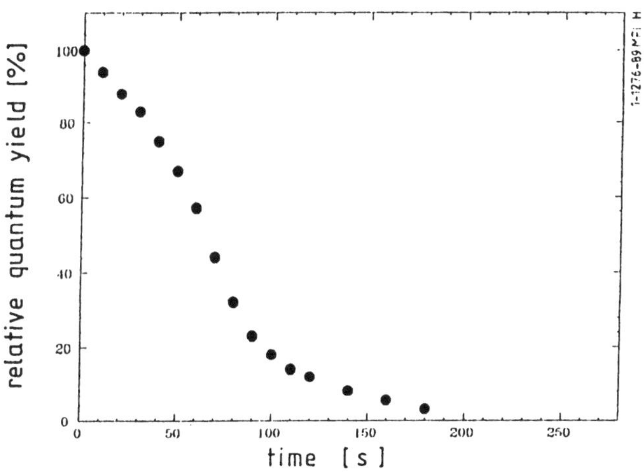

Figure 1: Relative decrease of the quantum yield as a function of the illumination time at a laser power of 100 mW and the initial current of 0.8 mA.

we conclude that laser-induced cesium desorption limits the stable operation of the photocathode at high current densities. A solution to this problem was found by continuous cesiation: at a low laser intensity and a Cs partial pressure of $6 \cdot 10^{-11}$ mbar in the test setup, an equilibrium between cesium desorption and adsorption was maintained such that a current of 0.1 mA could be extracted during several days.

3. THE ELECTRON ENERGY SPREAD

For determining the energy spread of the electron beam a retarding-field energy analyzer is situated at a distance of 4 cm in front of the cathode. By a 25 μm pinhole only a 10^{-4} fraction of the beam, originating from the cathode, is measured. The collector current I_{coll} measured behind the retarding field plate as a function of the retarding voltage U_{RAMP} gives the integrated 'electron distribution curve' (EDC). The differential signal is obtained by adding a small modulation (15 mV peak-to-peak, frequency 1000Hz) to the cathode voltage U_{cath} and by detecting the variations of I_{coll} with a lock-in-amplifier. It was the idea of this experiment to verify the measurements of [3],[4] that measured large energy broadenings (even more than 1eV)of the beam already at currents in the range of some μA, if the electron beam was generated by illuminating the cathode with a multimode laser. Furthermore we wanted to demonstrate that it was possible to produce high currents with small energy spread, necessary for the ultracold electron beam. The energy analyzer was tested with an electron beam produced by a conventional heated cathode. The EDC had a FWHM of 250meV for electron currents in the range of 0.5 and 100 μA. At very small currents and beam energies (e.g.1μA, 6eV) no relaxation processes uccured in the electron beam and the EDC had a half-Maxwellian shape; using this curve, the energy resolution could be determined [5] to be 50meV.

Figure 2, curve a) shows the EDC of a 500nA electron beam obtained by illuminating the cathode (cooled to liquid nitrogen temperature) with a HeNe-laser. Here U_{cath}=-6.5V and the two anodes were grounded. The asymmetric form of the EDC is due to the non relaxed electrons on the high-energy side [2]. When the anode voltage changed to +3V, the energy spread became extremely large for the same current(curve b). A further increase of the voltage between cathode and anode led to even larger broadenings, depending on the extracted current. When the two anodes were

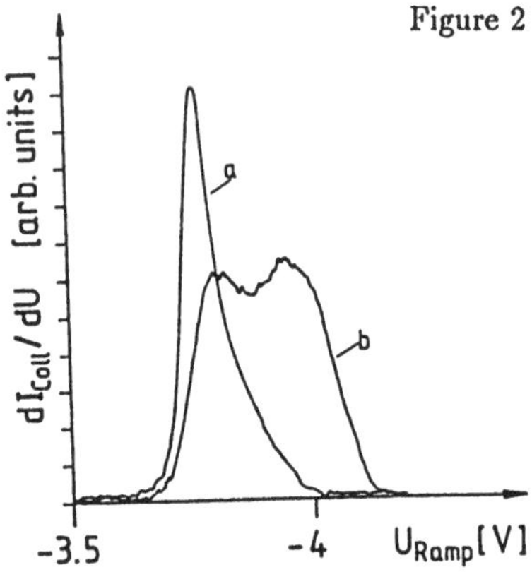

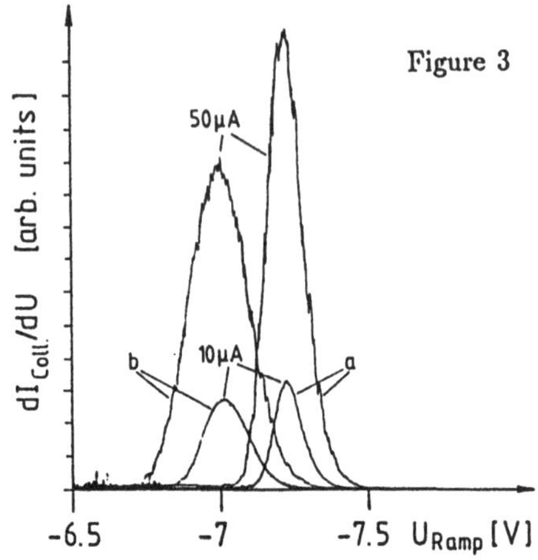

Figure 2: EDCs for 500nA electron current, obtained by illuminating the cathode with 0.5mW of the HeNe-laser. At a voltage of 6.5V between cathode and anode (curve a) the FWHM is 150meV. At 9.5V (curve b) the width of the EDC increases already to 400 meV.

Figure 3: EDCs obtained by illuminating the cathode with light from the laser diode. Curves a) were recorded at a small NEA, curves b) at a high NEA of the activated crystal. The influence of the NEA on the FWHM is important, whereas the total current does not have such an importance.

set to +20V, a similar situation to that measured in [3] was found: the energy spread was strongly correlated with the extracted current, the FWHM reached values of more than 1eV.
An explanation for this can easily be given: the HeNe-laser is a multimode laser and therefore has big power-fluctuations in the ns-region[6]. Therefore the current, generated by the photocathode, changes with a frequency that has the same value as the plasma oscillation frequency ω_p. Plasma waves in the beam are created, leading to these high energy broadenings[7]. If the electron gun is operated in or near to space charge limitation (curve a), such current fluctuations are strongly suppressed.
This situation changed when the crystal was illuminated with the single-mode GaAlAs-laser diode. In this case the voltage between cathode and anode did not have any influence on the energy spread, therefore U_{cath}=-9.5 and U_{anode}=+20V were fixed in the following measurements. The EDC of a 500nA electron beam, produced by 2mW of the laserdiode, yielded a FWHM of 80meV. This narrow energy distribution was obtained when the crystal was covered only with a small amount of Cs and NF_3, leading to a very small value of negative electron affinity (NEA) and quantum yield (smaller than 0.1%).
Figure 3 shows the influence of the NEA on the mean kinetic energy and the energy spread of the electron beam. The two curves a) were obtained at a small NEA (quantum efficiency 0.2%), while curves b) were recorded at a higher value of the NEA (quantum efficiency 0.8%). The EDCs show that the part of high energy electrons, that are able to leave the crystal at smaller values of NEA (curve a), decreases extremely if the photocathode is activated to a higher value of NEA (curve b). This leads to a smaller EDC that predicted in [8] for larger values of NEA. Possibly the high energy electrons lose some energy already at the surface of the GaAs or, more likely, a relaxation takes place in the beam, that can explane also the symmetrical shape of the EDCs. However, the EDCs at smaller NEA are narrower (for 10μA the FWHM is 110meV, for 50μA it is 130meV) than the

EDCs obtained at higher NEA (for 10μA the FWHM is 170meV, for 50μA it is 230meV). It is not yet understood if the slight increase of the energy spread with current is due to an intrinsic effect of the GaAs or if it is correlated with relaxation processes in the electron beam; in contrast to the measurements reported in [4], these curves demonstrate that it is possible to produce higher electron currents than 10μA with smaller energy widths than obtained by a thermocathode.
In order to reach the desired small energy spreads at high currents it thus will be necessary to obtain a high overall quantum efficiency at a small NEA. This will be the task of further experiments. The use of the multimode TiSapphire laser gave similar results as described in the beginning; with the help of an etalon the operation of this laser will be changed from multi-mode to single-mode.

4. OUTLOOK: THE BEAM TRANSPORT SYSTEM

In the beam transport system, a low energy spread of the electrons will be maintained by accelerating the electrons adiabatically with respect to the longitudinal plasma oscillations in the beam [9]. In the final design the electrons are accelerated non-adiabatically to 130 eV at the anode and will be decelerated again to 5 eV. This is the lowest energy possible in view of the space-charge forces in the beam.
Adiabatic acceleration then will take place in a NEC acceleration tube. Based on computer simulations it was decided to use a large diameter of 18 mm for all acceleration electrodes. This results in a very smooth electric field gradient along the beam axis except for the deceleration section (Figure 4a). In a magnetic field of 2 kG, the transverse electric field in the acceleration region creates a very small additional transverse electron energy of 3 meV at the entrance which relaxes to below 0.2

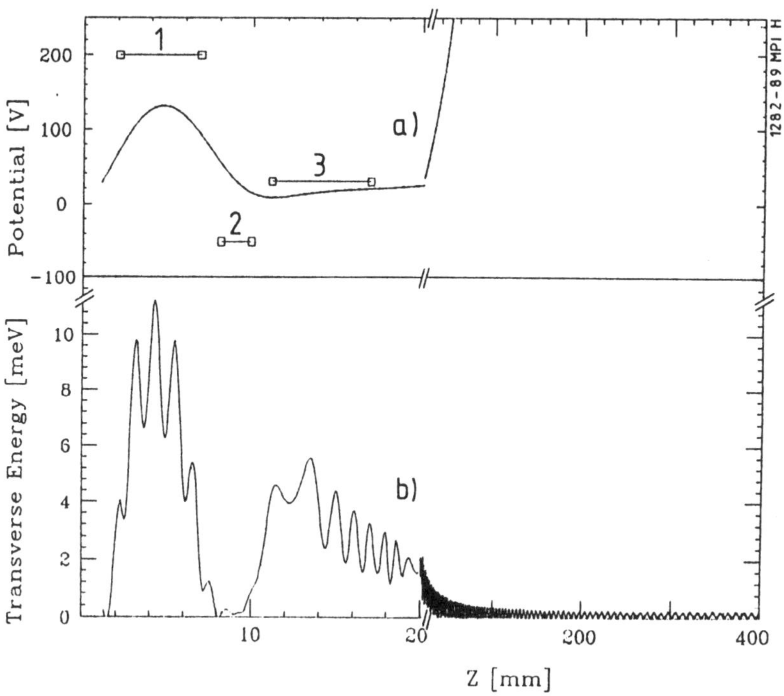

Figure 4: Computer modelling of an electron gun with smooth optics. (a) Potential on the axis, where the bars numbered 1-3 indicate the length and potential of the first three electrodes. (b) Transverse energy at the boundary of the beam, $I_e = 1$ mA, $E_{kin} = 7$ keV, $B = 2$ kG.

meV at the end of the acceleration section (Figure 4b). For comparison, the transverse temperature of the electrons at the cathode is expected to be 1 meV. The transverse energy already includes the space-charge effect. Therefore the decrease of the transverse energy during the acceleration is caused by the lowering of the electron density. Since the low transverse energy of this so called 'smooth optics' is achieved without using the conventional technique of resonant focusing, all parameters connected with electric and magnetic fields such as the final beam energy can be chosen independently from each other.
The beam transport and acceleration system will soon be tested in a linear test bench using a thermoionic cathode, which will later be replaced by the GaAs cathode. The longitudinal temperature will be measured by an energy analyzer similar to the one used in the photocathode studies.

ACKNOWLEDGEMENT
Special thanks are due to Valerie Lebedev and Vaino Kudelainen for the help in the development of the energy analyzing method and to Georges Lampel for the introduction in the lock-in technique.

REFERENCES
[1] D. Habs, J. Kramp, P. Krause, K. Matl, R. Neumann, and D. Schwalm, Physica Scripta T22, 269 (1988).
[2] H.J. Drouhin et al., Phys. Rev. B 31, 3859 (1985).
[3] K. Ertl et al., Rev. Sci. Instrum. 59, 1938 (1988).
[4] T. Dodt, Dissertation, Kernforschungsanlage Jülich (1988).
[5] H.Drouhin and M.Eminyan, Rev. Sci. Instrum. 57(6), 1052 (1986)
[6] C.Lecompte et al., Phys. Rev. Lett. 32, 265 (1983).
[7] V.A. Lebedev, Proceedings of ECOOL90, Legnaro (Italy), May 1990
[8] C.S. Pierce et al., Appl. Phys. Lett. 44, 866 (1984).
[9] V.A. Lebedev and A.N. Sharapa, Sov. Phys. Tech. Phys. 32(5), 594 (1987).

Spin-Polarized Photoelectrons from Optically Pumped p-$CdSiAs_2$(100)

F. Meier[1], *A. Vaterlaus*[1], *F.P. Baumgartner*[2], *M. Lux-Steiner*[2], *G. Doell*[2], *and E. Bucher*[2]

[1]Laboratorium für Festkörperphysik, ETH Hönggerberg, CH-8093 Zürich, Switzerland
[2]Institut für Physik, Universität Konstanz, W-7750 Konstanz, Fed. Rep. of Germany

Starting from photothreshold (1.5 eV) the polarization of the photoyield emitted from cesiated p-CdSiAs(100) by circularly polarized light at normal incidence has been measured as function of photon energy. The transitions from the only spin-degenerate top valence band levels have been identified. The observed spin-polarization of the individual transitions is significantly less than the theoretically expected $|P| = 100$ % which is attributed to shortcomings in the surface preparation of this ternary compound.

Optically pumped p-GaAs is the most widely used source of spin-polarized electrons [1]. Nevertheless, it suffers from the disadvantage of providing electrons which are at most 50 % polarized. Proposals for improving on the degree of polarization have been made already when the first GaAs source became operational [2].

The reason for the maximum 50 % polarization obtained by pumping GaAs is the symmetry of the electronic states involved in the optical transition [3]. With circularly polarized light two transitions are excited in GaAs from the 4-fold (including spin) degenerate top of the valence band at $\vec{k}=0$ to the bottom of the conduction band, see Fig. 1a.

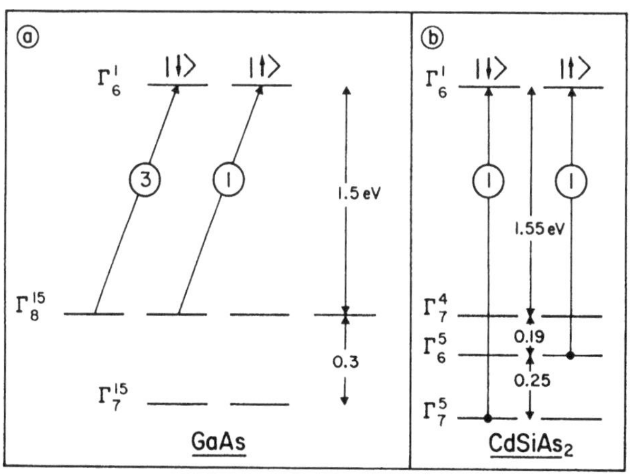

Figure 1

In this figure, the final states $|\uparrow\rangle$ and $|\downarrow\rangle$ indicate pure spin "up" and "down" states. Since one transition is 3 times more intense than the other - as shown by the encircled numbers in Fig.1 - a final state polarization P=(3-1)/(3+1)=0.5 or 50 % results. The ratio 3:1 of the intensities is determined entirely by the symmetry of the wave functions and it does not depend on the details, e.g.,of the atomic potentials.

The simultaneous excitation of electrons into oppositely polarized final states can be suppressed by lifting the orbital degeneracy of the valence band levels at $\vec{k}=0$. This can be done e.g. by applying an uniaxial stress [4] or by substituting GaAs by a material where one axis is inequivalent to the other ones as in the tetragonal chalcopyrite structure [5].

Ternary chalcopyrite semiconductors were investigated intensively many years ago [5]. To be potentially useful as a source of polarized photoelectrons the material must fulfill some minimum requirements:

1) In order to facilitate photoemission the vacuum level must be energetically lower than the polarized final states reached by the optically pumped electrons. These states are at or close to the bottom of the conduction band. Lowering of the vacuum level is achieved by a surface activation procedure consisting in the simplest case of depositing a submonolayer of an alkali metal, usually cesium, on top of the surface. In this way minimum photothresholds of the order of 1.5 eV can be obtained. This implies that the energy gap of semiconductor should not be smaller than 1.5 eV.

2) In order to allow for emission of electrons which were excited in the bulk semiconductor the band bending at the surface should be downward and its spatial extent should be small compared to the mean free path of the excited electrons. Then ballistic escape into the vacuum of the electrons excited in the interior of the crystal is possible. To this end, high p-doping levels $\geq 10^{18}$ cm^{-3} are required.

3) The separation of the non-degenerate valence band levels at $\vec{k}=0$ should preferably not be smaller than 0.1 eV in order to avoid overlap of neighbouring photoemissive transitions.

4) The material must be available in form of sufficently large single crystals which are properly oriented with the emitting surface perpendicular to the c-axis which is also the direction of the incident circularly polarized light. With cubic GaAs there are no orientational requirements for transitions at $\vec{k}=0$: one might as well use polycristalline material [3].

Conditons 1 and 3 point to $CdSiAs_2$ as the most desireable chalcopyrite material [5]: the energy gap is 1.55 eV and the separation between the valence band levels at $\vec{k}=0$ is 0.19 and 0.25 eV, respectively, see Fig.1b. In spite of numerous efforts, only recently the preparation of large, p-doped, and accurately oriented crystals has been accomplished [6].

Fig.1b shows the transitions excited at $\vec{k}=0$ for one type of circularly polarized light in $CdSiAs_2$(100) between levels of symmetry Γ_j^i [3]. The light is incident along the surface normal. The superscript i indicates the single group representation of the states and j the double group representation. For the relation of the level scheme of $CdSiAs_2$ to the one of GaAs, see Ref.[7]. The transitions scheme shown in Fig.1b is valid in the absence of any valence band hybridization between the levels.

The samples used were in form of thin, single-crystalline platelets with a surface area of about 12 mm^2 oriented perpendicular to the c-axis,i.e., along (001) [6]. They were Al-doped with $p_{Al}=2\times10^{18}$ cm^{-3} resulting in a room temperature resistivity of 45 $m\Omega/cm^2$. In the bulk, the Fermi level was 10 meV above the valence band maximum.

The surfaces were prepared by repeated scratching with a steel brush. In this way partially clean surfaces of presumably correct stoichiometry could be prepared. Ar^+-sputtering at 1000 V led to heavily damaged surfaces such that the photothreshold could not be sufficiently lowered by Cs-deposition. Evidently, scratching produces irregular surfaces, in particular because the {100}-planes are not the natural growth planes of $CdSiAs_2$. These are {112} and {011} [7]. Therefore, on a microscopic level the orientation of the surface with respect to the c-axis is not perfect. The optimum surface preparation of $CdSiAs_2$ with its large unit cell is still an open problem.

The photothreshold of a scratched surface could be lowered to 1.5 eV by depositing a submonolayer of cesium. Fig.2a shows the square root of the photoyield $Y^{1/2}$ vs. photon energy. A state of negative electron affinity is at least very closely achieved.

Fig.2b shows the spectrum of the spin-polarization $P(h\nu)$ of the total photoyield. At threshold, the electrons are negatively polarized, at higher photon energies they are predominantly positively polarized and finally P becomes zero above $h\nu$ = 2.5 eV.

The interpretation of the experimental results shown in Fig.2 is straightforward but requires - apart from the symmetry properties of the wave functions exploited in Fig. 1b - an additional ingredient: hybridization.

According to Fig.1b the transition $\Gamma_7^4 \rightarrow \Gamma_6^1$ occuring at 1.55 eV is forbidden for the sample and light geometry used. However,Fig.2a shows that photoemission is already observed at exactly this energy. The reason is that the topmost valence band Γ_7^4 is slightly hybridized with

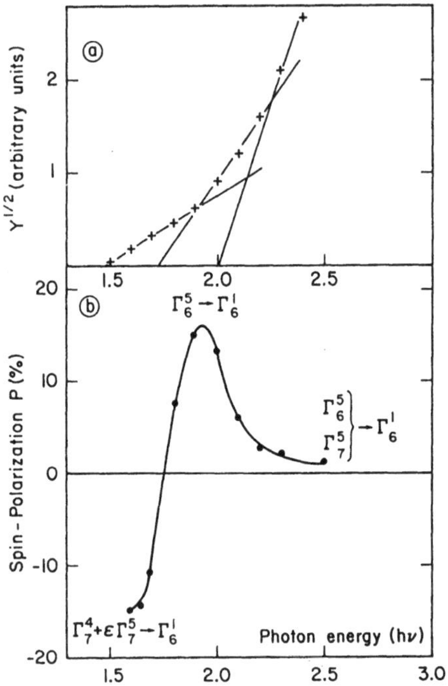

Figure 2

the lowest valence band Γ_7^5. Note that hybridization - i.e. superposition of wave functions - is only possible between states of the same double group symmetry [3]. Since Γ_7^5 produces $|\downarrow\rangle$ electrons, the transition from the hybridized top valence band level does the same: this explains the negative part of the polarization spectrum in Fig.2b close to photothreshold. Hybridization of the top valence band level has been observed already in the electroreflectance measurements reported in Ref.[7].

Above 1.74 eV the photoyield becomes considerably more intense because the transition $\Gamma_6^5 \to \Gamma_6^1$ starts contributing. This becomes evident by a change of slope of $Y^{1/2}(h\nu)$ in Fig. 2a. The polarization, according to Fig.1b, is positive. Since the intensity far exceeds the one due to the hybridized top valence band level the polarization of the total photoyield becomes positive, too. At still higher photon energies the lowest valence band level Γ_7^5 produces negatively polarized photoelectrons - see Fig.1b - of comparable intensity as Γ_6^5 such that the total polarization vanishes above 2.5 eV. With increasing photon energy the transitions excited between a given valence band and the conduction band move away from $\vec{k}=0$ into the outer part of the Brillouin zone. For sufficiently small increments of the photon energy, however, the polarization is affected only very little [8].

Similar experiments as shown in Fig.2 have been performed at higher photothresholds. The polarizations have always been << 100 % which is believed to be mainly due to the less than perfect technique of surface preparation. However, the experiment shows that the expected qualitative features of the spectrum of the spin polarization are present. No problems were found keeping the photothreshold stable at 1.5 eV just by Cs deposition without additional oxygen treatment.

We thank H.C.Siegmann for his interest in this work and K.Brunner for his expert technical assistance. The financial support by the Schweizerische Nationalfonds is gratefully acknowledged.

References.

1) D.T.Pierce and R.J.Celotta, Modern Problems in Condensed Matter Sciences: "Optical Orientation" 8, 259 (1984)

2) D.T.Pierce, F.Meier, and H.C.Siegmann, United States Patent 3,968,376 (July 6, 1976)

3) F.Meier and D.Pescia, Modern Problems in Condensed Matter Sciences: "Optical Orientation" 8, 295 (1984)

4) P.Zorabedian,"Optical pumping and photoluminescence detection of spin-polarized electrons in uniaxially stressed gallium arsenide", SLAC-Report 248, Stanford Linear Accelerator Center, Stanford University, Stanford California 94305, USA (March 1982)

5) J.L.Shay and J.H.Wernick, Ternary Chalcopyrite Semiconductors: Growth, Electronic Properties and Applications, Pergamon Press, Oxford (1975)

6) F.P.Baumgartner, M.Lux-Steiner, G.Doell, E.Bucher, F.Meier, and A.Vaterlaus, J.Cryst.Growth (to be published)

7) J.L.Shay and E.Buehler, Phys. Rev. B3, 2598 (1971)

8) R.Allenspach, F.Meier, and D.Pescia, Phys. Rev. Lett. 51,2148 (1983)

Development of Polarized Electron Source of GaAs-AlGaAs Superlattice and Strained GaAs

T. Nakanishi and S. Nakamura

Department of Physics, Nagoya University, Nagoya 464, Japan

1. Our apparatus

At Nagoya University,we have continued the development of GaAs polarized electron source(PES) for several years. Our test apparatus (NPES-1) is shown in Fig.1, which consists of a gun producing (1-5) KeV polarized electrons and a standard 100 KeV Mott polarimeter.[1] By this apparatus, we extracted electrons with the spin electron polarization(ESP) of (30 ~ 33)% and the quantum efficiency(QE) of ~ 4% from the bulk GaAs. We obtained also the beam with ESP of (35 ~ 40)% and QE of ~ 0.5% from the thin(0.5μm thick) MOCVD GaAs layer whose surface was covered by the oxide film for ambient transfer.

At present,we have been trying to develop the new material which may give the higher ESP than the bulk GaAs,those are, GaAs-AlGaAs superlattice(SL), strained GaAs and chalcopyrite crystal,in the collaboration with KEK,NEC,Osaka Pref.Univ. and Daido Steel groups for the Japan Linear Collider(JLC) project and with Bonn Univ. group for the ELSA experiment. Here, we report our approach to prepare SL and strain GaAs samples and preliminary results of ESP and QE measurement, although the studies are not yet completed,rather in progress.

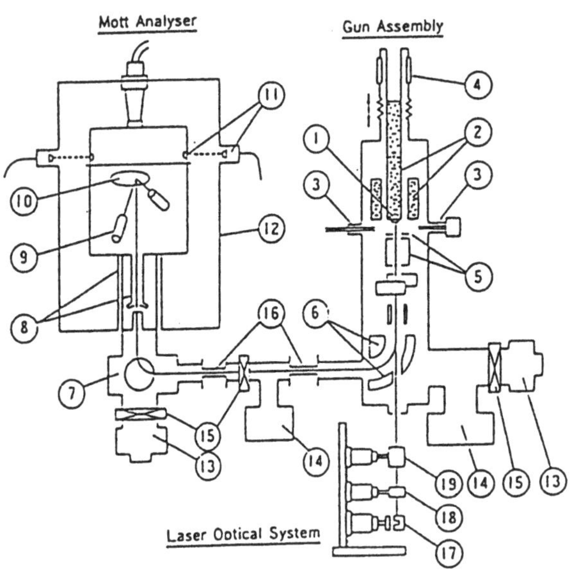

1) GaAs Crystal
2) Liq. N_2 Container
3) Activator (O_2 and C_s)
4) 5 KV Ceramic Bushing
5) Extraction Electrode
6) Spherical Condenser
7) Helmholtz Coil
8) 100 KV Ceramic Bushing
9) Surface Barrier Detector
10) Target Foil and Wheel
11) LED and Photodiode
12) Freon Gas Tank
13) Turbo-Molecular Pump
14) Ion Pump
15) UHV Gate Valve
16) Conductance Reducing Tube
17) Laser Diode
18) Linear Polarizer
19) Pockels Cell

Fig.1 Test Apparatus NPES - I

2. GaAs-AlGaAs Superlattice

The superlattice is a candidate to overcome the ESP limitation of 50% ,lifting the degeneracy of heavy-hole (*hh*-) and light-hole(*lh*-) band by the quantum size effect. But, the past experimental results did not exceed the limit of 50% and this reason is explained by depolarizations which are accumulated by multi-collisions with potential barriers before electrons can escape to the vacuum.[2)]

This depolarization should be reduced by use of the narrower layers of GaAs and AlGaAs than those($\geq 50\AA$) of previously used. However,on the other hand,the tunneling effect brings the energy spread(mini-band) of each *hh*- and *lh*-band and make it difficult to select only the *hh*- band excitation. Thus, optimum thickness must be determined from such two contradict requirements and a trial of such optimization was done by KEK group.[3)]

Sample preparation

The first two kind of MBE samples(SL#1 and SL#2) were made by NEC group. The 0.4 μm thick GaAs-AlGaAs SL layers were grown on the GaAs substrate with an insertion of 1 μm thick layer of AlGaAs. For SL#1,the thickness of GaAs and AlGaAs layer are $21\AA$ and $24\AA$, the resultant energy separation(δ_s) is estimated to be $\sim 20\,meV$. **For SL#2,those parameters are $21\AA$,$33\AA$ and $\sim 43\,meV$,respectively.** The surface of the SL was covered by $\sim 1\mu m$ thick As layer and it was removed by heat cleaning at $\sim 400°C$,before the NEA activation. The photoluminescence(PL) spectrum from **SL#2** taken at room temperature is shown in Fig.2(a),where a large peak ($\lambda \sim 750nm$) comes from the SL and a small peak ($\lambda \sim 675nm$) corresponds to the AlGaAs layer behind the SL.

Preliminary ESP results and its interpretation

The preliminary results of ESP and QE measurement for **SL#2** at room temperature are shown in Fig.2(b) and 2(c), where the wave-length was changed by Ti:sapphire tunable laser whose available wave-length is limited to $\lambda \geq 700\AA$. Data points indicated by (•) were taken 10 days after the others(□),although no re-activation of NEA surface was done during two measurements.

In Fig.2(b),the ESP seems to step up from the lower level of $\sim$ 30% to the higher level of $\sim$ 50% and this feature was well reproduced by several separate measurements, independent on the surface state which gave different quantum efficiencies.

Such behavior can be interpreted as that the $\sim$ 50% level of ESP at the range of $\lambda = (775 - 800)nm$ is brought by the selective *hh*- band excitation, while both excitations of *hh*- and *lh*-band make the ESP to decrease to the $\sim$ 30% level in the region of shorter wave-length. This interpretation seems to be well supported by the both QE and PL data which show that electrons with ESP of $\sim$ 50% were produced by the threshold energy excitation.

The ESP averaged over the flat-region($\Delta\lambda = 25nm$) was (**52.4.0** $\pm 0.3 \pm 2.5$)% for **SL#2**, while it was (**45.2** $\pm 0.3 \pm 2.3$)% for **SL#1**,where the first(second) error notes statistical

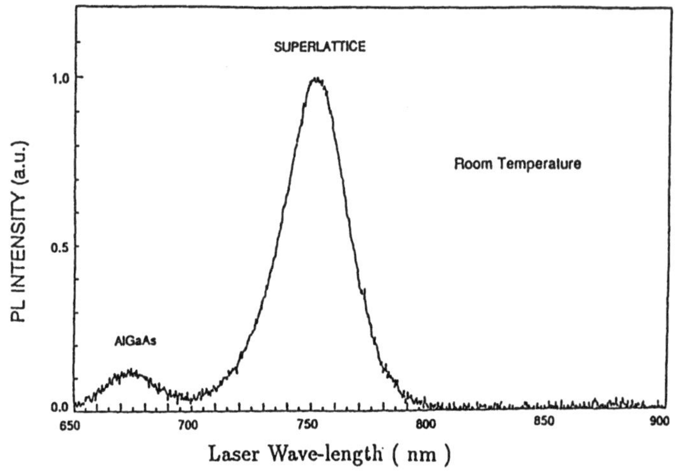

Fig.2(a) Photoluminescence(PL) data for SL#2

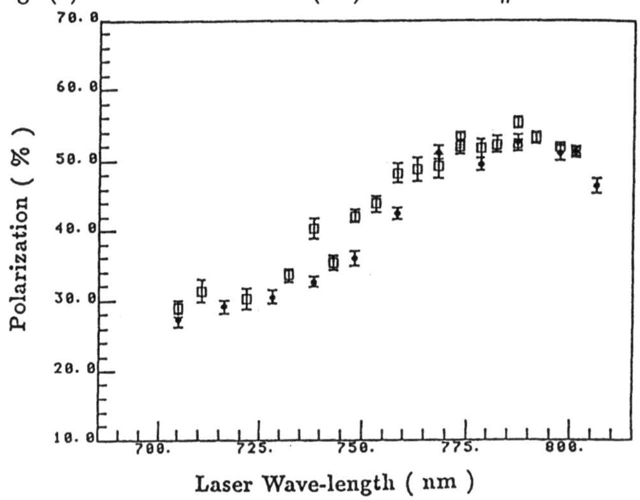

Fig.2(b) The ESP results for SL#2

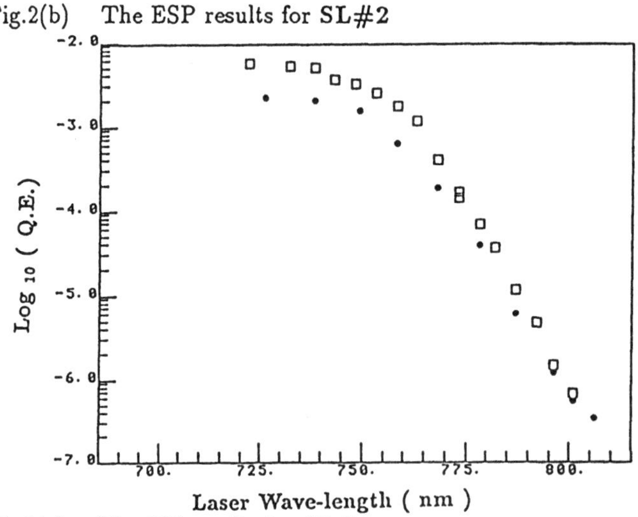

Fig.2(c) The QE results for SL#2

(systematic) one. This difference seems to suggest the importance of δ_s parameter. The systematic works to examine other SL samples which have different energy splittings and mobilities are scheduled in order to get the much higher ESP.

3. Strained GaAs

The introduction of an uniaxial stress to GaAs is another approach to overcome the ESP limitation of 50%. The enhancement of circular polarization of PL from GaAs with a tensile stress produced mechanically along [001] axis was observed by P.Zorabedian.[4] The similar enhancement was also observed by H.Horinaka et al. from InP whose strain was introduced through the interface stress caused by an *artificial* lattice-mismatch(LM) heterojunction.[5] Motivated by the latter experiment, we started the study of the strained GaAs layer formed on the substrate of a mixed-semiconductor with an *artificial*-LM heterojunction. Another type of interface stress (which we define as *thermal*-LM) is also possible, using the difference of expansion coefficients between the film and substrate.[6]) Details of our study is given in ref.7) and here,the very short explanation is described.

Designing of strained GaAs samples

Following knowledges are required to prepare the strained GaAs layer with the *artificial* LM heterojunction.

(1) Energy splitting – Strain relation Energy splitting(δ_s) is given as a linear function of the LM ($\Delta a/a_0$), where Δa is the difference between the unstrained lattice constants of film(a) and substrate(a_0). For GaAs,it becomes $\delta_s = 7.6\,(\Delta a/a_0)(eV)$.

(2) Critical thickness – Strain relation Coherent growth of the strained epitaxial layer on the substrate is allowed within a critical thickness(t_c),but the dislocations or micro-cracks begin to arise to relax the strain over this thickness. The critical thickness can be evaluated by Matthews'relation[8] and the result becomes,for example, $t_c = 0.1\mu m$ for $\Delta a/a_0 = 0.2\%$.

This small t_c limits the available range of LM to be $0.1\% \leq \Delta a/a_0 \leq 0.3$ %, for which $t_c = (0.22 \sim 0.07)\mu m$ and $\delta_s = (8 \sim 23)meV$ are given.

(3) Choice of the substrate Two conditions are required for the substrate; (a) The lattice constant of substrate must be smaller than that of GaAs(5.653 Å at $300K$) to apply the tensile strain or *vice versa* for the compressive strain. (b) The band gap energy of the substrate must be larger than that of GaAs(E_g = 1.42 eV at $300K$),to insure that no electron is extracted from the substrate. There exist some candidates in ternary or in quaternary mixed-semiconductors; for example,$Ga_xIn_{1-x}P$, GaP_xAs_{1-x} or $Al_xGa_{1-x}P_yAs_{1-y}$, $Ga_xIn_{1-x}P_yAs_{1-y}$ etc.

Sample preparation

The first sample of GaP_xAs_{1-x} type(STP#1) was made by a MOCVD apparatus of Daido Steel Co. It has the structure shown in Fig.3, where the oxide film was employed for surface protection. The ESP and QE measurement is just under way and soon first preliminary results will be obtained.

Oxide Film for Surface Passivation
$Zn - doped \ \ GaAs \ \ (t = 1000\,\mathring{A}, \ \ p \geq 5 \times 10^{18} cm^{-3})$
$Zn - doped \ \ GaP_xAs_{1-x} \ \ (x = 0.06 \pm 0.01)$ $(t = 2\,\mu m, \ \ p \geq 5 \times 10^{18} cm^{-3})$
$Zn - doped \ \ GaAs \ \ Substrate$ (t=350 μm, $p \geq 5 \times 10^{18} cm^{-3}$)   Orientation : (1 0 0)

Fig.3 Structure of STP#1 sample

This work was supported partially by the Grant-in-Aid for International Joint Research Project of JSPS and the Kurata Research Grant of Kurata Foundation.

References

1) T.Nakanishi et al.; Jpn.J.Appl.Phys.25(1986)766-767
2) S.F.Alvarado et al.;Appl.Phys.Lett.39(1981)615-617
3) Y.Kurihara et al.;KEK Preprint 90-77
4) P.Zorabedian; Stanford Linear Accelerator Center Report-248 (1982)
5) H.Horinaka et al.; Jpn.J.Appl.Phys.27(1988)765-770
6) G.Lampel; Contribution to this workshop
7) T.Nakanishi et al.;Nagoya Univ.Preprint DPNU-90-16
8) J.W.Mathews et al.;J.Cryst.Growth 27(1974)118-125

Spinrotator for MAMI

K.H. Steffens and H.G. Andresen

Institut für Kernphysik, Johannes Gutenberg Universität,
Postfach 3980, W-6500 Mainz, Fed.Rep.of Germany

A source of polarized electrons based on the photoemission of a GaAsP-cathode [1] has been set up to inject polarized electrons in the Mainz microtron MAMI. In order to compensate the spin precession in MAMI a spinrotator system has to be integrated in the 100 keV beamline from the source to the injector linac to achieve the required longitudinal spin direction at target position.

1 Introduction

A numerical analysis of the depolarization effects and changes of the polarization direction showed that for an accelerator as MAMI in contrast to storage rings or synchrotrons it is not necessary to orient the electron polarization vertically to the acceleration plane in order to avoid depolarization effects. The "in-plane"-component of the polarization precesses in the magnetic field of MAMI by a factor of $(1+\gamma a)$ faster than the electron momentum, where "a" stands for the g-factor anomaly of the electron. Since the spin precession angle accumulated during the acceleration process exceeds that of the momentum by about $118*2\pi$, it is difficult to predict the spin direction with sufficient accuracy. However it is possible to monitor the polarization direction of the high energy electrons (Møllerpolarimeter [2]) and to optimize the required longitudinal polarization component at the target position by providing an appropriate spin rotation at the low energy injection side. This can be accomplished with minor technical effort.

2 Spinrotator

The spinrotator consists of two spherical electrical condensors and two systems of solenoids (fig. 1). This system seems to be more complicated than another possible system realized by a wienfilter and a solenoid, but it has the advantage of almost fixed optical beam properties for all spin angle adjustments in particular it is advantageous that the focal length can be kept constant.

The source produces a beam with a longitudinal polarization. During the deflection in the first spherical electrical condensor the spin precesses very slowly, so that after a momentum deflection of 107.6° the longitudinal polarization is transformed to a transverse polarization. The polarization can now be rotated azimuthally by the two pairs of solenoids as shown in fig. 1. The best location to measure the polarization at 100 keV via mott scattering is between the two condensors, because for mott scattering transverse polarized electrons are needed. The next electrical deflection bends the beam back to its

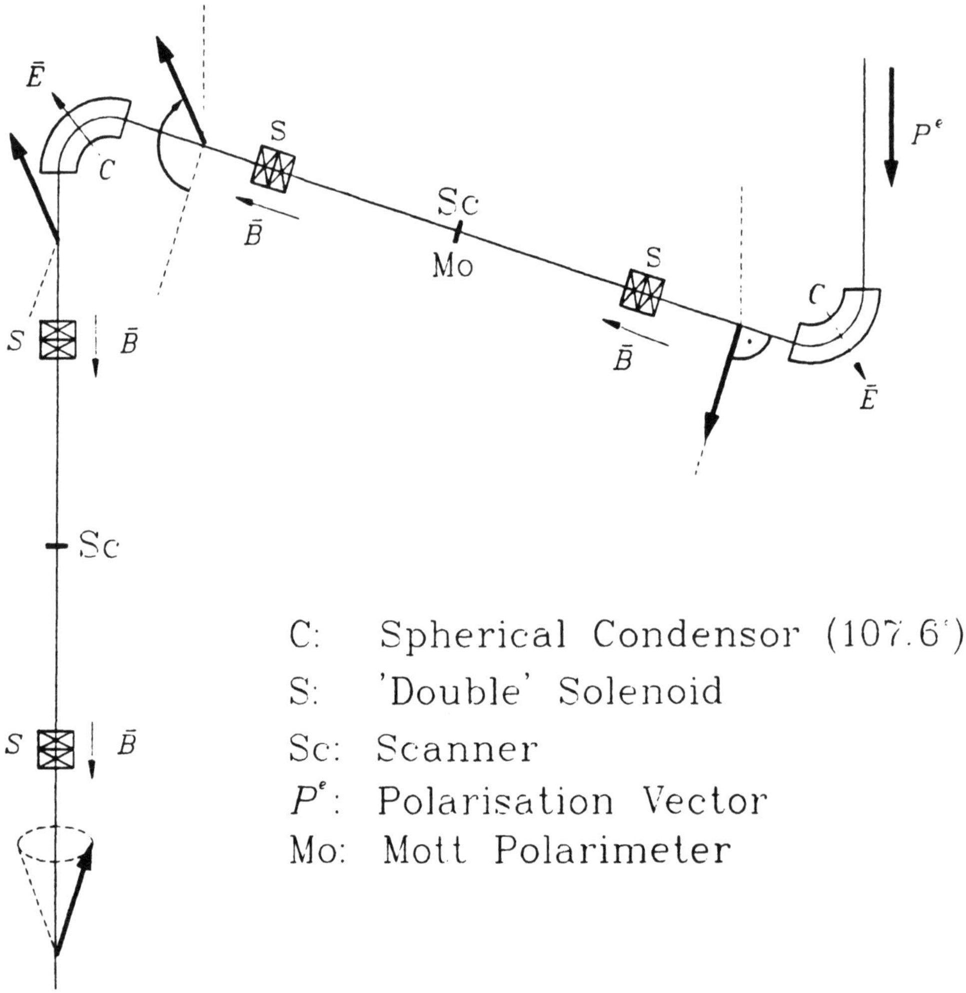

Figure 1: Spinrotator with two spherical condensors and solenoids

original direction which is accompanied by only a small change in the spin orientation. The last two solenoids allow another azimuthal rotation of the polarization vector. The overall system may be used to orient the polarization vector of the beam in any desired direction. For a spin rotation alone only one solenoid would be needed, however in order to achieve fixed optical beam properties each solenoid is realized by two symmetrically arranged twin-solenoids. In each twin-solenoid the two coils are driven back-to-back. Since the spin rotation angle is proportional to the $\int B_z dz$ and the focal length is proportional to the reciprocal of the $\int B_z^2 dz$, the spin orientation may be adjusted independently from the beam focussing, which can be kept constant.

For all rotation angles the envelope of the beem varies only very slightly as shown in fig. 2, depicting the beam envelopes for the spin rotation angles 0°, 90° and 180° respectivly. This slight variation of the beam extend is caused by the shift of the principal planes of a thick, asymmetrically driven twin-solenoid. The effect is so small, however, that it does not have to be corrected.

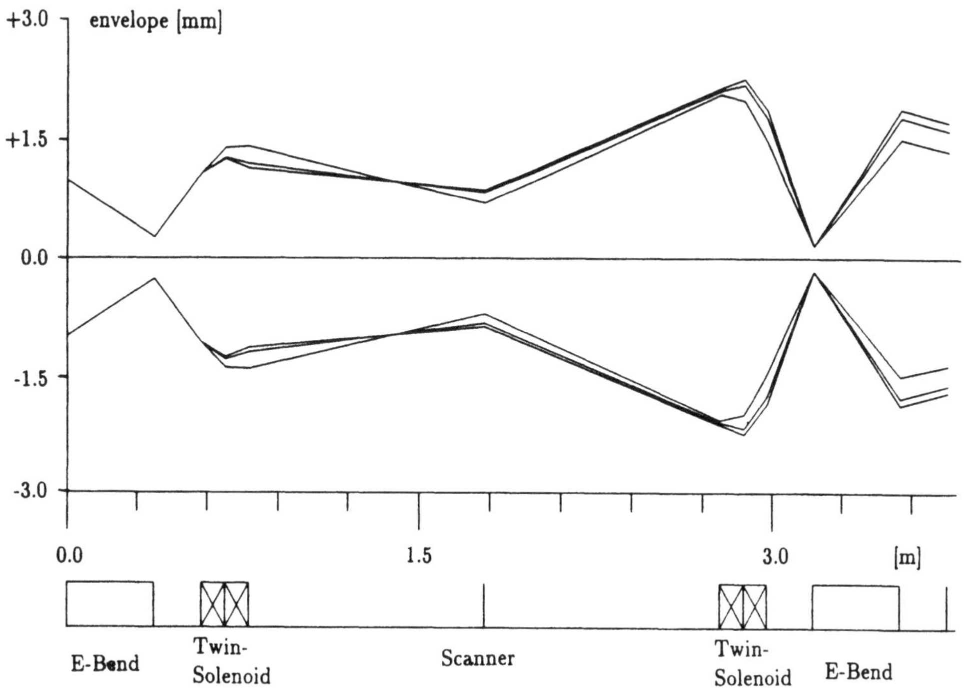

Figure 2: Beamoptic: Spin rotationangle = 0°, 90°, 180°

In order to avoid transverse phase space rotation effects by solenoids the input beam has to exhibit rotational symmetry and thus stigmatic focussing properties of the electrical condensors are required. A spherical electrical condensor achieves stigmatism only for nonrelativistic energies. In the case of 100 keV electrons ($\gamma = 1.2$) the spherical electrical condensor has to be modified to a toroidal condensor with slight different radii of curvature in order to achieve stigmatic focussing. This result in an electric field index $n = r_o/R_o = 0.85$ where r_o stands for the bending radius.

The spin rotator is under contruction now and will be tested under beam condition at the end of 1990.

References

[1] K. Aulenbacher: *HIGH ENERGIE SPIN-PHYSICS CONFERENCE Bonn, Sep. 10-15, 1990, Workshop: POLARIZED ELECTRON SOURCES AND ELECTRON POLARIMETERS*
Source of polarized electrons for MAMI B

[2] M. Ertel: *HIGH ENERGIE SPIN-PHYSICS CONFERENCE Bonn, Sep. 10-15, 1990, Workshop: POLARIZED ELECTRON SOURCES AND ELECTRON POLARIMETERS*
Møllerpolarimeter for MAMI B

Status of the HERA Electron Polarimeter

M. Lomperski [1]

University of Wisconsin, Madison, WI 53706, USA

The status of the design and construction of the polarimeter for the HERA electron storage ring are presented.

Introduction

The electron and proton storage rings of HERA are nearing completion. The physics program of the facility includes the possibility of colliding longitudinally spin polarized electrons on protons. An early step in achieving this is the optimization of the transverse (vertical) electron spin polarization; Spin rotators are then used to rotate the polarization direction. A laser polarimeter for measuring vertical polarization is now under construction, and I wish to describe its present status. More information about the HERA facility and the prospects for polarization there can be found in D.P. Barber's contribution to these proceedings.

The Method

Laser polarimeters, widely used at e^+e^- storage rings (LEP[2], TRISTAN[3], etc.) utilize the spin dependent cross section for Compton scattering of polarized photons on electrons: The scattering of circularly polarized photons (from a laser) on vertically polarized electrons is asymmetric with respect to the orbital plane of the electrons, and the magnitude of the asymmetry is proportional to the degree of polarization of the electrons [5].

The measurement of the asymmetry of the cross section is sensitive to the synchrotron radiation and beam gas bremsstrahlung backgrounds produced by the storage ring. With large backgrounds it becomes inefficient (if not impossible) to measure single Compton photons (the *single-photon* method). This problem can be overcome by producing many ($\approx 10^4$) backscattered photons in each laser-electron bunch interaction using a high power laser (the *multi-photon* method). This technique is less sensitive to backgrounds but does not allow for detailed analysis of the backscattered distributions (e.g. as a function of the Compton photon energy) which provides consistency checks and enables one to optimize the measured asymmetry with data cuts.

The position chosen for the electron-laser photon interaction point (I.P.) is ideally located. The short (4 m) straight section, at the beginning of the arc after the West Hall,

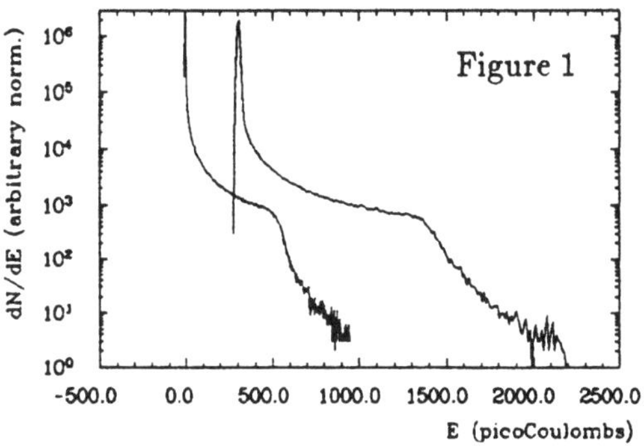

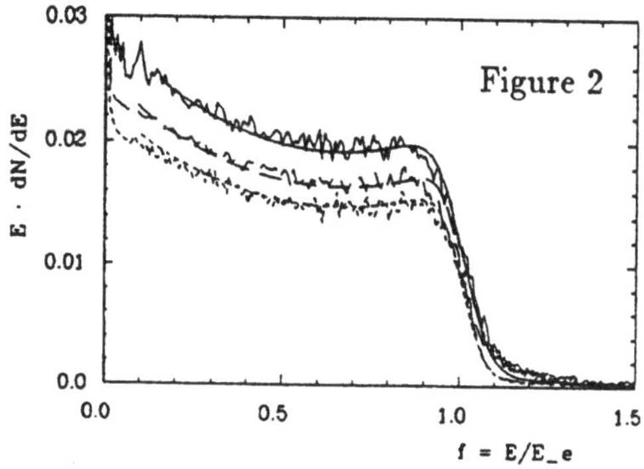

Figure 1: Energy spectra measured with 13 and 25 GeV electrons, showing the effective pedestal shift due to synchrotron radiation.

Figure 2: Measured and predicted bremsstrahlung spectra for 13, 20 and 22 GeV electrons.

is isolated from the long straight sections (360 m) and from the strong bending magnets of the arc ($\rho = 608$ m) by two weak bending magnets ($\rho = 3215$ m), greatly reducing the beam gas bremsstrahlung and synchrotron radiation in the detector aperture. During the electron machine run in September 1989 we made background measurements [4]. Plotted in Figure 1 are bremsstrahlung spectra taken with 13 and 25 GeV electrons. The relative shift is due to synchrotron radiation. In Figure 2 the bremsstrahlung spectra (with synchrotron shifts subtracted) measured at various beam energies are plotted with our predictions. The measured synchrotron radiation and beam gas bremsstrahlung backgrounds were in excellent agreement with our expectations, and on the basis of these results, we are building a single-photon Compton polarimeter.

The Laser

Simplicity and reliability were important features of the laser for the polarimeter. We have chosen an off-the-shelf continuous-wave argon ion laser with 10 Watts at 514 nm. With realistic parameters (crossing angle, bunch currents) we expect a luminosity of about 5 Compton photons per 100 laser-bunch interactions. The laser is installed and is in operation.

The laser is located in the West Hall and a system of 4 mirrors will direct the light to the I.P. The total distance is 200 m, and so the beam must be transported in a partially evacuated channel. The system of monitors for alignment of the beam must be carefully designed to enable re-alignment of the beam inside the channel without access to the tunnel. The laser light enters the electron vacuum chamber from above, through a special window in the chamber and exits after passing through the I.P. The circular polarization of the laser light is generated by a Pockels cell in the West Hall, and must be preserved during its passage through the transport system. For this reason we are using dielectric mirrors and have designed the vacuum chamber so the last mirror is outside the chamber. Detectors positioned after the exit window will measure the power and polarization of the beam.

The backscattered photons leave the I.P. travelling with the electron beam. The two beams are separated by the second weak bend and the first strong bend of the arc. The photons exit the electron vacuum chamber through a 0.5 mm thick Al window. The detector is 65 m from the I.P., as far back as possible to take advantage of the energy dependent Compton opening angle to increase the size of the Compton spot at the detector. The optics of the storage ring near the I.P. will be optimized to focus the backscattered photons on the detector. To better define the photon beam, collimators will be placed 16 m upstream from the detector, followed by sweeping magnets to help remove any secondary electrons which are produced.

The Detector

Kinematics. In the lab system, the initial electron energy is 30 GeV, the photon energy is 2.41 eV, and the collision is nearly head-on (the vertical crossing angle is 3 mrad). Boosting to the electron rest system, the photon has an initial energy of 280 keV and its final energy is between 280 keV ($\theta = 0$) and 130 keV ($\theta = \pi$). Boosting back into the lab system, the final photon energy range is between 2.41 eV ($\theta = 0$) and 15.7 GeV ($\theta = \pi$), and the scattering angles are focussed into a narrow cone about the electron direction (e.g. $\theta_{REST} = \pi/2$ boosts to $\theta_{LAB} = 17\mu$rad). In summary, the photons to be measured have very high energies and at the detector are in a spot of radius 1-2 mm.

Measuring the Asymmetry. The term in the Compton cross section which is proportional to the electron vertical polarization depends on the azimuthal angle ϕ as $\sin\phi$. The amplitude also depends on the energy of the scattered photon. One way to find the polarization is to calculate the net asymmetry of the scattering about the median plane. This asymmetry can be written

$$A = \frac{N_U - N_D}{N_U + N_D} = P_e P_\gamma \Pi, \tag{1}$$

where N_U and N_D are the numbers of events above and below the machine plane and P_e and P_γ are the degrees of electron vertical and photon circular polarizations. The constant Π, called the analyzing power, depends on the data cuts used to define N_U and N_D and can be calculated by Monte Carlo. The dN/dy distributions from three representative situations are shown in Figure 3: unpolarized electrons and photons (broken curve), fully polarized ($P_e = P_\gamma = 1$) electrons and photons (solid curve), and full polarization including the divergence of the electron beam (dotted curve). The curves have been calculated by Monte Carlo, and only the energies between 7.8 and 14.2 GeV are included.

The Calorimeter. The detector must measure the energy and the y-impact parameter of high energy photons. The photon detector is a sampling calorimeter with 12 layers of tungsten (6.2 mm) and scintillator (2.5 mm); see Figure 4. Conceptually, it can be considered to consist of two calorimeters. one on top of the other, with each layer consisting of a single absorber plate and two optically decoupled scintillator plates. The

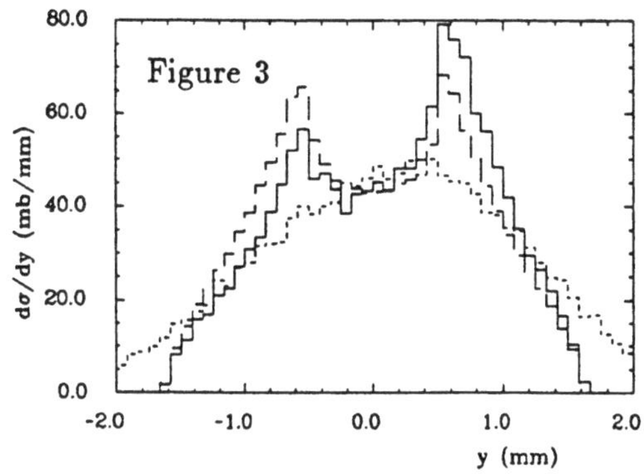

Figure 4

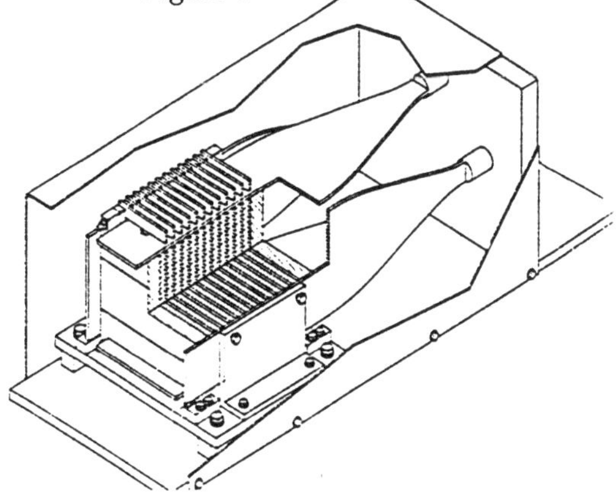

Figure 3: dN/dy distributions with $P_e = P_\gamma = 0$ (broken curve), $P_e = P_\gamma = 1$ (solid curve), and $P_e = P_\gamma = 1$ with the smearing effects of the electron beam divergence included.

Figure 4: Perspective view of the calorimeter, showing the sampling layer structure and the four wavelength shifters.

scintillation light from each half is collected with wavelength shifters (WLS) along the top and bottom of the detector. The vertical position measurement is made by comparing the energy measured in the two channels. For calibration and redundancy, there are also WLSs on the right and left sides which read the upper and lower halves together. Photomultiplier tubes collect the light from the WLSs. The detector has been tested in a DESY test beam with electrons of energy 3 GeV. Summing the two halves, the energy resolution is $22\%/\sqrt{E(GeV)}$ and the sampling fraction is 2.5%. The vertical position measurement has a resolution in the center of the detector of about 0.5 mm at 3 GeV. Test have been made at higher energies at CERN, and the data is now being analyzed.

Data Acquisition. The data acquisition system has special requirements. To use the polarimeter to guide the machine operators, we need results in real time. But the analyzing power defined above is typically 10-20% and to achieve an absolute statistical error in the value of the polarization of 1% one sees that one needs approximately 10^6 photons. Therefore, the data-acquisition system must accumulate data as rapidly as possible (up to about 200 kHz), reduce the size of the data (to save space), and analyze the data in real-time. Tapping the experience of the ZEUS Data Acquisition Group, we have found a solution in the "Nevis Digital Card"[6], a VME module designed for the ZEUS calorimeter DAQ system. The card has four ADCs, large input, output, and storage memories, and a digital signal processor for on-board computations. We are basing our DAQ system around this card.

Conclusions

The third run of the electron ring and the first run of the proton ring of HERA are scheduled to begin in Spring '91. We are hoping to have the polarimeter in operation at that time, to begin parasitic "de-bugging", and to begin optimizations of the machine orbit parameters for maximum transverse electron polarization.

Acknowledgments

I would like to thank the entire ZEUS collaboration for the help I have received in working on the polarimeter. For lack of space, I will mention only a few people (and apologize to many more): A. Dwurazny, G. Jahnen, B. Machowski, S. Ritz, W. Vogel, and K. Westphal. And I would also like to thank B. Dehning and M. Placidi for many educational discussions about building polarimeters.

References

[1] For the HERA Polarimeter Group: D.P. Barber, W. Bialowons, H.-D. Bremer, U. Camerini, H. Goetschel, H. Gressmann, H. Kaiser, R. Kaiser, R. Klanner, U. Koetz, H.-C. Lewin, E. Lohrmann, M. Lomperski, G. Meyer, P.M. Patel, T. Romanowski.

[2] M. Placidi, Proceedings from this workshop.

[3] K. Nakajima, Proceedings from this workshop.

[4] D.P. Barber, et al., Background Measurements for the HERA Electron Polarimeter, DESY internal note.

[5] U. Fano, Remarks on the Classical and Quantum-Mechanical Treatment of Partial Polarization, J. Op. Soc. Am., 39 (10) 1949 859-863.

[6] The Zeus Front End Digital Card: Architecture, Nevis Labs, Columbia University, ZEUS note, May 1989

The Møller Polarimeter for MAMI

M. Ertel, H.G. Andresen, R. Loos, and K.H. Steffens

Institut für Kernphysik, Johannes Gutenberg Universität,
Postfach 39 80, W-6500 Mainz, Fed. Rep. of Germany

1 Design Considerations

The polarization of high-energy electrons can conveniently be measured by means of Møller scattering from polarized electrons in a magnetized ferromagnetic target. This process exhibits in the limit of high energies ($E \geq 100$ MeV) still relatively large, energy independent and well known analysator strengths. Furthermore, the laboratory cross section in the kinematical range suitable for polarimeter applications is large

$$\frac{d\sigma}{d\Omega}(\Theta_{cm} = 90^\circ) = Z \cdot 0.18 \ barn/sr$$

and energy-independent.

Fig. 1 depicts the analysator strengths a_{zz}, a_{xx} and a_{yy} in the high energy limit, where the indices refer to the polarization components of the beam and the target, respectively. The z-axis coincides with the direction of the primary beam; the y-axis is chosen to be

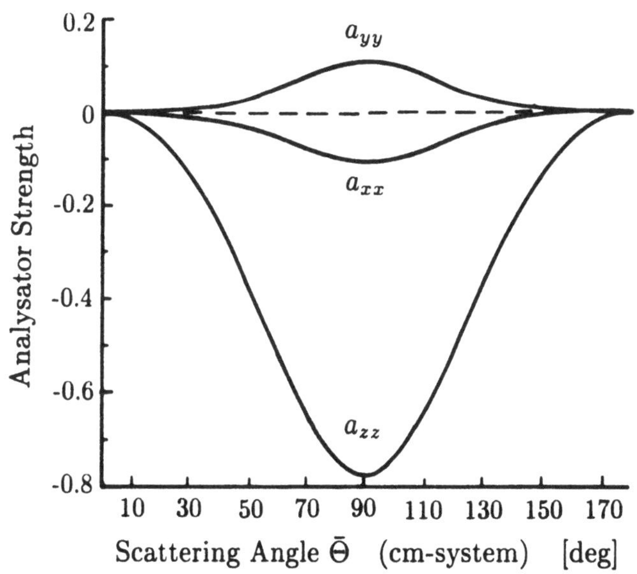

Figure 1: High-energy limit of the analysator strengths a_{zz}, a_{yy} and a_{xx}

(in the Madison convention) perpendicular to the scattering plane. A scattering angle of 90° in the cm-system will result in optimum polarimeter performance since the analysator strengths in this case exhibit their maxima resulting in a first order insensitivity to small variations of the scattering angle. General design considerations for high energy electron Møller polarimeters have been discussed in detail in references [1, 2, 3].

Since the polarized target electrons will be provided by magnetized ferromagnetic foils, the dominating background for the Møller scattering stems from the elastic Mott-scattering on Fe-nuclei. For "symmetric" Møller scattering with $\Theta_{cm} = 90°$ the Mott-scattered background will be 12 times larger than the Møller signal. Since the energies of the Møller- and the elastically scattered Mott-electrons differ for $\Theta_{cm} = 90°$ by a factor of two, a momentum analysis with modest resolution will already considerably decrease the Mott background; the background can be further reduced by the coincidence detection of the Møller scattered recoil electron. For the energy of MAMI between 180 and 850 Mev the scattering angle in the laboratory system for "symmetric" scattering varies between 4.3° and 2.0°. The required very forward scattering angles, the possibility of a moderate momentum analysis and the coincidence detection of the symetrically scattered recoil electron can be conveniently accomplished with the quadrupole magnet arrangement, shown in fig. 2.

The quadrupole magnet will not deflect the primary beam and faciliates beam dump requirements. The electrons, which are scattered by the magnetized ferromagnetic target foils will be separated from the primary beam and are momentum analyzed in the defocusing plane of the quadrupole magnet. The variation of the angle Θ_{sym} for the "symmetric" Møller scattering can be accomplished by moving the target along the beam axis, while keeping the detector and collimator geometry fixed. The relatively large scattering chamber is due to the long way of the target movement required to cover the energy range between 150 and 900 MeV.

2 Target Mounting for Beam Polarization Measurements

For foils only an "in-plane" magnetization can be achieved. Therefore it is necessary to incorporate target foils with different orientations in the target area of the polarimeter, schematically indicated in fig. 2 by T_1, T_2 and T_3. The required longitudinal polarization component of the target polarization P_z^T for the measurement of the longitudinal beam polarization P_z^B can only be provided by tilting the plane of the target foil with an angle α relative to the beam axis. For the target orientations T_1 and T_2 the angle α in the y-z plane was chosen to be $\pm 30°$. This allows to determine the longitudinal beam polarization P_z^B as well as the transverse beam polarization P_y^B. The target foils for both orientations T_1 and T_2 will be magnetized by a pair of Helmholtz-coils with a magnetic field in beam axis direction. In order to determine the transverse beam polarization P_y^B or P_x^B the plane of the target foil T_3 was aligned perpendicular to the beam axis and can be magnetized in the y- or the x-direction by two other pairs of Helmholtz-coils. The components of the beam polarization can be determined by means of the following equations, where the symbols A^{T_ν} specify the counting rate asymmetries for a reversal of the electron beam polarization for the target foil orientations T_ν. Furthermore, it has been assumed that for

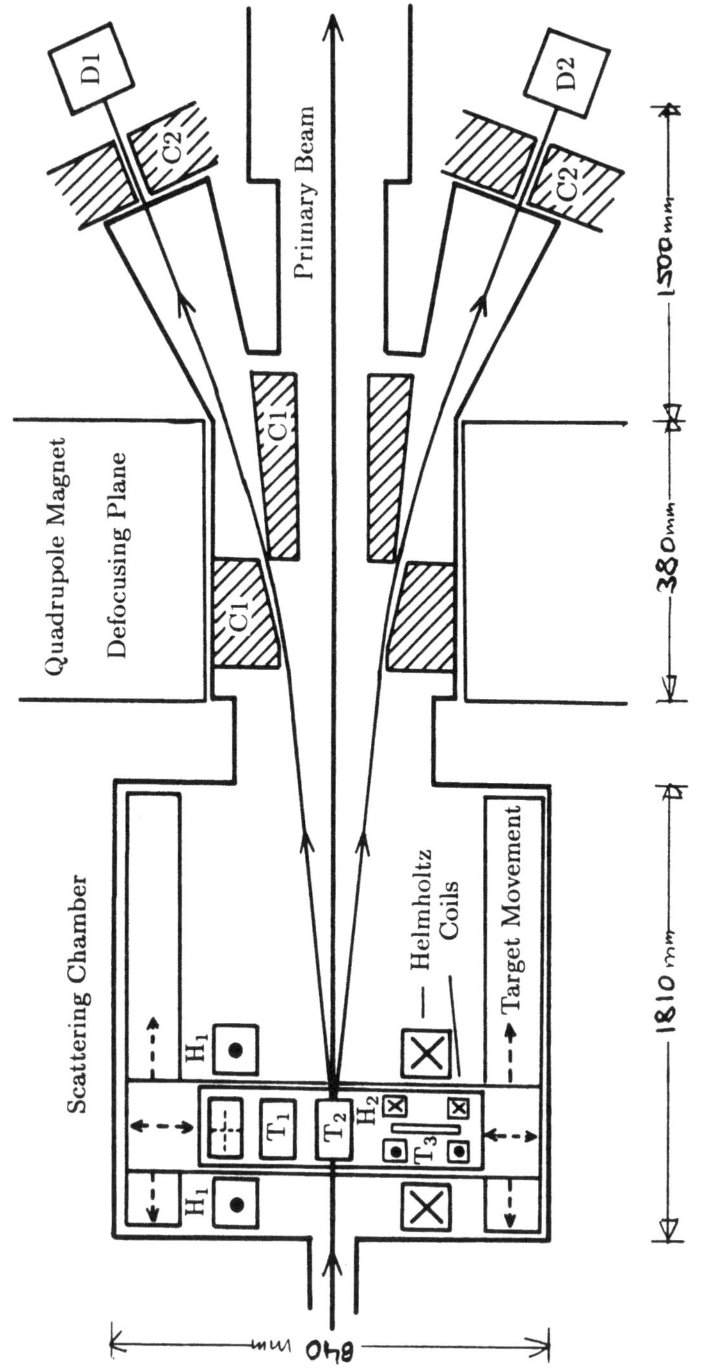

Figure 2: Schematic picture of the Møller Polarimeter for MAMI

both target foil orientations T_1 and T_2 the "in-plane" magnetization P^T will be equal.

$$\begin{aligned} P_z^B &= \frac{A^{T_1} + A^{T_2}}{2a_{zz} P^T \cos\alpha} \\ P_y^B &= \frac{A^{T_1} - A^{T_2}}{2a_{yy} P^T \sin\alpha} \\ P_y^B &= \frac{A^{T_3}}{a_{yy} P_y^{T_3}} \\ P_x^B &= \frac{A^{T_3}}{a_{xx} P_x^{T_3}} \end{aligned}$$

3 Performance Specifications

A polarimeter of similar design, but scaled for the energy range between 25 and 180 MeV was already in operation and has been tested with the Mainz-Linac and with the low energy stage MAMI A. The results of these tests, which are described in detail in reference [1, 3], allow to extrapolate that the new polarimeter should meet the following performance specifications:

With a target foil thickness of 6 μm or 4.5 mg/cm^2 and an electron beam current of 10 μA the longitudinal and transverse polarization components of the electron beam can be determined with a counting statistic limited absolute error of $\Delta P = \pm 0.01$ within a measurement time of 6 sec or 4 min, respectively. It should be possible to determine the longitudinal polarization with a total relative error of 2%, which will mainly be caused by the uncertainty of the target polarization. For transverse beam polarization measurements a total relative error of about 6% can be expected, which will mainly be caused by the target orientation uncertainty.

References

[1] B. Wagner, H.G. Andresen, K.H. Steffens, W. Hartmann, W. Heil, and E. Reichert, Nucl. Instr. and Meth. A294 (1990) 541

[2] B. Wagner, Masters Thesis, Inst. f. Kernphysik, Universität Mainz, FRG (1981) KPH 15/81

[3] B. Wagner, Ph. D. Thesis, Inst. f. Kernphysik, Universität Mainz, FRG (1986) KPH 1/87

Monte Carlo Studies of Compton Scattering for the LEP Laser Polarimeter

G. Barbagli

IROE-CNR and INFN, I-50127 Florence, Italy

Monte Carlo calculations of the Compton scattering of circularly polarized laser photons from transverse polarized electrons have been performed using the design parameters of the LEP laser polarimeter.
Energy and spatial distributions of the backscattered photons are obtained and the influence of some effects, such as spurious components of linear polarization in the laser beam, is considered.
Backscattered photons are propagated through a model of detector consisting of silicon strips shielded by a variable thickness lead absorber.
The expected pulse distribution in the silicon strips is finally predicted.

1 Introduction

Electron and positron beams, circulating in a storage ring, acquire transverse polarization due to the so-called Sokolov-Ternov mechanism [1]: the different transition rates for emission of synchrotron radiation with spin-flip in the two opposite senses lead to an alignment of the electron spin anti-parallel (parallel for positron) to the guide vertical magnetic field. A standard physical process that allows to measure the degree of vertical polarization is Compton scattering of circularly polarized photons from the electron beam. It has been used to build polarization monitors at SPEAR [2], DORIS and PETRA [3], and it is is being used for the LEP laser polarimeter [4] and for HERA too [5].
The Compton cross section contains a spin dependent term that, for vertically polarized electrons and circularly polarized photons, reflects in an up-down asymmetry in the projected spot of backscattered photons. What is experimentally done is to reverse the helicity of the laser beam with some optical device and to compare the two profiles obtained with two different helicities, to extract an asymmetry curve. The peak value of this curve - as well as the shift of the center of gravity between the two profiles - is proportional to the polarization of the electron beam.
In this work I simulate this scattering processes in the physical conditions and with the design parameters of the LEP laser polarimeter [4], in order to predict the experimentally measured distributions.

2 Basic kinematics and cross sections

Compton scattering is the scattering of energetic photons from free electrons. Let E, $\mathbf{p}$ and ω, $\mathbf{k}$ be the energy and momentum of the incoming electron and photon, and E', $\mathbf{p}'$, ω', $\mathbf{k}'$ the energy and momentum of the scattered particles.
From the conservation laws one finds:

$$\frac{\omega'}{\omega} = \frac{1-\beta cos\theta}{1-\beta cos\theta' + \frac{\omega}{E}(1-cos(\theta'-\theta))} \tag{1}$$

β (and γ) have the ususal relativistic meaning.

θ, θ' are the polar angles of the incoming and the scattered photon. From now on let $\hbar = 1$ and $c = 1$.
In the electron rest frame:

$$\frac{\omega'}{\omega} = \frac{1}{1 + \frac{\omega}{m}(1 - cos\theta^*)} \tag{2}$$

In the laboratory frame, choosing $\theta = \pi$:

$$\frac{\omega'}{\omega} = \frac{1+\beta}{1 + \frac{\omega}{E} + (\frac{\omega}{E} - \beta)cos\theta'} \tag{3}$$

The maximum value of the energy is obtained when $\theta' = 0$ (backscattered photons). Assuming $\omega/E << \beta$ this value is given by the equation:

$$\frac{\omega'}{\omega} = \frac{1+\beta}{1 - \beta + \frac{2\omega}{m\gamma}} \simeq \frac{4m\gamma^2}{m + 4\omega\gamma} \tag{4}$$

At LEP: E= 46.525 GeV, ω=2.38 eV, ω'=28.875 GeV.

From the lowest order of QED the cross section for the Compton scattering of polarized photons from polarized electrons follows. In the electron rest frame one gets [6]:

$$\begin{aligned} \frac{d\sigma}{d\Omega} = \frac{1}{2} r_o^2 (\frac{\omega'}{\omega})^2 [\quad & (1 + cos^2\theta) - (\omega' - \omega)(1 - cos\theta) \\ + \quad & \xi_o^1 sin^2\theta \\ - \quad & \xi_o^3 (1 - cos\theta) \zeta \cdot (\mathbf{k} cos\theta + \mathbf{k}')] \end{aligned} \tag{5}$$

θ is the scattering angle. r_o is the electron classsical radius. $\boldsymbol{\zeta}$ the electron polarization vector. ξ_o^1 (ξ_o^2) ξ_o^3 are the Stokes parameters, which enter in the description of polarized photon beams. The density matrix of a photon beam is:

$$\rho = \frac{1}{2} \begin{pmatrix} 1 + \xi^1 & \xi^3 + i\xi^2 \\ \xi^3 - i\xi^2 & 1 - \xi^1 \end{pmatrix} \tag{6}$$

Particularly:
1) ξ_o^1 describes the preponderance of linear polarization along the horizontal axis over the vertical linear polarization;
2) ξ_o^3 describes the circular polarization (helicity).

Integrating over the solid angle one finds the Klein-Nishina formula:

$$\sigma = 2\pi r_o^2 \left\{ \frac{1+\epsilon}{\epsilon^2} \left[\frac{2(1+\epsilon)}{1+2\epsilon} - \frac{1}{\epsilon} ln(1+2\epsilon) - \frac{1+3\epsilon}{(1+2\epsilon)^2} \right] \right\} \tag{7}$$

where $\epsilon = \omega/m$.

3 Calculations and results

The calculations refer to the following parameters, describing the horizontal and vertical betatron motion, and concerning the L.I.R. (Laser Interaction Region) where the collision between electrons and photons takes place:

$\alpha_x=$	0.588			$\alpha_y=$	3.278		
$\beta_x=$	14.995		m	$\beta_y=$	156.576		m
$\epsilon_x=$	36.000	$\cdot 10^{-9}$	m	$\epsilon_y=$	2.160	$\cdot 10^{-9}$	m
$\sigma_x=$	0.735		mm	$\sigma_y=$	0.581		mm

The electron energy is 46.525 GeV. The photon energy is 2.38 eV, corresponding to the Nd-Yag laser chosen for the polarimeter. The distance between the interaction point and the polarimeter is 247 m. Fig. 1 shows a set of pictures of the expected spatial distributions simply got by integrating the scattering cross section in the electron rest frame and boosting to the laboratory frame.They describe, for different cases of electron and photon polarization, the predicted vertical profiles of the backscattered photons and the asymmetry. The two profiles correspond to reversed light polarizations. The asymmetry is defined for each value of the vertical coordinate z by taking the difference of the two profiles and dividing by the sum. Circular polarization reflects in an

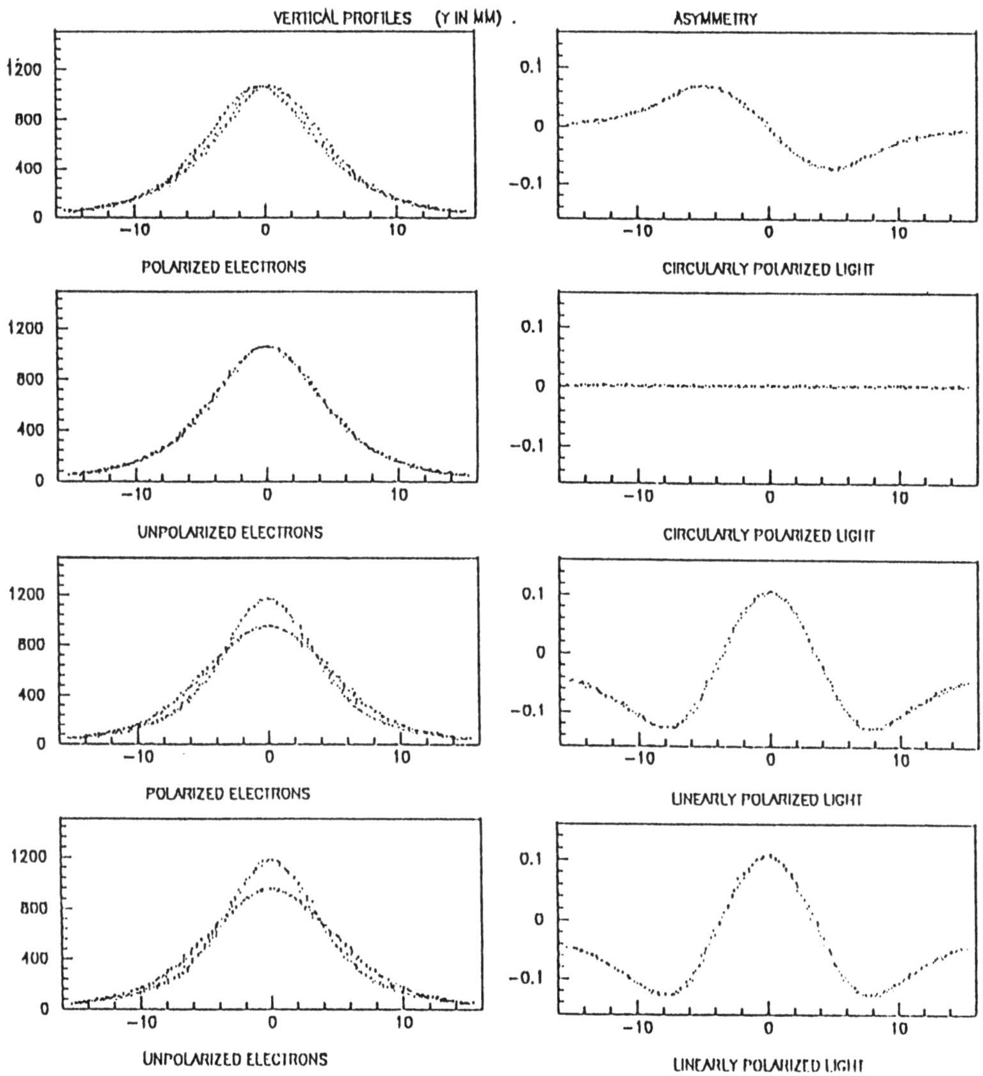

Figure 1: Vertical profiles and asymmetry curves

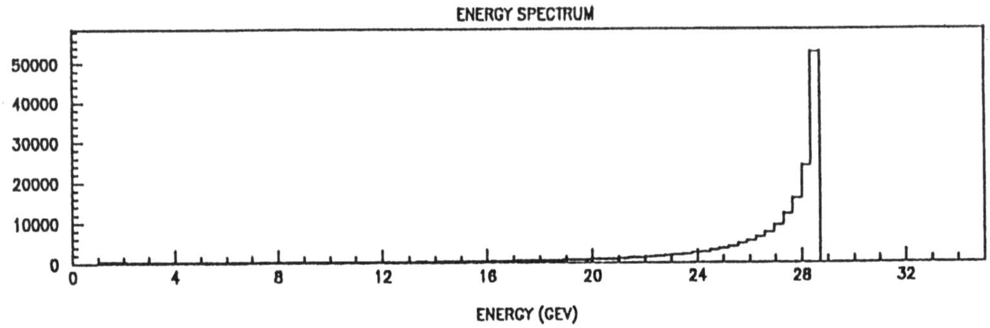

Figure 2: Energy spectrum of scattered photons

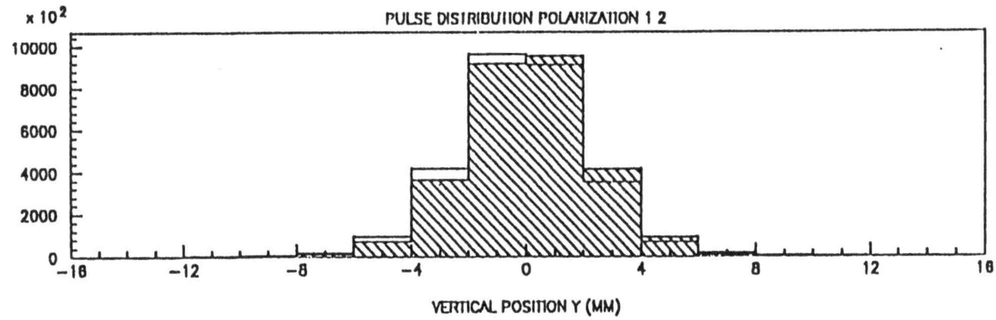

Figure 3: Monte Carlo prediction for the vertical profiles. Empty histogram (filled histogram) correspond to positive (negative) helicity

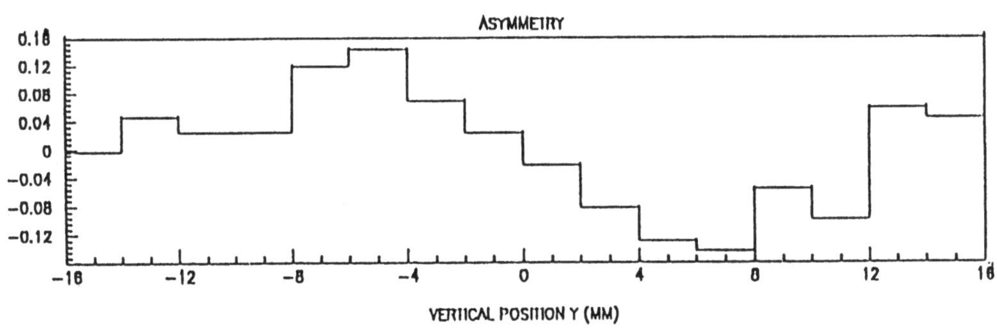

Figure 4: Monte Carlo prediction for the asymmetry

up-down asymmetry whose amount(peak values and also center of gravity shift) is proportional to the vertical electron polarization and disappears for an unpolarized beam. Linear polarization gives a change in the r.m.s. without shifting the mean value and this effect is not dependent on the electron polarization. What happens passing from purely circular polarization to purely linear polarization is that the two profiles move simultaneously towards the origin and their sizes become different, while the asymmetry curve acquires 3 peaks instead of 2, the intermediate being just in the center, where the helicity asymmetry vanished.

Fig 2, 3 and 4 show Monte Carlo distributions, obtained generating scattering events. Fig.2 shows the energy spectrum of the scattered photons, with the kinematical cut at about 28.9 GeV. For each photon of given energy a shower profile, taken from a fit to EGS data, is used in order to weight the vertical distribution with the energy released in the silicon strips. The binning (2 mm) is chosen in agreement with the size of the silicon strips used to sample the electromagnetic tracks distribution after a certain depth of absorber (2.5 radiation lengths of lead). In this way the expected pulse distribution is obtained. Fig. 3 and 4 show, respectively, the predicted pulse distribution in the detector and the corresponding asymmetry. The maximum value of the asymmetry (analyzing power) is about 12%. The shift of the center of gravity is about 420 μm for full electron (and photon) polarization (the expected value from the cross section is about 390 μm).

4 Acknowledgements

I would like to thank M. Placidi who proposed to me this work and gave me hints and suggestions. I am also indebted to B. Dehning and R.Schmidt for stimulating discussions.

5 References

[1] A.A. Sokolov and I.M. Ternov, Sov. Phys. Dokl. 8 (1964) 1203.
[2] D.B. Gustavson et al., Nucl. Instr. and Meth. 165 (1979) 177.
[3] R. Schmidt, Diplom Thesis, DESY Internal Report M-80-04 (1980).
R. Schmidt, Doktor Thesis, DESY Internal Report M-82-22 (1982).
R. Rossmanith and R. Schmidt,DESY Report 84-105 (1984).
[4] M. Placidi and R. Rossmanith, Nucl. Instr. and Meth. 274 (1989) 79.
[5] D. P. Barber et. al., Polarimeters at HERA, DESY-HERA 86-05 (1986).
[6] F.W. Lipps and H.A. Tolhoek, Physica 20 (1954), 85, 395.

List of Participants

Andresen,H.G.	Mainz /Germany
Arianer,J.	Orsay /France
Aulenbacher,K.	Mainz /Germany
Avakian,R.	Yerevan /USSR
Barbagli,G.	Florence /Italy
Barber,D.P.	DESY /Germany
Baum,G.	Bielefeld /Germany
Brissaud,I.	Orsay /France
Clendenin,J.E.	SLAC /U S A
Cords,D.	SLAC /U S A
Drachenfels,W.von	Bonn /Germany
Ertel,M.	Mainz /Germany
Essabaa,S.	Orsay /France
Ester,T.	Muenster /Germany
Fieguth,Th.	SLAC /U S A
Gellrich,A.	Muenster /Germany
Hanne,G.F.	Muenster /Germany
Hartmann,P.	Mainz /Germany

Hillebrecht,U.	Duesseldorf /Germany
Hopster,H.	Irvine /U S A
Kimmel,M.	SLAC /U S A
Kirschner,J.	Berlin /Germany
Kisker,E.	Duesseldorf /Germany
Lampel,G.	Palaiseau /France
Lomperski,M.	DESY /Germany
Maruyama,T.	SLAC /U S A
Mayer,S.	Muenster /Germany
Meier,F.	Zuerich /Switzerland
Merz,H.	Muenster /Germany
Moffeit,K.	SLAC /U S A
Mueller,N.	Bielefeld /Germany
Nakamura,S.	Nagoya /Japan
Nakanishi,T.	Nagoya /Japan
Oestreich,G.	Mainz /Germay
Placidi,M.	CERN /Switzerland
Pluetzer,St.	Mainz /Germany
Prescott,C.Y.	SLAC /U S A
Reichert,E.	Mainz /Germany
Schoch,B.	Bonn /Germany
Schultz,D.	SLAC /U S A
Sinclair,C.K.	CEBAF /U S A
Sorokin,P.V.	Kharkov /USSR

Steffens, K.H.	Mainz /Germany
Voigt, S.	Bonn /Germany
Wehrmeister, D.	Bonn /Germany
Zwickler, S.	Heidelberg /Germany

Part II

Siberian Snakes and Polarization in Circular Machines

Experiments with Siberian Snakes*

A.D. Krisch

Department of Physics, The University of Michigan,
Ann Arbor, MI 48109, USA

Our experimental study of Siberian Snakes began at a 1985 workshop in Ann Arbor on how to have polarized protons at the SSC/1/. The Workshop concluded that it would be practical to have polarized protons at the SSC, but only if the Siberian Snake Concept worked as everyone hoped. Fig. 1 is a diagram of how a polarized SSC might appear with Siberian Snakes installed in the various rings. Thus, the Workshop's first conclusion was that polarized beam acceleration would be possible at the SSC. The second conclusion was that one must test the Siberian Snake Concept, which had been invented around 1974 by Derbenev and Kondratenko/2/. While most of the world's distinguished accelerator theorists were quite convinced that the concept would work, we felt that it would be difficult to convince the SSC Director and the funding people to spend tens of millions of dollars on Siberian Snakes, if it was an untested theoretical concept.

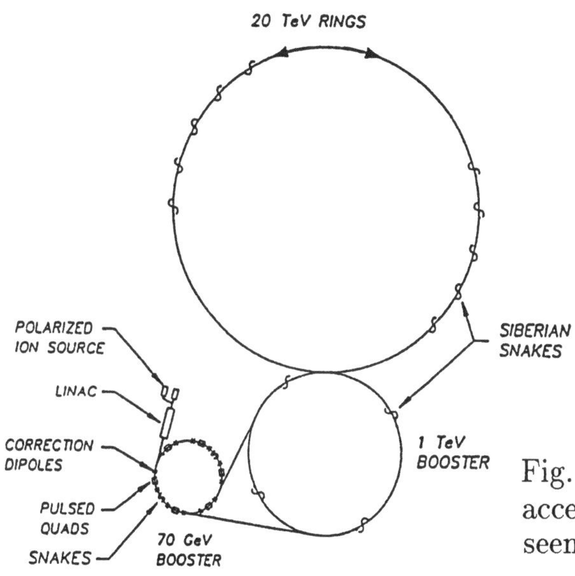

Fig. 1 Polarized proton acceleration at the SSC as seen in 1985.

/ * / Supported by research grants from the U.S. Department of Energy and the U.S. National Science Foundation.

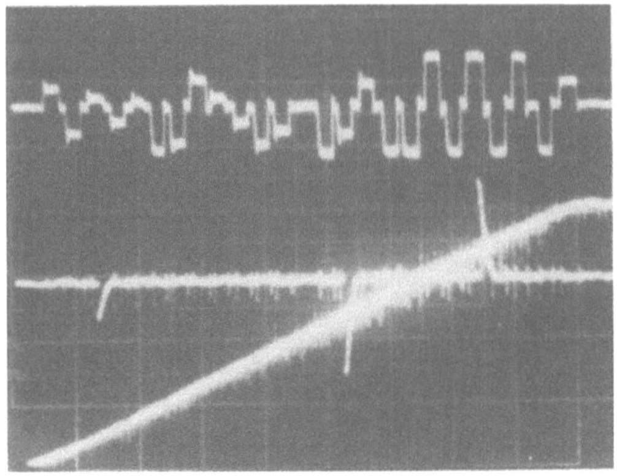

Fig. 2 AGS main control room oscilloscope trace showing the acceleration of polarized protons to 16.5 GeV.

After the Ann Arbor Workshop a number of participants began looking around the world for some accelerator where we could test the Siberian Snake concept. As I recall, we looked seriously at the AGS, at Saturne, at KEK, and at Indiana. We settled on the Indiana Cooler Ring mostly because its straight sections were long enough for a Siberian Snake. We then formed a collaboration and began to study Siberian Snakes. The first results were published during the past year in two Physical Review Letters/3/4/. The collaboration includes people from Michigan, Indiana, and Brookhaven who are listed in reference 3.

The importance of Siberian Snakes is illustrated by the difficulty of accelerating polarized protons to 22 GeV at the AGS, which was the world's highest energy polarized proton beam. One had to individually overcome each of the many intrinsic and imperfection depolarizing resonances. Fig. 2 shows the oscilloscope trace from the AGS main control room; the bottom curve is the AGS magnet cycle during the acceleration of polarized protons to 16.5 GeV. The correction currents in the 96 correction dipoles for some 25 imperfections resonances are shown in the upper curve; the middle curve shows the pulsed quadrupoles firing three times to overcome three intrinsic resonances. To reach 22 GeV we had to overcome about 45 resonances; this required seven weeks of total dedication of the AGS to these polarized beam studies which cost a million dollars each week. At the SSC each 20 TeV ring will have 36,000 depolarizing resonances; it seems completely impractical to try to overcome them individually. It was clear that one needed a new technique to accelerate polarized protons to TeV energies.

A diagram of the Indiana Cooler Ring is shown in Fig. 3. IUCF had the potential for injecting polarized beam into the Cooler Ring from the two stage cyclotron injector; the maximum energy of the second Cyclotron is 200 MeV. The Cooler Ring itself is a synchrotron storage ring, which can accelerate protons up to 500 MeV. The study of Siberian

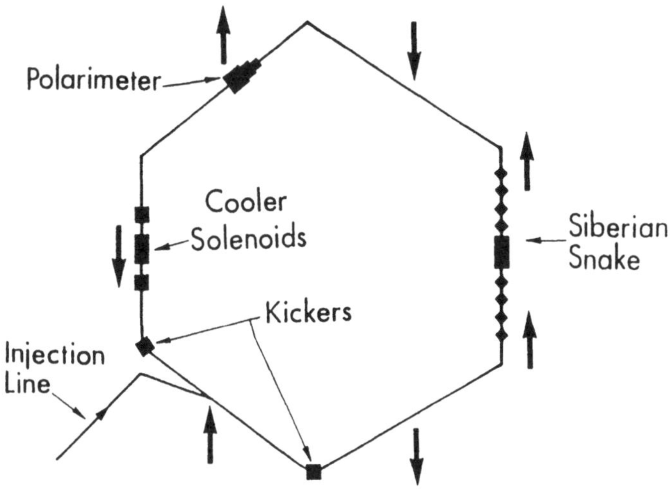

Fig. 3 The IUCF Cooler Ring with the Michigan-Indiana Siberian Snake.

Snakes at the Cooler Ring required several new hardware items. We used the electron cooling solenoids to create imperfection fields to study the depolarizing resonances. To inject polarized protons into the Cooler Ring, we installed some kicker magnets. The Cooler Ring normally used stripping injection of H^- ions; however the polarized source emitted normal polarized protons, which made stripping injection impossible. Two kicker magnets were built quickly and inexpensively at Michigan using some spare ferrite that remained after building the pulsed quadrupoles for the AGS. These kickers had a risetime of about 100 ns and effectively injected the polarized beam into the Cooler Ring.

The polarimeter shown in Fig. 3 was built by Indiana as a spectrometer for various nuclear physics experiments. It had azimuthal symmetry and contained a carbon target, wire chambers, and scintillation counters. We used it as a polarimeter by accepting the $\Delta\theta$ region of about 5° to 17° and splitting it into four $\Delta\phi$ bins of 90° each. The left and right quadrants measured the up-down polarization, while the up-down quadrants measured the sideways polarization. Near the $G\gamma = 2$ depolarizing resonance, which occurs at 108 MeV, the average analyzing power was about 25%. Near the $G\gamma = -3 + \nu_y$ depolarizing resonance at 177 MeV, the cross-section was smaller but the analyzing power was larger. It normally took about 10 minutes to make a 3% measurement of the beam polarization with about 20 nA of stored beam.

The most important piece of hardware was of course the Snake which is shown in Fig. 4. This was not a conventional high energy Snake where 8 transverse dipole magnets produced the 180° spin rotation. Such snakes were not very practical at these low energies; instead we used the superconducting solenoid magnet, shown at the center, which easily rotated the spin by 180°. It was a rather strong solenoid magnet, with an $\int B \cdot d\ell$

Fig. 4 The Michigan-IUCF Siberian Snake.

of about 2 Tesla-meters. Unfortunately, this solenoid caused considerable optical distortion to the low energy beam. This orbit distortion focused the beam and rotated its phase space by about 30° which produced considerable xy-coupling. To compensate for this orbit distortion we used a series of eight quadrupole magnets; the four magnets on the outside were standard quadrupoles, the four on the inside were identical quadrupoles which could be rotated to become skew quadrupoles at various angles. By properly adjusting the currents and the rotation angles of these eight quadrupoles we could eliminate all of the orbit distortions introduced by the solenoid magnet and thus make this straight section optically transparent. The quadrupoles were rather strong and they had to be carefully adjusted, but when they were all properly adjusted the Snake was optically transparent and there was no beam loss.

Our first experiment was a study of the beam polarization at a fixed energy of 120 MeV. This energy was some distance from the $G\gamma = 2$ imperfection resonance at 108 MeV, but the resonance was certainly still strong. With the injection of vertically polarized protons, the measured vertical polarization had a clear peak which looked similar to the imperfection depolarizing resonance correction curves at the ZGS[5] and AGS[6]. The radial polarization was fairly large; it also had a zero crossing and changed sign. We later came to understand that we were rotating the stable spin direction as we varied the longitudinal imperfection field. The imperfection field came from the cooling solenoid which was itself a weak partial Snake that could rotate the stable spin direction.

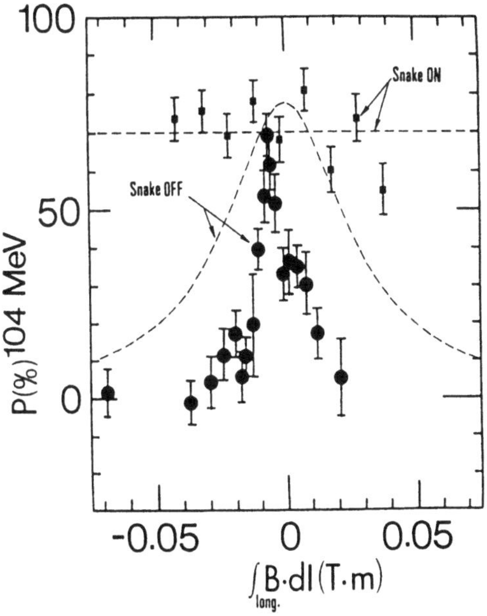

Fig. 5 The 104 MeV polarization in the stable spin directions with the Snake on and the Snake off.

We next did a similar study at 104 MeV with the Snake off as shown in Fig. 5; the curve was now about 10 times sharper than at 120 MeV. We could only maintain full polarization with the Snake off if we exactly corrected the imperfection fields. Any slight imperfection would quickly change the polarization. We next turned the Snake on and injected protons polarized in the radial direction, which was the stable spin direction when the Snake was on. We then found that the polarization remained large and constant as we varied the imperfection field over a wide range. With the Snake on, the polarization appeared totally insensitive to the imperfection fields; the beam remained horizontally polarized even with a strong imperfection field. We considered Fig. 5 to be the first clear indication that the Siberian Snake concept worked.

Our calculation of the polarization at 104 MeV is shown in Fig. 5 as a dashed curve. This calculation, which assumed that the resonance was at 108 MeV, did not agree very well with the Snake-off data. At first we were not sure if the calculation was correct, so we did not take this disagreement too seriously. We were quite pleased that the Snake-on calculation agreed with the Snake-on data which showed that the Snake worked.

During 1989 and 1990 we had eight different running period each lasting about a week. We did many different experiments; I will discuss some of these experiments in an order which is more logical than temporal.

As time progressed our measurement ability improved considerably. We, therefore, remeasured the polarization at 104 MeV in much finer detail as shown in Fig. 6. This curve should have been similar to Fig. 5, but on a much expanded scale. However, with this fine detail, we found

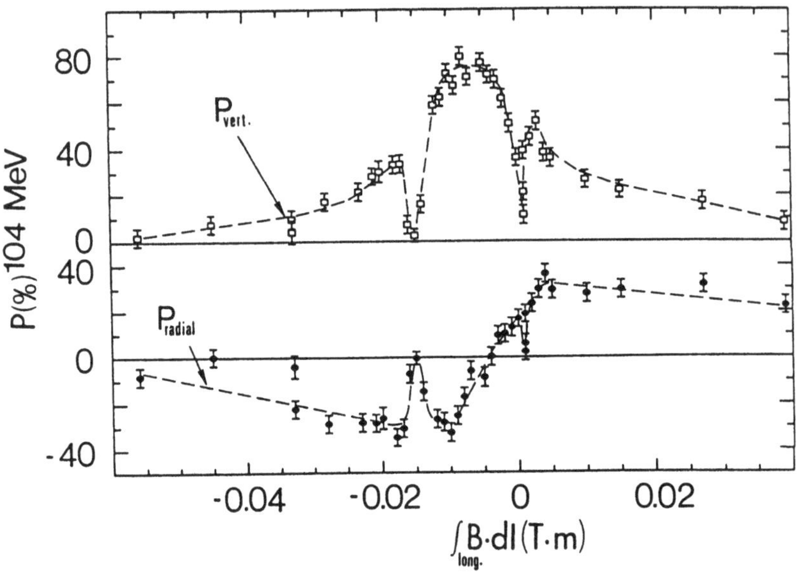

Fig. 6 Observation of synchrotron depolarizing resonances.

a surprise; at first we saw one low data point which we thought was an incorrect measurement. But when we remeasured the curve several more times, there were clear dips on both sides of the central peak. The existence of these dips was further confirmed by identical dips in the radial polarization. Note that the beam was not spin-rotated in these dips; it was fully depolarized because both the vertical and the radial polarizations were zero. We thought that these dips must be due to some type of depolarizing resonance. Eventually we decided that each dip might be a synchrotron depolarizing resonance, which would occur whenever $G\gamma = 2 \pm \nu_s$, where ν_s was the synchrotron tune.

We next tried to prove that each dip was a synchrotron depolarizing resonance by measuring the polarization while changing the synchrotron tune, ν_s. We thought that the easiest way to vary ν_s might be to vary the RF voltage. Therefore, we set $\int B \cdot d\ell$ in the center of the dip on the left and then measured the polarization while varying the RF voltage. As shown in Fig. 7 this experiment worked very well; there was clearly a strong relationship between the polarization and the RF voltage. A synchrotron depolarizing resonance seemed the only explanation for this relationship. It now seems highly probable that each dip in Fig. 6 was due to a synchrotron depolarizing resonance.

We then turned the Snake on with $\int B \cdot d\ell$ again set in the center of the dip; we again varied V_{rf} and thus ν_s. As shown in Fig. 7 the Snake now maintained full polarization over the entire range of V_{rf}. We had not originally considered synchrotron depolarizing resonances, but Fig. 7 showed that a Snake can certainly overcome these resonances.

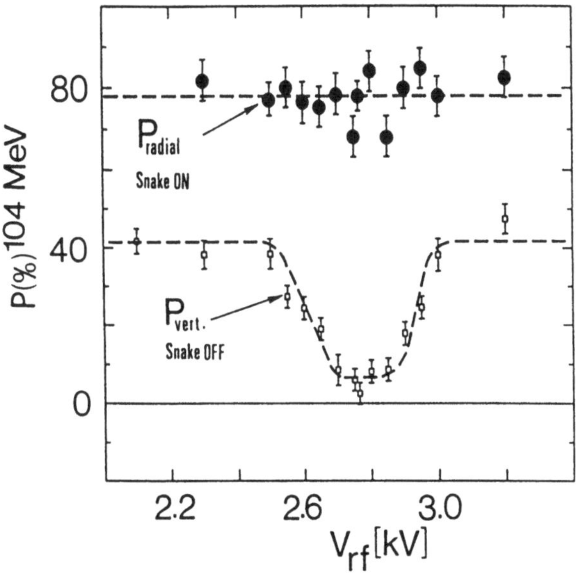

Fig. 7 The Snake overcoming a synchrotron depolarizing resonance

Next we studied the effect of the $G\gamma = 2$ resonance on the polarization at some nearby energies. Recall that the calculation did not quite agree with the data in Fig. 5; we thought that perhaps the energy of the Cooler Ring was not properly calibrated. We recalled that at two previous accelerators, the ZGS/5/ and AGS/6/, the polarized beam studies had discovered an energy miscalibration of about 1%. Our IUCF colleagues were rather concerned about this possibility. Bob Pollock and Hans-Otto Meyer believed that the calibration was correct. Fig. 8 shows a study at 108 MeV which should be near the resonance if the energy calibration was correct; but the width of the resonance curve was quite similar to the width at 104 MeV shown in Fig. 6. This increased the concern about the energy calibration. [This curve was less detailed than Fig. 6, but one could again see the synchrotron resonance dips.]

After studying the 104 and 108 MeV curves we thought that the resonance might be near 106.5 MeV. Therefore, we studied the polarization at 107 MeV which is shown in Fig. 9. This curve was very narrow; it was about a factor of 2 sharper than the 108 MeV curve. Also note that one never reached full polarization in the vertical direction, apparently because 107 MeV was so close to the resonance. Moreover, the synchrotron depolarizing resonances now apparently either disappeared or became very narrow. Thomas Roser and S.Y. Lee believed that as one approached the resonant energy, the two synchrotron resonances should move together, eventually merge, and then cease to exist. [The 107 MeV data was certainly consistent with no synchrotron resonances; but the existence of very narrow dips can never be completely ruled out. However, we had about reached the stability limit of the Cooler Ring; probably we cannot reliably get points much closer together than in Fig. 9.]

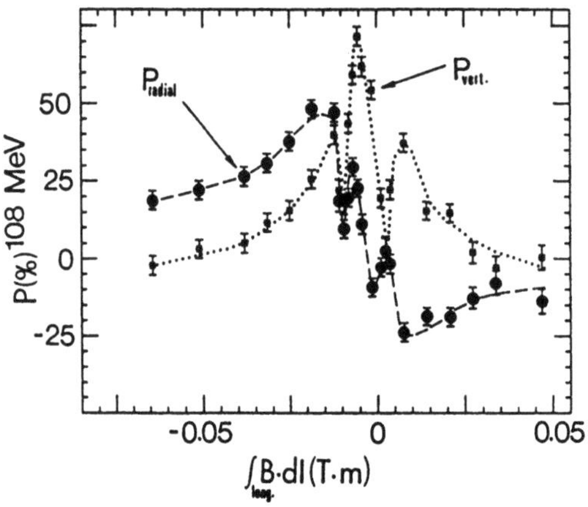

Fig. 8 The $G\gamma = 2$ resonance as seen at 108 MeV.

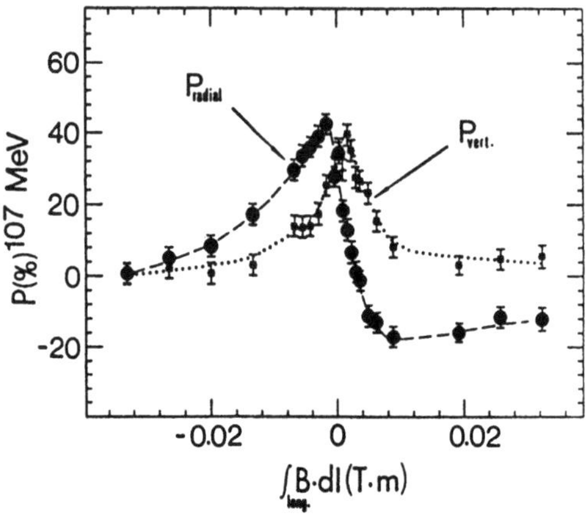

Fig. 9 The $G\gamma = 2$ resonance as seen at 107 MeV.

One of our original goals was to determine if a Snake could overcome an intrinsic depolarizing resonance; Fig. 10 shows the polarization at 177 MeV plotted against the vertical betatron tune, ν_y. The correction quads were not always used; however for this study of the $G\gamma = -3 + \nu_y$ intrinsic resonance we used the correction quads to insure that the vertical betatron tune, ν_y, was exactly right. It was difficult to ramp the Snake because the superconducting solenoid and the quads must then exactly match the Ring ramp. Therefore, S.Y. Lee proposed ramping the tune instead. Normally, during acceleration one holds the tune fixed and ramps the energy; here we held the energy fixed and changed ν_y in discrete steps. We did a measurement at one tune and then we changed the tune for the next measurement. Notice in Fig. 10 that with no Snake, the polarization was zero near the resonance. However, with the Snake on, full polarization was maintained over the entire ν_y range. This clearly demonstrated that the Snake was capable of overcoming a quite strong intrinsic depolarizing resonance.

I will now discuss partial Siberian Snakes. Many people have considered these partial Snakes; probably Derbenev and Kondratenko were first. Courant has been quite involved with them and Roser and Ratner have recently been actively studying partial Snakes. You will soon hear more about them from Ratner and Roser. We made a study of partial Snakes by measuring the polarization at 104 MeV, while lowering the spin rotation angle in the Snake. As shown in Fig. 11, the Snake was able to maintain full polarization from a 100% Snake, down to about a 5% Snake. Near a few percent, the Snake was no longer strong enough and the polarization dropped rapidly. The broad flat region of full polarization seems to be clear evidence that partial Snakes can overcome imperfection resonances.

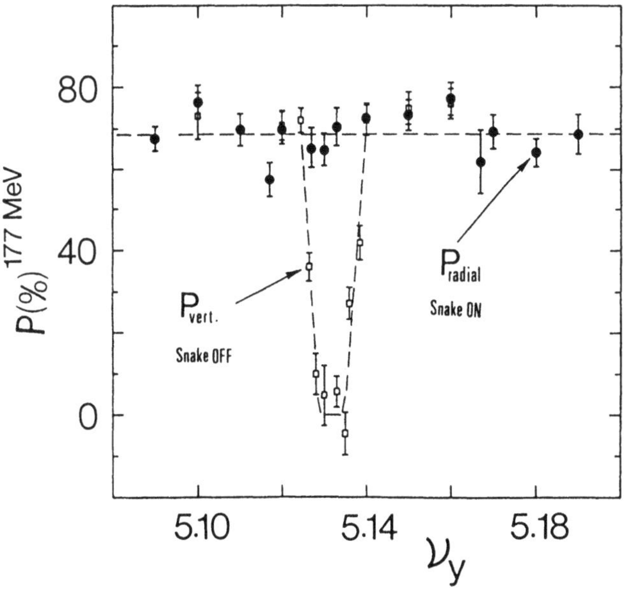

Fig. 10 The Siberian Snake overcoming the $G\gamma = -3 + \nu_y$ intrinsic depolarizing resonance.

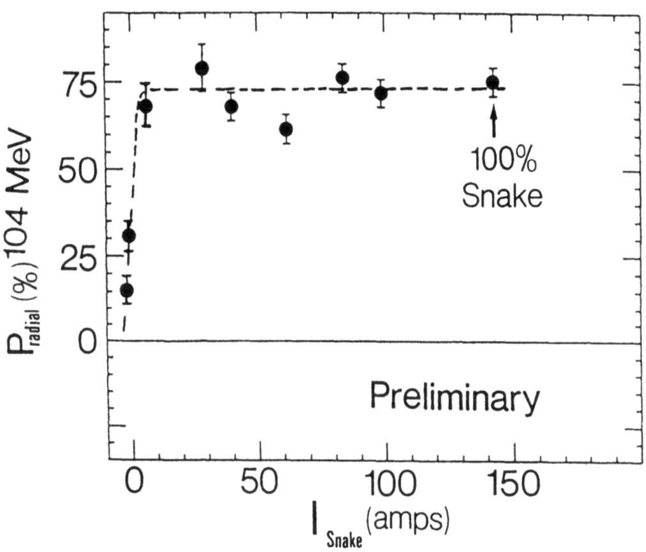

Fig. 11 A partial Snake overcoming the $G\gamma = 2$ intrinsic depolarizing resonance.

We have many more studies of Siberian Snakes and their interactions with various depolarizing resonances. We have about 50 data curves of one sort or another, but many of them are not yet organized. Our thesis student, Michiko Minty, is organizing all of this data. She has compiled

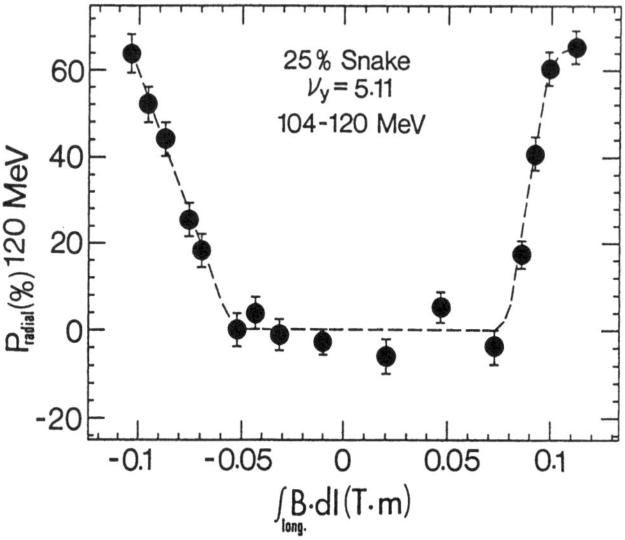

Fig. 12 The 25% partial Snake moving the intrinsic depolarizing resonances into the 104 to 120 MeV range.

a book about one inch thick with many cross references. It is important to keep track of all this data; several times we were almost ready to start some new measurement when we discovered that we had already done it some months earlier.

At this point we had shown that a Siberian Snake was capable of overcoming an imperfection depolarizing resonance, an intrinsic depolarizing resonance, and a synchrotron depolarizing resonance. We had also shown that a partial Snake can overcome intrinsic depolarizing resonances.

Partial Snakes were only predicted to be able to overcome imperfection depolarizing resonances; a full Snake is needed for an intrinsic resonance. I will not give a detailed theoretical discussion; however a partial Snake could apparently shift the spin tune and thus drive an intrinsic resonance to a different energy. We searched for this shift by using a 25% Snake while ramping the Cooler Ring energy from 104 to 120 MeV; we measured the radial polarization while injecting horizontally polarized protons. We changed the total percentage of Snake in the Ring by varying the correction solenoids; they were weak partial Snakes which could vary the percentage of total Snake by a few percent. As shown in Fig. 12 there was a strong depolarization in the predicted region. This seemed to be a clear indication that the spin tune shift in the partial Snake shifted the energy of the $G\gamma = -3 + \nu_y$ intrinsic resonance from 177 MeV to the 104 to 120 MeV ramp. There was another intrinsic resonance, $G\gamma = 7 - \nu_y$, which normally occurred well below injection. However the partial Snake apparently made both these intrinsic resonances occur at the same energy. In any case, we were apparently moving some intrinsic

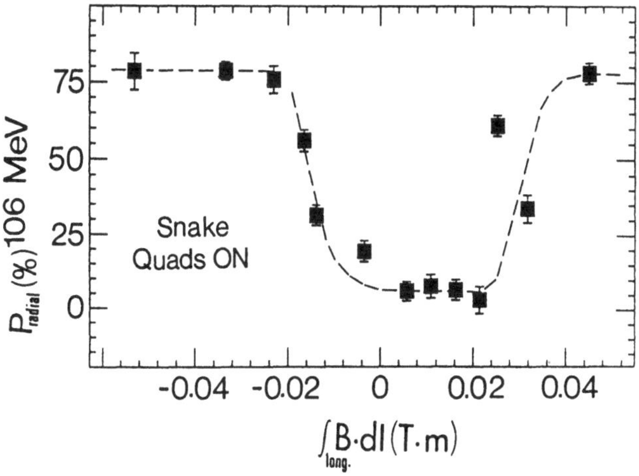

Fig. 13 A possible Snake resonance.

depolarizing resonance into the ramp region of 104–120 MeV; there was certainly a strong depolarization.

We observed a surprising result during a recent run. We found for the first time a depolarizing resonance that the Snake did not overcome. This result is shown in Fig. 13 where we varied the current in the correction solenoids while sitting at 106 MeV near the $G\gamma = 2$ intrinsic resonance. There was a very clear dip in this curve.* We had a 100% Snake turned on, but this was the first time that we ran with the Snake quadrupoles turned on near the $G\gamma = 2$ imperfection resonance.† We had been looking for Snake resonances for some time, but never before saw any depolarization when the Snake was on. Derbenev and I now believe that the depolarization in Fig. 13 was probably due to a Snake resonance which was driven by the Snake quadrupoles.

To understand a Snake resonance, let us first recall how a Snake works. Consider a vertically polarized proton circling a ring containing a Snake. Normally the spin gets slightly rotated during each turn around the ring by the horizontal imperfection fields. On one turn the vertical spin may get rotated from 0° to perhaps 3°; then it passes through the Snake which rotates it by 180° to 183°. On the next turn around the Ring, the imperfections act in the same direction as on the first turn; therefore, they now rotate the spin by 3° back to 180°. Finally the Snake again rotates the spin by 180° bringing it back to 0° after two full

* Do not yet take too seriously the two points on the right which may not agree; one may just be due to an error in recording the data.

† Recall that at 104 MeV the Snake easily overcame the $G\gamma = 2$ resonance, as was shown in Fig. 5. We later made a similar Snake-on study at 108 MeV; there was again no depolarization. However these studies were done with the Snake quadrupoles turned off. Often we did not use the quadrupoles because they were so strong that they had to be finely adjusted or they would kill the beam. The quadrupoles shifted the horizontal tune, ν_x, by a full unit of one.

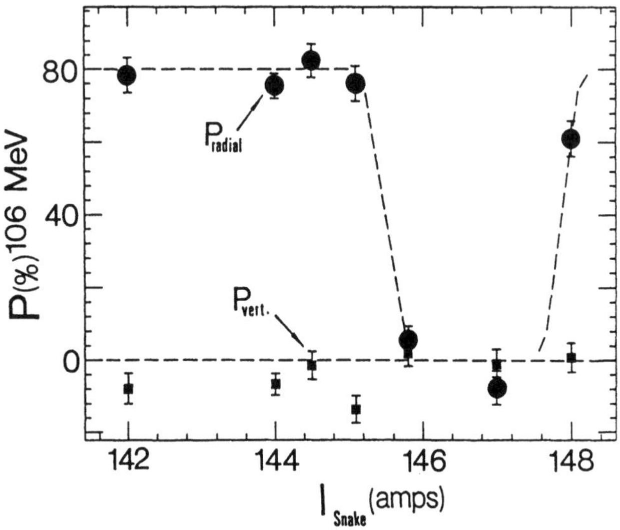

Fig. 14 Studying a Snake resonance by varying the Snake Solenoid.

turns around the ring. Thus, the Snake makes each imperfection field cancel itself. The only exceptions to this cancellation are apparently the magnetic fields inside the Snake itself.

To see how the Snake quadrupoles might drive a Snake resonance, recall that our Snake was not a normal high energy Snake constructed of 8 dipoles; it contained a solenoid and eight quadrupoles. There was a sharp focus at the center of the solenoid so that if a proton in the upstream quadrupole was above the beam axis, then in the downstream quadrupole it would be below the axis. Moreover, the proton's spin was rotated through 180° by the Snake. Finally, the quadrupoles were mirror symmetric. These three factors combined to force the two quadrupoles to have a constructive depolarizing effect on the spin. If in the upstream quadrupole the field pointed to the left and the spin was up, then in the downstream quadrupole the field pointed to the right and the spin was down. The two quadrupoles would therefore interact coherently with the proton's spin and cause a coherent depolarization inside the Snake. We think that the depolarization in Fig. 13 was probably a Snake resonance which was being driven by the Snake quadrupoles; these were much stronger than the normal Cooler Ring quadrupoles. The Snake apparently could not overcome the depolarizing quadrupole fields because they existed inside the Snake itself. The Snake could cure other problems, but it could not cure its own problems. This type of Snake resonance should not occur with a high energy Snake made of eight dipoles.

We tried to study this Snake resonance by slightly varying the strength of the Snake solenoid. As shown in Fig. 14, there was clearly a dip in the polarization as the solenoid current was varied near 144.2

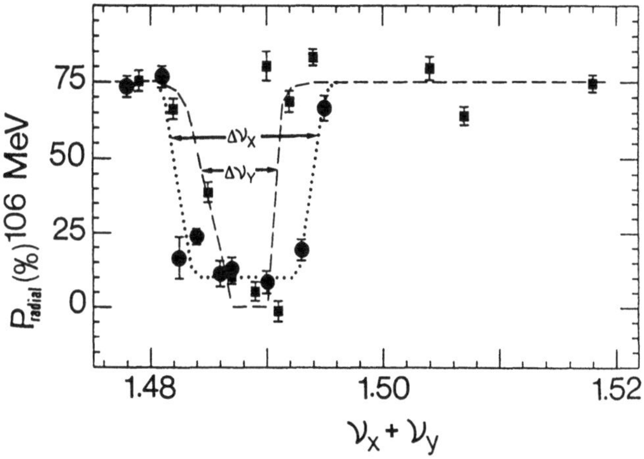

Fig. 15 Studying a Snake resonance by varying ν_x and ν_y.

amps which gives a 100% Snake at 106 MeV. Notice that the width of the dip was about 2%; this may provide information about the nature of the Snake resonance.

We then tried to study the Snake resonance by varying the horizontal and vertical betatron tunes. As shown in Fig. 15, we again saw clear dips in the polarization. During the first data run, where we were changing the vertical tune, we were not careful enough about keeping fixed the horizontal tune. We did obtain good data where we held fixed the vertical betatron tune and varied the horizontal tune. Fig. 15 clearly shows that the resonance depended strongly upon the horizontal betatron tune, ν_x, and probably upon the vertical betatron tune, ν_y. Notice that the dip occurred near $\nu_x + \nu_y \approx 1.50$ for both curves. Since the spin tune, ν_s, was close to one-half, this data showed a strong depolarization near $\nu_x + \nu_y + \nu_s \approx 2$ which is an integer. The shift from an exact integer was probably due to the Type-3 Snake which we will soon discuss. This $\nu_x + \nu_y + \nu_s =$ integer behavior supports the Snake resonance hypothesis. Notice that the width of the dip in the horizontal curve, $\Delta\nu_x$, was considerably larger than the vertical width, $\Delta\nu_y$. These widths may provide important information about the nature of the Snake resonance. In other similar curves with the Snake solenoid on, there was no evidence of depolarization. Thus the unusually strong driving term due to the Snake quadrupoles may be essential for this Snake resonance.

Recall the serious concern about the possibility that the Cooler Ring energy might be miscalibrated. Our data indicated that the $G\gamma = 2$ resonance was occurring near 106.5 MeV, while much data was based upon a well established 108.4 MeV energy calibration for both the Cooler Ring and the Cyclotron. Pollock believed that this 108.4 MeV calibration was correct. Pollock, with some advice by Roser, then proposed that there

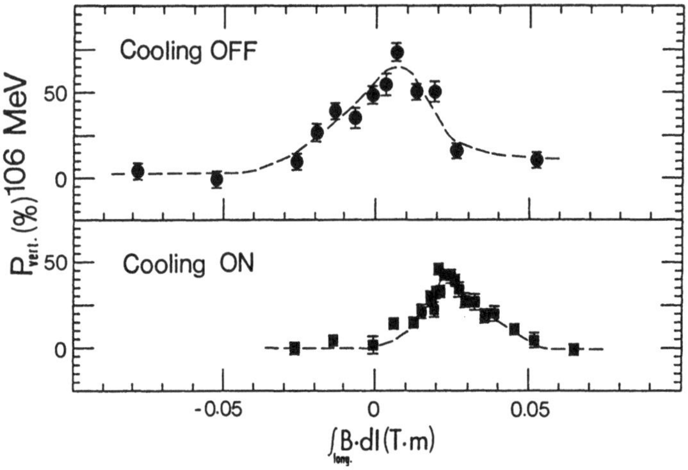

Fig. 16 Measuring the resonance width at 106 MeV with a Type-3 Snake on and off.

Just before our August run, Pollock submitted a paper/7/ proposing that a Type-3 Snake would explain our apparent shift in the energy calibration. He suggested that the Electron Cooler magnet system was a Type-3 Snake, because it contained several solenoids and toroids. Spin operators do not commute, so these magnets could produce a weak Type-3 Snake which could shift the spin tune and thus the resonant energy. I afterwards found a paper by Shatunov and Skrinsky/8/, which never referred to Type-3 Snakes, but stressed that a depolarizing resonance energy calibration is only reliable provided there are no longitudinal fields.

We tested the Type-3 Snake proposal by first measuring the vertical polarization at 106 MeV as a function of $\int B \cdot dl$ with the cooling magnets on; this data is shown in Fig. 16. As expected, we saw a sharp peak. This peak at 106 MeV had a width almost identical to the peak at 107 MeV shown in Fig. 9; this equality gave a calibration of the resonance energy at about 106.5 MeV. Then we turned the cooling system completely off. This made the measurements somewhat more difficult because the beam intensity and life-time both deteriorated; however, we were able to make some measurements. Notice that with the cooling magnets turned off and thus any Type-3 Snake turned off, the peak was much broader; in fact it was quite consistent with the resonance having shifted back up to near 108 MeV. This shift seems to be rather direct evidence for the existence of a weak Type-3 Snake which was accidentally built into the Ring's cooler section. This seems a clever bit of detective work by Pollock and Roser.

At multi-TeV facilities there will be a new problem called overlapping depolarizing resonances; these will occur as the various resonances get stronger and thus wider, while the 523 MeV spacing between the imperfection resonances remains fixed. This overlapping was no problem

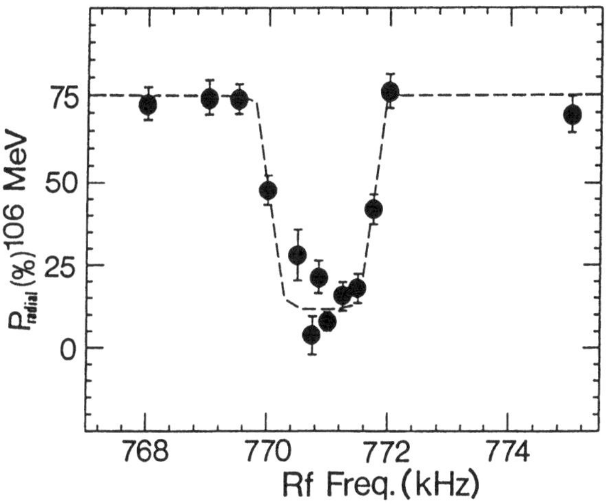

Fig. 17 A depolarizing resonance induced by an rf dipole field.

up to the maximum energy of 22 GeV reached at the AGS. However, at multi-TeV energies, overlapping depolarizing resonances should be common. Experts, such as Courant and Derbenev, believe that Siberian Snakes should be able to overcome the overlapping resonances at the SSC. We plan to experimentally study these overlapping resonances by first creating an induced depolarizing resonance at the Cooler Ring as was done at Novosibirsk/8/. We will then move this induced depolarizing resonance into the region of either the $G\gamma = 2$ or the $G\gamma = -3 + \nu_y$ depolarizing resonance. We are now building a high power RF solenoid to create a strong induced depolarizing resonance that will be made to interfere with each of these two resonances.

During our August run we made a simple first study of an induced depolarizing resonance. We disconnected the RF knock-out system which was normally used for measuring the Cooler Ring betatron tunes; we then connected its power supply to one of our two kicker magnets. This produced a rather weak RF vertical field. With 106 MeV radially injected polarized protons we then turned on the Snake and thus forced the spin tune to be about 1/2. The spin tune was not exactly 1/2 because the cooling was on, which presumably shifted ν_s slightly. Then we varied the RF frequency in an attempt to produce a weak induced depolarizing resonance. As shown in Fig. 17. this study was quite successful. We saw a clear depolarizing resonance when we varied the frequency around the calculated resonant frequency. We estimate that the center of the dip was about 771.0 ± 0.2 kHz; thus the relative error was about $3 \cdot 10^{-4}$. This seems a very precise way to directly measure the spin tune with a precision of about $3 \cdot 10^{-4}$.

Recall that the long term goal of our Siberian Snake studies was to accelerate polarized protons to TeV energies at some facilities such as the SSC. Our SPIN collaboration recently submitted a preliminary proposal to install Siberian Snakes in the SSC and then accelerate polarized protons to 2 TeV and then to 20 TeV. At the June PAC meeting at SSC our collaboration contained 38 people from Michigan, Protvino and Dubna; recently 28 additional accelerator physicists and experimenters from Indiana, Moscow, KEK, and Kyoto joined our SPIN collaboration. We have had considerable interaction with the SSC people and they now seem rather interested in polarized proton beams but certainly not yet committed. This SSC project could be the first significant application of Siberian Snakes to a very high energy facility. Other possibilities include using Siberian Snakes to accelerate polarized protons at the Fermilab Main Ring, at RHIC, or at the 400 GeV and 3 TeV UNK rings.

/1/ Proc. of the 1985 Ann Arbor Workshop on Polarized Protons at the SSC, eds. A. D. Krisch, A.M.T. Lin, and O. Chamberlain, AIP Conf. Proc., **145** (New York, 1986).

/2/ Ya. S. Derbenev and A. M. Kondratenko, JETP, **35**, 230 (1972); Part. Accel., **8**, 115 (1978).

/3/ A. D. Krisch, S. R. Mane, R. S. Raymond, T. Roser, J. A. Stewart, K. M. Terwilliger, J. E. Goodwin, H-O. Meyer, M. G. Minty, P. V. Pancella, R. E. Pollock, T. Rinckel, M. A. Ross, F. Sperisen, E. J. Stephenson, E. D. Courant, S. Y. Lee, and L. G. Ratner, Phys. Rev. Letts., **63**, 1137 (1989).

/4/ J. E. Goodwin, et al., Phys. Rev. Lett., **64**, 2779 (1990).

/5/ T. Khoe et al., Part. Accel., **6**, 213 (1975).

/6/ F. Z. Khiari et al., Phys. Rev., **D39**, 45 (1989).

/7/ R. E. Pollock, Nucl. Instrum. and Meth., (to be published).

/8/ Yu. M. Shatunov and A. N. Skrinsky, Part. World, **1**, 35 (1989).

A Partial Snake for the AGS*

L.G. Ratner

Brookhaven National Laboratory, Associated Universities, Inc.,
Upton, Long Island, NY 11789, USA

Based on snake experiments at the Indiana University Cyclotron Facility and computer simulations at Brookhaven National Laboratory, as well as the conclusions of a BNL mini-workshop, we feel that a partial Siberian snake is a practical device for the AGS. It is anticipated that such a device could reduce the polarized beam tune-up time from 2-3 weeks to 2-3 days.

Requirements:

1. Fit into a 10′ straight section with 6" i.d.
2. Have sufficient strength to correct all imperfection resonances up to $G\gamma = 48$ (above RHIC transition and a "magic" energy which preserves beam polarization in the AGS to RHIC transfer line).
3. Turn on before or at $G\gamma = 6\text{-}1/2$ and off at $G\gamma = 47\text{-}1/2$ on $\nu_s = 1/2$ flattops or ramp between flattops. Nominally take less than 10% of duty cycle (300 msec) for manipulations. Not an absolute requirement. Other scenarios are possible depending on how much duty cycle one is willing to give up.

Solution:

1. Use an air-core solenoid. No orbit distortions and easily rampable in short times. Yields good duty cycle and low power consumption.
2. Choose $L_{eff} = 90"$ (leave about 10" on each side for fittings and another 5" for flanges).
3. $$\int_{\theta_p=\pi} B d\ell = 3.752\ (P_{GeV/c}) = 93.2 \text{ T-M for } 24.84 \text{ GeV/c.}$$
4. Determine resonance strength from previous polarized proton runs at the AGS and select suitable precession angle.

From Table I we see that a 5° snake can work up to 22 GeV/c with $P_f/P_o \sim 0.76$ and a 6° snake would give $P_f/P_o \sim 1.0$. To get to RHIC injection energy requires an 8° snake. We feel that it would be useful to

*Work performed under the auspices of the U.S. Department of Energy.

TABLE I

AGS Resonance Strengths and (P_f/P_o) for Various Strength Snakes

Gγ	P(GeV/c)	AGS ε	P_f/P_o 4°	5°	6°	7°	8°	NOTE: π = product of all preceeding P_f/P_o
7		.0008	.99	1.00	1.00	1.00		
8		.0027	.93	1.00	1.00	1.00		
9		.0045	.78	.965	1.00	1.00		
10		.0011	.98	1.00	1.00	1.00		
11		.0013	.97	1.00	1.00	1.00		
12		.0045	.78	.965	1.00	1.00		
13		.0013	.98	1.00	1.00	1.00		
13-1/2	7.00	--	π=.52		1.00	1.00		
14		.0008	.0008	1.00	1.00	1.00		
15		.0007	.0007	1.00	1.00	1.00		
16		.0010	.0010	1.00	1.00	1.00		
17		.0007	.0007	1.00	1.00	1.00		
18		.0005	.0005	1.00	1.00	1.00		
19		.0005	.0005	1.00	1.00	1.00		
20		.0030	π=.40		1.00	1.00		
20-1/2	10.69	--		1.00	1.00	1.00		
21		.0009		1.00	1.00	1.00		
22		.0010		1.00	1.00	1.00		
23		.0020		1.00	1.00	1.00		
24		.0012		1.00	1.00	1.00		
25		.0026		1.00	1.00	1.00		
26		.0009		1.00	1.00	1.00		
27		.0026		1.00	1.00	1.00		
27-1/2	14.36	--		π=.93		1.00		
28		.0059		.895	.995	1.00		
29		.0030		1.00	1.00	1.00		
30		.0004		1.00	1.00	1.00		
31		.0007		1.00	1.00	1.00		
32		.0009		1.00	1.00	1.00		
33		.0011		1.00	1.00	1.00		
34		.0029		1.00	1.00	1.00		
35	18.30	.0024		1.00	1.00	1.00		↑ 1988 values for ε
36		.0040		.98	1.00	1.00		↓ 1986 values for ε
36-1/2	19.08	--		π=.82				
37		.0022		1.00	1.00	1.00		
38		.0025		1.00	1.00	1.00		
39		.0054		.93	1.00	1.00		
40		.0024		1.00	1.00	1.00		
41		.0008		1.00	1.00	1.00		
42		.0029		1.00	1.00	1.00		
42-1/2	21.96	--		π=.76	π=.995	π=1.00		↓ Comparison of ε
43		.0055		.92	1.00	1.00	1.00	measured from Gγ=
44		.0134		0	0	.655	.94	36-42 with a
45		.0126		0	.10	.79	.96	Courant calcula-
46		.0059		.92	.995	1.00	1.00	tion for RMS=3.0
47		.0108		0	.64	.96	1.00	mm gave ϵ_M=.37 ×
47-1/2	24.84	--		π=0	π=0	π=.46	π=.90	ϵ_{calc}.

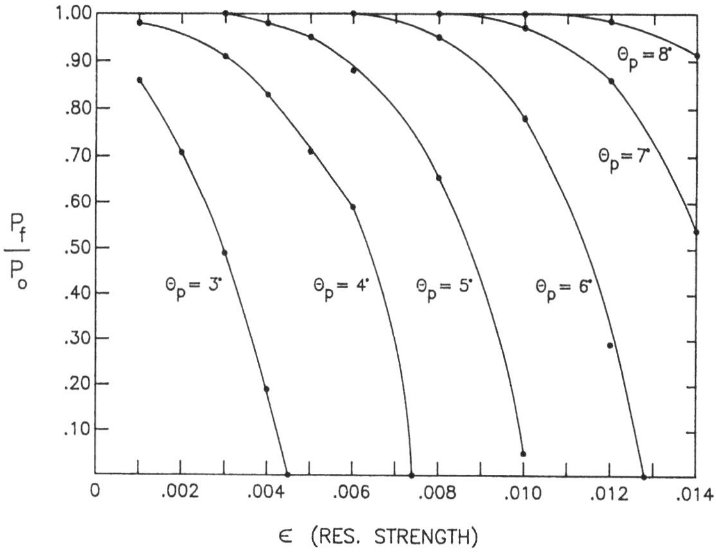

Figure 1: Resonance strengths vs. polarization survival (P_f/P_o) from computer simulations for several different strength snakes.

design a solenoid that can reach 8° or more, but to power it to lower levels from readily available power supplies. This is cost effective and provides an opportunity to do machine studies with partial snakes at energies which are not particularly desirable for physics experiments and/or RHIC.

At the AGS, there are several 4.8 kA 125 V power supplies and after a few iterations we have come up with the following design making use of them.

$$\int_{\theta_p=\pi} B d\ell = 3.752 \text{ P T-M} \qquad \int_{\theta_p=8°} B d\ell = 4.142 \text{ T-M for P} = 24.84 \text{ GeV/c}$$

Choose L = 90" = 2.286 M, B = 1.812 T.
Then, NI ≈ 3.4 x 10^6 amp-turns.

Using OTOKUMPO METALS #6810 29 mm x 29 mm x 16.2ϕ with 0.02" insulation wrap, we can get 76 turns per layer in 90". Use four layers, 304 turns.

$\ell_{\text{Longest layer}} = 260.4'$ $\ell_{\text{Shortest layer}} = 147.9'$ $\bar{\ell} = 204.15'$
$\bar{\ell}_{\text{Total}} = 817'$

$R_L = 2.25$ mΩ $R_S = 1.28$ mΩ $\bar{R} = 1.76$ mΩ $\bar{R}_{\text{Total}} = 7.04$ mΩ

Ramp power supply in 0.6 sec to 9.6 kA.
Average current 4.8 kA.

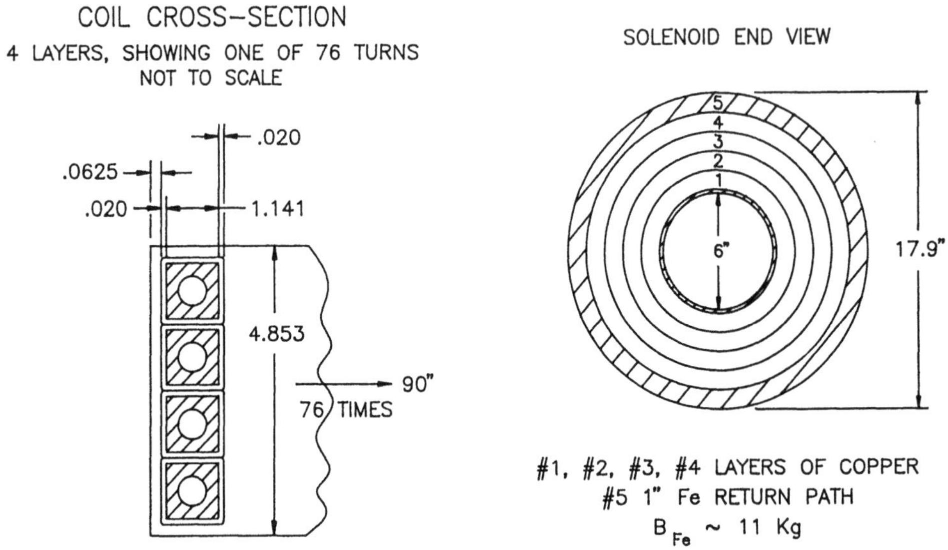

Figure 2: Schematic of coil and end view--not to scale.

$E_{Longest} = 10.8$ V $\quad P_L \approx 52$ kW $\quad$ Duty Cycle = 0.6/3 = 0.2.

$\overline{P}_L = 10.4$ kW in longest circuit ⇒ 40 GPM/°C rise.

Allow 8° rise then need 5 GPM = q.

Total length = 817′for four layers, so pressure required:

$$\Delta P = \frac{817}{100} \times 4.84 \times 10^{-2} \; \frac{q^{1.85}}{d^{4.8655}}, \quad d = 16.2 \text{ mm} = 0.638''$$

$$\Delta P = 3.53\, q^{1.85} = 69.3 \text{ psi for all four layers in series.}$$

Total power $\overline{P} = \overline{R}_T\, \overline{I}^2 \times 0.2 = 7.06\ \text{m}\Omega \times 4.8^2\ \text{kA} \times 0.2 \approx 32.5$ kW and 5 GPM gives an approximately 25°C rise.

Inductance $L = \dfrac{0.32\, a^2 n^2}{6a + 9b + 10c}$ μH, where a = 13.943 cm; n = 304 turns;

b = 228.5 cm; c = 12.327 cm.

L = 5.653 mH.

With one power supply, could operate the AGS to 16.9 GeV/c passing through many resonances as well as transition. This is lower than desired for physics runs. AGS physics runs likely at 18.5 and 22 GeV/c. So, with two power supplies at 6.3 kA could run program. RHIC would probably want a new supply.

TABLE II

Current Required for Different AGS Conditions

I	Number of Power Supplies	θ_p	P_{AGS}	P_f/P_o
4.8 kA	1	5°	20.3 GeV/c	0.82
4.8 kA	1	6°	16.9 GeV/c	1.00
6.3 kA	2 in parallel	6°	22.0 GeV/c	1.00
8.3 kA	2 in parallel	7°	24.8 GeV/c	0.46
9.5 kA	2 in parallel	8°	24.8 GeV/c	0.90
10.7 kA	Different P.S.	9°	24.8 GeV/c	1.00

At 6.3 kA $E_{max} = I_R + L\, dI/dt = 104$ V with a 0.6 sec rise time. At 8.3 kA for RHIC, would be voltage limited at about a 1 sec rise time.

We feel that this design will enable us to do a lot of machine studies and to prepare for future physics runs and RHIC at a modest cost. The solenoid and power supply could be commissioned for less than $100,000.

An Operational Description of the Compact and Tunable LEP Spin Rotator

T. Fieguth[1,*] *and G. Guignard*[2]

[1]Stanford Linear Accelerator Center, Stanford University, Stanford, CA 94309, USA
[2]CERN, CH-1211 Geneva 23, Switzerland

Special configurations of horizontal and vertical bending magnets can rotate the spin vector from the vertical to the longitudinal direction. The compact and tunable LEP spin rotator discovered recently fits neatly into the list of such configurations that generate a group of 24 rotation operators. It is shown that any of these configurations can be achieved with a string of 8 simple dipoles. Results can, in principle, be applied to electron or proton beams.

1. Introduction

It has been shown[1] that it is possible to find special configurations of horizontal and vertical bending magnets which will rotate the spin polarization vector from the vertical (or horizontal) direction to the longitudinal direction and at the same time have the following important attributes.

1. The configuration can be inserted into a drift region and will not change the overall direction of the beam.
2. The operation performed on the spin polarization vector is independent of the energy of the beam.
3. The configurations do not induce dispersion in the regions outside of the region of the configuration itself.

In Ref. 1, it was also shown that these configurations and combinations of them could generate a group of 24 operators isomorphic to the group of rotational symmetries of a cube (Octahedral Group).

Recently an investigation[2] of configurations of this type lead to the discovery of one previously unreported which also is the most economic in that it requires less integrated field than those previously known. When the operation which it performs on the three components of the spin vector was examined it was found that it neatly provided the two operations needed to *complete the list of simple magnet configurations* which rotate vertical spin into longitudinal spin. This new configuration is called the Compact LEP Rotator.

⋆ Work supported in part by Department of Energy contract DE–AC03–76SF00515 while visiting CERN.

2. Operation of the Compact LEP Rotator

To describe the operation of the new Vertical Compact LEP Rotator (**VLR**) on the components of the spin polarization vector we use the same conventions described in Ref. 1 where $\mathbf{V}(+45°)$ and $\mathbf{H}(+90°)$ denote the *orthogonal (3 × 3) matrix operators* representing precession of +45° or +90° in vertical or horizontal bending magnets. The new VLR configuration of magnets is shown in Fig. 1, where only its action on the vertical component of the initial spin vector is shown. The overall operation of this configuration on *all of the components of the spin vector* **p** is obtained by the sequential application of the operator for each bend magnet.

$$\mathbf{p}' = \mathbf{V}(+45°)\mathbf{H}(+90°)\mathbf{V}(-45°)\mathbf{H}(-180°)\mathbf{V}(-45°)\mathbf{H}(+90°)\mathbf{V}(+45°)\ \mathbf{p} \quad . \tag{2.1}$$

We will represent this configuration by the notation, $\mathbf{VLR}(+45°, +90°)$ and rewrite Eq. (2.1) as

$$\mathbf{p}' = \mathbf{VLR}(+45°, +90°)\ \mathbf{p} \quad . \tag{2.2}$$

which after the matrix calculation becomes,

$$\mathbf{p}' = \begin{pmatrix} 1 & 0 & 0 \\ 0 & 0 & -1 \\ 0 & 1 & 0 \end{pmatrix} \begin{pmatrix} H \\ V \\ S \end{pmatrix} \quad \text{or} \quad \mathbf{p}' = \begin{pmatrix} H \\ -S \\ V \end{pmatrix} \quad . \tag{2.3}$$

As expected, we note that the vertical component V of the initial vector has now been rotated to become the longitudinal component. The matrix operator is described in Ref. 1 and given an identifying label $\mathbf{E}_3$. We could now write

$$\mathbf{p}' = \mathbf{E}_3\ \mathbf{p} \tag{2.4}$$

The operator $\mathbf{E}_3$ is but one of 24 operators representing the group of rotational symmetries of a cube. In Ref. 1, all these operators are described in both matrix form and in a second more compact form whereby the eigenvector (rotational axis) is calculated and specified along with a rotational angle about that axis. Thus the operator $\mathbf{E}_3$ represents a rotation of +90° about the unit vector $+\mathbf{i}$ which is along the horizontal axis. The complete list of all such operators is given in Table I.

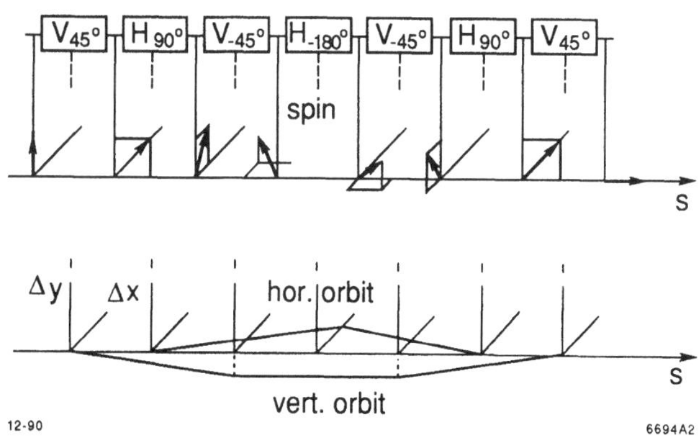

Figure 1: The 7 **VLR** magnets, orbit and action on vertical spin component.

Table I: Summary of the 24 operators of the Octahedral Group with respective rotation angles and eigenvectors (rotation axes).

$\mathbf{A}_1, \mathbf{A}_1^{-1}$	$+120°, -120°$	$\mathbf{a}_1 = 1/3(+\sqrt{3}\mathbf{i} + \sqrt{3}\mathbf{j} + \sqrt{3}\mathbf{k})$
$\mathbf{A}_2, \mathbf{A}_2^{-1}$	$+120°, -120°$	$\mathbf{a}_2 = 1/3(+\sqrt{3}\mathbf{i} - \sqrt{3}\mathbf{j} - \sqrt{3}\mathbf{k})$
$\mathbf{A}_3, \mathbf{A}_3^{-1}$	$+120°, -120°$	$\mathbf{a}_3 = 1/3(-\sqrt{3}\mathbf{i} + \sqrt{3}\mathbf{j} - \sqrt{3}\mathbf{k})$
$\mathbf{A}_4, \mathbf{A}_4^{-1}$	$+120°, -120°$	$\mathbf{a}_4 = 1/3(-\sqrt{3}\mathbf{i} - \sqrt{3}\mathbf{j} + \sqrt{3}\mathbf{k})$
$\mathbf{C}_1$	$180°$	$\mathbf{c}_1 = \pm\mathbf{k}$
$\mathbf{C}_2$	$180°$	$\mathbf{c}_2 = \pm\mathbf{i}$
$\mathbf{C}_3$	$180°$	$\mathbf{c}_3 = \pm\mathbf{j}$
$\mathbf{I}$	$0°$	
$\mathbf{D}_1$	$180°$	$\mathbf{d}_1 = 1/2(\pm\sqrt{2}\mathbf{i} \mp \sqrt{2}\mathbf{k})$
$\mathbf{D}_2$	$180°$	$\mathbf{d}_2 = 1/2(\pm\sqrt{2}\mathbf{i} \pm \sqrt{2}\mathbf{k})$
$\mathbf{D}_3, \mathbf{D}_3^{-1}$	$+90°, -90°$	$\mathbf{d}_3 = +\mathbf{j}$
$\mathbf{E}_1$	$180°$	$\mathbf{e}_1 = 1/2(\pm\sqrt{2}\mathbf{j} \mp \sqrt{2}\mathbf{k})$
$\mathbf{E}_2$	$180°$	$\mathbf{e}_2 = 1/2(\pm\sqrt{2}\mathbf{j} \pm \sqrt{2}\mathbf{k})$
$\mathbf{E}_3, \mathbf{E}_3^{-1}$	$+90°, -90°$	$\mathbf{e}_3 = +\mathbf{i}$
$\mathbf{S}_1$	$180°$	$\mathbf{s}_1 = 1/2(\pm\sqrt{2}\mathbf{i} \pm \sqrt{2}\mathbf{j})$
$\mathbf{S}_2$	$180°$	$\mathbf{s}_2 = 1/2(\pm\sqrt{2}\mathbf{i} \mp \sqrt{2}\mathbf{j})$
$\mathbf{S}_3, \mathbf{S}_3^{-1}$	$+90°, -90°$	$\mathbf{s}_3 = +\mathbf{k}$

By permuting the field directions and recalculating we obtain:

$$\mathbf{VLR}(+45°, \pm 90°) = \begin{pmatrix} 1 & 0 & 0 \\ 0 & 0 & -1 \\ 0 & 1 & 0 \end{pmatrix} = \mathbf{E}_3 \qquad \begin{pmatrix} H \\ V \\ S \end{pmatrix} \Longrightarrow \begin{pmatrix} H \\ -S \\ V \end{pmatrix}$$
$$\mathbf{VLR}(-45°, \pm 90°) = \begin{pmatrix} 1 & 0 & 0 \\ 0 & 0 & 1 \\ 0 & -1 & 0 \end{pmatrix} = \mathbf{E}_3^{-1} \qquad \begin{pmatrix} H \\ V \\ S \end{pmatrix} \Longrightarrow \begin{pmatrix} H \\ S \\ -V \end{pmatrix} \tag{2.5}$$

where the operation $\mathbf{E}_3^{-1}$ is also shown in Table I.

The entire ensemble of magnets in Fig. 1 can be rolled about the beam axis by 90°. We describe this action by calling this configuration the Horizontal Compact LEP Rotator **HLR** and note that the operation upon the spin polarization vector is given by the operators $\mathbf{D}_3$ and $\mathbf{D}_3^{-1}$ in Table I.

3. Generation of the remaining symmetry operators

It was shown in Sec. 2 how the $\mathbf{VLR}(\pm 45°, \pm 90°)$ and $\mathbf{HLR}(\pm 90°, \pm 45°)$ configurations are associated with the operators $\mathbf{E}_3$, $\mathbf{E}_3^{-1}$, $\mathbf{D}_3$, and $\mathbf{D}_3^{-1}$. Reference 1 also describes similar correspondences between the serpents denoted as $\mathbf{V}(\pm\psi, \pm\theta)$ or $\mathbf{H}(\pm\psi, \pm\theta)$ with the 8 operators identified by the letter **A** and the two snake operators $\mathbf{S}_3$ and $\mathbf{S}_3^{-1}$, the Left-Right Snakes $\mathbf{LR}(\pm\psi, \pm\theta)$ with the operators $\mathbf{D}_1$ and $\mathbf{D}_2$, and

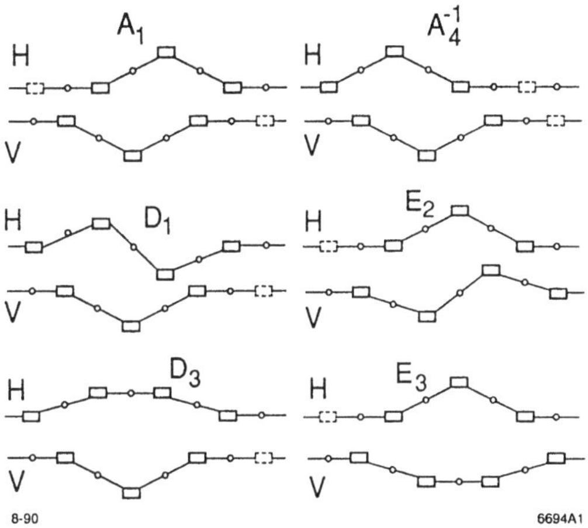

Figure 2: Examples of eight magnets configured to perform different operations.

half of an Up-Down snake termed $\mathbf{UD}(\pm\psi, \pm\theta)$ with the corresponding operators $\mathbf{E}_1$ and $\mathbf{E}_2$. By combining some of these configurations in sequence the remaining operators $\mathbf{C}_1$, $\mathbf{C}_2$, $\mathbf{C}_3$, $\mathbf{S}_1$ and $\mathbf{S}_2$ can also be generated along with $\mathbf{E}_3$, $\mathbf{E}_3^{-1}$, $\mathbf{D}_3$, and $\mathbf{D}_3^{-1}$, the four operators which we now find can be obtained more simply without the need of sequential configurations.

A brief summary of these configurations and the corresponding operators cited in Ref. 1 is recalled here:

$$\mathbf{V}(\pm 45^\circ, \mp 45^\circ) = \mathbf{H}(\pm 45^\circ, \pm 45^\circ) = \mathbf{S}_3 \qquad \mathbf{V}(\pm 45^\circ, \pm 45^\circ) = \mathbf{H}(\mp 45^\circ, \pm 45^\circ) = \mathbf{S}_3^{-1} \tag{3.1}$$

$$\begin{aligned}
\mathbf{V}(+90^\circ, \pm 90^\circ) &= \mathbf{A}_{1,2} & \mathbf{V}(-90^\circ, \pm 90^\circ) &= \mathbf{A}_{3,4} \\
\mathbf{H}(+90^\circ, \pm 90^\circ) &= \mathbf{A}_{4,3}^{-1} & \mathbf{H}(-90^\circ, \pm 90^\circ) &= \mathbf{A}_{2,1}^{-1} \\
\mathbf{LR}(\pm 90^\circ, +45^\circ) &= \mathbf{D}_1 & \mathbf{LR}(\pm 90^\circ, -45^\circ) &= \mathbf{D}_2 \\
\mathbf{UD}(+45^\circ, \pm 90^\circ) &= \mathbf{E}_2 & \mathbf{UD}(-45^\circ, \pm 90^\circ) &= \mathbf{E}_1 \\
\mathbf{VLR}(+45^\circ, \pm 90^\circ) &= \mathbf{E}_3 & \mathbf{VLR}(-45^\circ, \pm 90^\circ) &= \mathbf{E}_3^{-1} \\
\mathbf{HLR}(\pm 90^\circ, +45^\circ) &= \mathbf{D}_3 & \mathbf{HLR}(\pm 90^\circ, -45^\circ) &= \mathbf{D}_3^{-1}
\end{aligned} \tag{3.2}$$

where the last two lines are the configurations reported here and have now been added. There are 18 operators listed here and we can add the identity operator by setting the fields of all magnets to zero. Next, we note that all 19 operators can be obtained using a single string of 8 magnets (4 horizontally bending magnets alternating with 4 vertically bending magnets). Examples of how this is done are shown in Fig. 2 and a complete list of operators and corresponding configurations is given in Table II.

Table II: Configurations using 8 magnets showing the operators generated.

Operator	Magnet Type and Precession Angle								Sum of Angles
	H	*V*	*H*	*V*	*H*	*V*	*H*	*V*	
$\mathbf{A}_1$	—	90	90	−180	−180	90	90	—	720°
$\mathbf{A}_2$	—	90	−90	−180	180	90	−90	—	
$\mathbf{A}_3$	—	−90	90	180	−180	−90	90	—	
$\mathbf{A}_4$	—	−90	−90	180	180	−90	−90	—	
$\mathbf{A}_4^{-1}$	90	90	−180	−180	90	90	—	—	
$\mathbf{A}_3^{-1}$	−90	90	180	−180	−90	90	—	—	
$\mathbf{A}_2^{-1}$	90	−90	−180	180	90	−90	—	—	
$\mathbf{A}_1^{-1}$	−90	−90	180	180	−90	−90	—	—	
$\mathbf{D}_1$	45	90	−90	−180	90	90	−45	—	630°
	45	−90	−90	180	90	−90	−45	—	
$\mathbf{D}_2$	−45	90	90	−180	−90	90	45	—	
	−45	−90	90	180	−90	−90	45	—	
$\mathbf{E}_1$	—	−45	90	90	−180	−90	90	45	
	—	−45	−90	90	180	−90	−90	45	
$\mathbf{E}_2$	—	45	90	−90	−180	90	90	−45	
	—	45	−90	−90	180	90	−90	−45	
$\mathbf{D}_3$	45	90	−45	−180	−45	90	45	—	540°
	45	−90	−45	180	−45	−90	45	—	
$\mathbf{D}_3^{-1}$	−45	90	45	−180	45	90	−45	—	
	−45	−90	45	180	45	−90	−45	—	
$\mathbf{E}_3$	—	45	90	−45	−180	−45	90	45	
	—	45	−90	−45	180	−45	−90	45	
$\mathbf{E}_3^{-1}$	—	−45	90	45	−180	45	90	−45	
	—	−45	−90	45	180	45	−90	−45	
$\mathbf{S}_3$	45	45	−90	−90	45	45	—	—	360°
	−45	−45	90	90	−45	−45	—	—	
		45	−45	−90	90	45	−45	—	
		−45	45	90	−90	−45	45	—	
$\mathbf{S}_3^{-1}$		45	45	−90	−90	45	45	—	
		−45	−45	90	90	−45	−45	—	
	45	−45	−90	90	45	−45	—	—	
	−45	45	90	−90	−45	45	—	—	
BL max (T-m)	2.309	2.309	4.618	4.618	4.618	2.309	2.309	1.155	

4. Conclusions

Of the operators shown in Table II, there are 8 which rotate the vertical component of the initial spin polarization vector into the longitudinal direction. These 8 operators are $\mathbf{A}_1$, $\mathbf{A}_2$, $\mathbf{A}_3$, $\mathbf{A}_4$, $\mathbf{E}_1$, $\mathbf{E}_2$, $\mathbf{E}_3$, and $\mathbf{E}_3^{-1}$. Either examination of Table I or Ref. 1 shows that these 8 are the only operators of the set of 24 which have this property. In Ref. 1, only the first 6 of this list were known to be obtainable from a short simple string of magnets (less than 8 magnets) whereas we now know how to obtain all 8 operators in this manner. By placing a string of 8 magnets on either side of an Interaction Point (IP) it is possible to provide longitudinal polarization at the IP while at the same time generating an overall operation which is desired such as that of a rotator (**I**), a snake or another of the 24 possibilities. A multiplication table which shows the product of two successive operations whereby the first is one which causes the polarization to be longitudinal can be found in Ref. 1 but is not shown here due to lack of space.

References

1. T. Fieguth, "Snakes, Rotators, Serpents and the Octahedral Group," *Particle Accelerators*, **22**, No. 3, 183–203 (1987).
2. G. Guignard and J. P. Gourber, "Feasibility Study of Compact and Tunable Spin Rotators," CERN SL/90–04 (AP).
3. B. W. Montague, "Polarized Beams in High Energy Storage Rings," *Physics Reports*, **113**, No. 1, 8–13 (1984).

First Evidence of Transverse Polarization in LEP

J. Badier[1], *A. Blondel*[1], *M. Crozon*[2], *B. Dehning*[3], *L. Knudsen*[4], *J.P. Koutchouk*[4], *M. Placidi*[4], *and R. Schmidt*[4]

[1]LPNHE Ecole Polytechnique, IN2P3, France
[2]Collège de France, Paris, France
[3]Max-Planck-Institut, Munich, Fed. Rep. of Germany
[4]CERN, CH-1211 Geneva 23, Switzerland

Results on first experimental evidence of transverse polarization in LEP at $E = 46.5\,\mathrm{GeV}$ ($\nu_s = 105.55$) are reported. The polarization level associated to the significant asymmetry observed in the Compton back-scattered photon distribution when illuminating the electron beam with circularly polarized laser light is estimated to be $9\% \pm 2\%$ (systematic) $\pm 2\%$ (statistical). The validity of the polarization signal was assessed by exciting a depolarizing resonance. The measurement is consistent with the predicted polarization degree with the especially prepared polarization optics.

1 INTRODUCTION

A fast laser polarimeter has been designed [1] and installed [2] as an essential tool to optimize orbit correction strategies expected to improve the polarization level in LEP, presently limited by optics imperfections like betatron coupling and residual vertical dispersion [3]. The procedures adopted in the commissioning of the polarimeter are described in [4] and briefly recalled here. Experimental results on the first measurements of transverse polarization at LEP are presented together with the optics manipulations developed to check the consistency of the observed polarization signal.

2 SPIN-DEPENDENT COMPTON SCATTERING

The differential Compton scattering cross section expressed in terms of the electron and photon polarization states in the $e^{\pm}$-rest frame [5] [6] is:

$$\frac{d\sigma_c(\vec{\xi},\vec{P}_e)}{d\Omega} = \frac{1}{2}\left(r_e\frac{k'}{k'_0}\right)^2 \left[\Phi_0 + \Phi_1\left(\vec{\xi}\right) + \Phi_2\left(\vec{\xi},\vec{P}_e\right)\right] \tag{1}$$

where $r_e = e^2/m_ec^2$ = classical electron radius
k'_0, k' = incident and scattered photon momenta
$\vec{P}_e \equiv (P_x, P_y, P_z)$ = $e^{\pm}$ Polarization vector
$\vec{\xi} \equiv (\xi_1, \xi_2, \xi_3)$ = Photon Polarization vector in terms of the normalized Stokes parameters ($\sum_{i=1}^{3} \xi_i = 1$).

In the general case of transverse ($P_x = 0, P_y \equiv P_\perp$) and longitudinal ($P_z \equiv P_\parallel$) beam polarization, the differential cross section (1) reads:

$$\begin{aligned}\frac{d\sigma_c(\vec{\xi},\vec{P}_e)}{d\Omega} &= \frac{1}{2}\left(r_e\frac{k'}{k'_0}\right)^2\Big\{1+cos^2\theta'+(k'_0-k')\,(1-cos\,\theta') \\ &+\xi_1\,sin^2\,\theta' \\ &-\xi_3\,P_\perp\,[(1-cos\,\theta')\,k'\,sin\,\theta']\,sin\,\phi' \\ &-\xi_3\,P_\parallel\,[(1-cos\,\theta')\,(k'_0+k')\,cos\,\theta']\Big\}\end{aligned} \tag{2}$$

Circular light ($\xi_3 \neq 0$) produces asymmetry via the Φ_2 term in presence of transverse or longitudinal beam polarization. In both cases an asymmetry function $A(y)$ can be defined:

$$A(y) = \frac{n_R(y)-n_L(y)}{n_R(y)+n_L(y)} = \frac{\Phi_2(\vec{\xi},\vec{P}_e)}{\Phi_0} \tag{3}$$

where $n_{R,L}(y)$ are the γ-rates at a vertical position y for right and left circular light, or horizontal and vertical linear light.

3 THE LEP POLARIMETER

The LEP polarimeter, based on spin-dependent Compton scattering of circularly polarized photons from polarized particles [7], is installed in the LEP LSS1 straight section to monitor the transverse polarization of the electron beam. A Nd-YAG laser in the Optical Laboratory (Fig. 1) generates 12 ns long, 50 mW light pulses at a $\sim$ 30 Hz repetition rate.

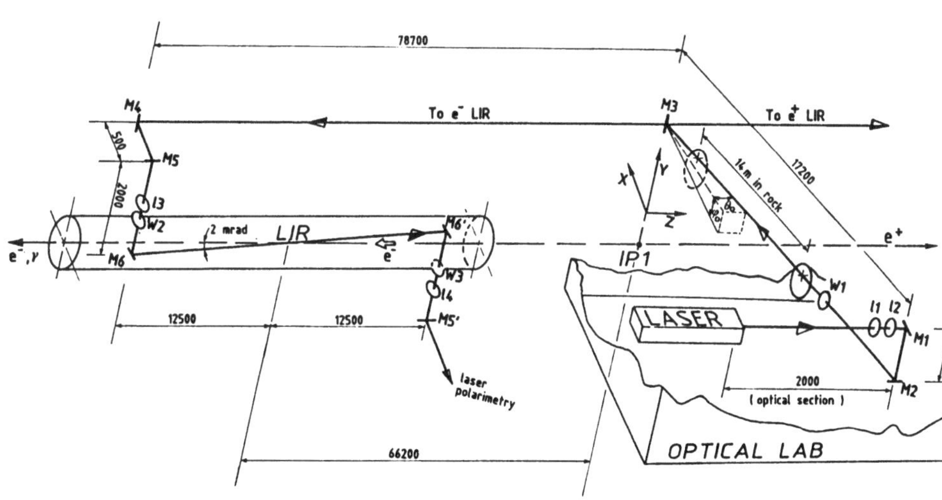

Figure 1: Optical layout of the LEP laser polarimeter.

The 532 nm wavelength light is guided over 115 m to the Laser Interaction Region (LIR) by five multilayer dielectric mirrors in a roughly evacuated transport line, steered onto the incoming electron beam by the in-vacuum Ag-coated Cu mirror M_6 and finally analyzed in the optical box outside the LIR. Photons in an energy range of 5-28 GeV are backscattered in the direction of the incoming electrons and their transverse profiles recorded $\sim$ 247 m downstream in a detector constituted of silicon strip planes behind a remote controlled variable-thickness lead absorber.

Changes in the shape of the vertical γ-distributions originating from asymmetry properties of the Compton cross section provide the observable beam polarization signal. The associated centre-of-gravity shift (*mean-shift*) of the distribution is proportional to the photon and electron degree of polarization :

$$\Delta\langle Y\rangle = \kappa\ \xi_3\ P_\perp \tag{4}$$

In our case the "analyzing power" κ (total centre-of-gravity shift for unit electron and photon polarization) evaluated at 46.5 GeV is :

$$\kappa = 500 \pm 30\,\mu\text{m}.$$

4 RESULTS

The polarimeter was commissioned and optimized mainly running in parasitic mode on physics runs by adjusting the mirror positions at the LIR to avoid any perturbation to the accelerated beams. Progress in the strategies adopted to optimize the laser-e^- beam relative alignment for the best overlap, to steer the electron orbit at the LIR for optimum transmission of the recoil photons and to define the detector parameters could be achieved saving the allocated MD time for polarization studies.

The optical section installed on a bench at the laser output is used to control the light polarization. A rotating $\lambda/2$ plate, a $\lambda/4$ plate and a "push-pull" $\lambda/2$ plate produce any elliptical light state, from linear to circular. Linear light has proved fruitful to the setting up of the polarimeter since for ($\xi_1 \neq 0$) the Compton cross section does not depend on beam polarization (Equ. (2)). The push-pull $\lambda/2$ plate introduces an additional π phase-shift, thus reversing once more the handedness. This provides a simple way to correlate the observed sign reversal in the measured mean-shift to the polarization signal (Fig. 3).

The ellipticity of the light at the LIR was controlled in the optical section by timing the angular position of the rotating plate w.r.t the laser pulse to compensate for depolarizing effects from elements in the transport line (reflections, birefringence etc.) which would spoil an initially perfect circular light.

In Fig. 2 an example is shown of measured asymmetries for linear and circular light. Good agreement with simulation is found in both cases: the predicted

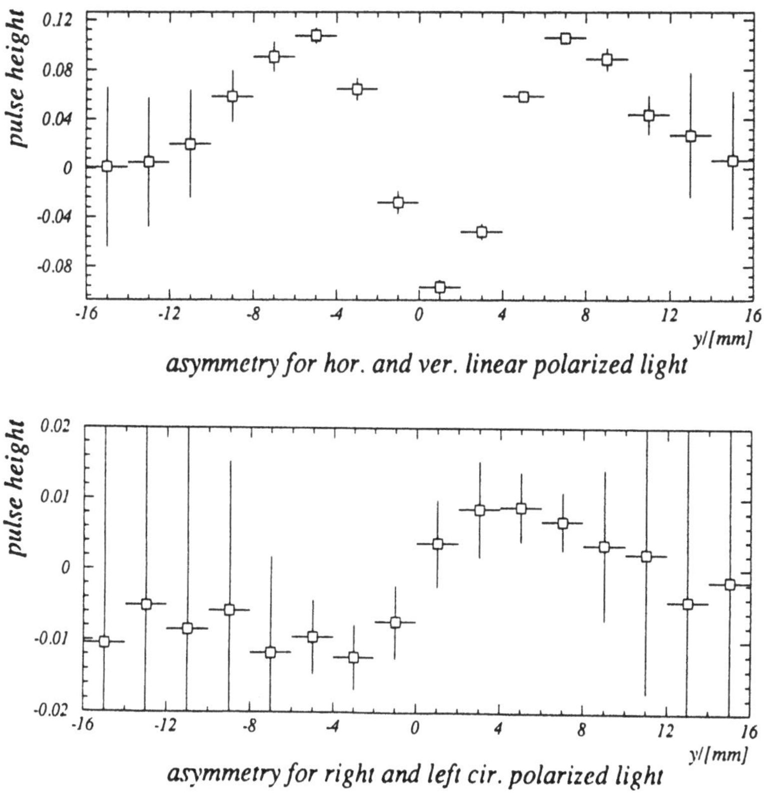

Figure 2: Measured asymmetry for linear and circular laser light.

$\pm 11\%$ maximum asymmetry for linear light is obtained and measurements with circular light are compatible with a transverse beam polarization of $\sim 9\%$.

With an effective 85% degree of circular polarization, measured at the optical box outside the LIR, a centre of gravity shift $\Delta\langle Y\rangle$ of $\sim 40\mu$m was detected when switching from left to right circular light. The shift vanished when using linearly polarized light. The polarization data collected at a beam energy of 46.5 GeV ($\nu_s = 105.55$) are shown in Fig. 3 in terms of the mean shift $\Delta\langle Y\rangle$. Negative $\Delta\langle Y\rangle$ values correspond to measurements with the push-pull $\lambda/2$ plate in "IN" position. The associated polarization levels from (3) are collected in Table I and subdivided into 9 periods of time according to different optics manipulations.

5 CONCLUSIONS AND FUTURE PLANS

The commissioning and the studies of the polarimeter with linearly and circularly polarized light produced the expected results. The centre-of-gravity shift measurements with right and left circularly polarized light indicated an asymmetry compatible with a degree of polarization of $\sim 9\%$. According to simulations, the shift vanished when illuminating the beam with linear light and no changes

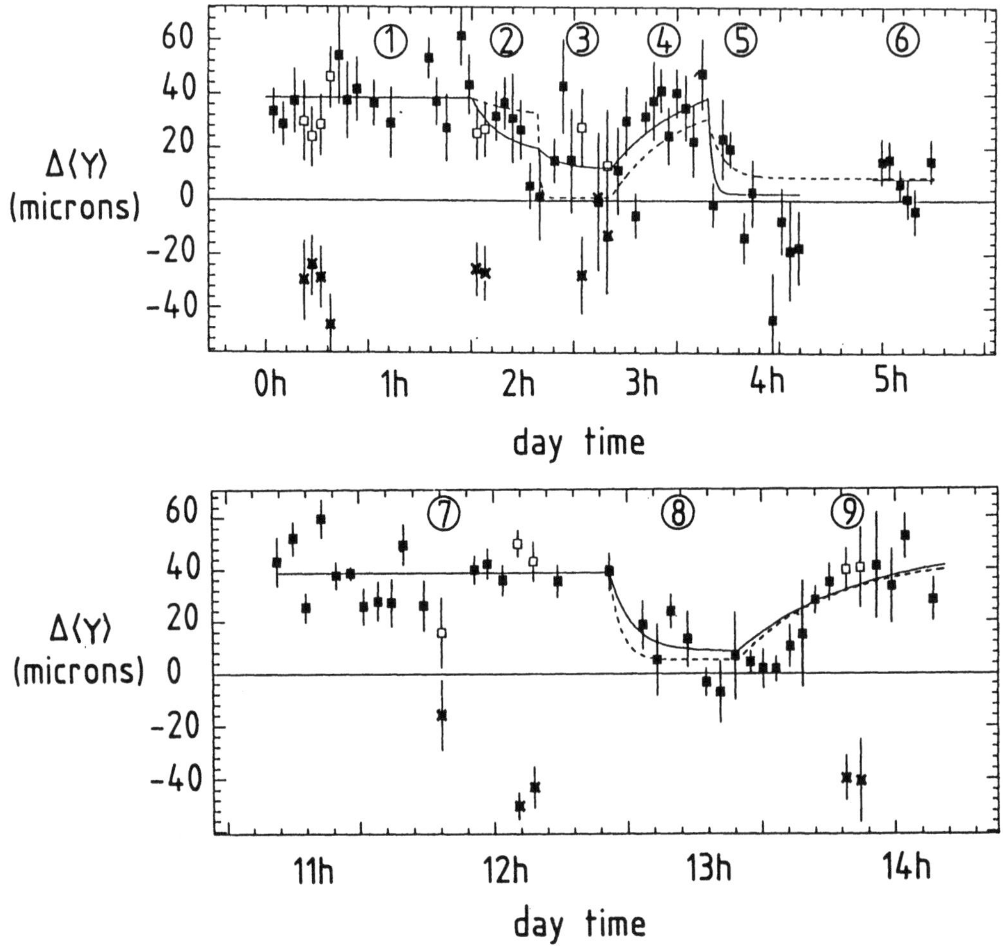

Figure 3: Experimental results in terms of mean shift $\Delta\langle Y\rangle$. The level of transverse polarization is evaluated from (4) for $\kappa = 500\,\mu\text{m}$, as resulting from simulations, and $\xi_3 = 85\%$ measured circular laser light. The expected behaviour from applied depolarizing resonances (dotted line) is compared to the experimental data fit. Negative $\Delta\langle Y\rangle$ values represent measurements with the push-pull half wave plate in "IN" position.

in the peak height and the rms width of the distributions were observed with circularly polarized light.

The LEP polarization program for 1991 contemplates the absolute beam energy calibration to an accuracy of some 10^{-5}. Improvements in the performance of the polarimeter are being implemented to cope with the polarization studies which will be devoted to improve the level of polarization on the present optics, to study the behaviour of the new wigglers expected to reduce the natural polarization time ($\sim$ 5.5 h at the Z^0 energy) to about 35 min and to develop a new optics

Table 1: Summary of fits to the polarization measurements for different optics conditions. The errors, statistical and systematic, include: a possible 5 μm systematic shift in $\Delta\langle Y\rangle$, uncertainties in the transition time between consecutive conditions (periods 2,3,4,9), and scale factor for the goodness-of-fit (periods 5,7,9). Expected asymptotic values [4] are also shown for comparison.

Period	χ^2/D.F. of the fit	P_∞ (%) (fit)	P_∞ (%) (expected)	Conditions and comments
1	0.99	9.1 ± 0.6 ± 1.2		asymptotic polarization
2	1.07	5.1 ± 0.6 ± 1.6	7.4	excitation of ν_s=106 (20 %)
3	0.74	2.5 ± 1.6 ± 1.3	0.3	excitation of ν_s=106 (100 %)
4	1.26	11.5 ± 3.1 ± 2.1	9.1	natural rise
5	4.4	0.6 ± 0.6 ± 1.7	2.1	RF trips, solenoid bumps ON
6	1.3	1.9 ± 0.7 ± 1.2	2.1	stable beam, solenoid bumps ON
7	2.36	9.1 ± 0.3 ± 1.3	9.1	asymptotic polarization
8	0.62	2.0 ± 0.6 ± 1.2	1.2	excitation of ν_s=106 (50 %)
9	2.4	11.7 ± 2.4 ± 2.7	9.1	natural rise

capable of producing appreciable polarization level at the Z^0 resonance and in physics conditions.

6 ACKNOWLEDGMENTS

The measurements reported here are the result of a close collaboration between IN2P3 (France), MPI (Munich) and CERN. The authors wish to acknowledge the encouragement and the strong support to the polarization program at LEP from the SL and PPE Divisions and the LEP users community. The realization of the polarimeter, its commissioning and the polarization studies could never have been accomplished in a relatively short time without the active participation and the skill of colleagues in the Beam Instrumentation, Accelerator Physics and Operation groups. Credit goes to the LEP users community for having granted to the measurements the last 24 hours period of the 1990 LEP physics program.

References

[1] M. Placidi, R. Rossmanith, NIM A274 (1989) 79-94.
[2] M. Placidi *et al.*, 2nd European Particle Acc. Conf., Nice, 1990.
[3] J.P. Koutchouk, this Conference.
[4] J. Badier *et al.*, LEP Performance Note 42, December 1990.
[5] U. Fano, J. Opt. Soc. Am., (1949) 39, 859.
[6] F.W. Lipps and H.A. Toelhoek, Physica XX (1954) 85, 395.
[7] V.N. Baier and V.A. Khoze, Sov. J. Phys. 9 (1969) 238.

Transverse Polarization in LEP

J.P. Koutchouk

CERN, CH-1211 Geneva 23, Switzerland

A significant asymmetry of the Compton back-scattered photons was observed when illuminating the LEP beam with circularly polarized laser shots. The corresponding polarization level is estimated to be about 9 %. The asymmetry could be modified by exciting a spin resonance. This experiment was made at dedicated energy and machine optics. The consistency of the measurement with the prediction of the polarization is satisfactory.

1 INTRODUCTION

The programme of transverse and longitudinal polarization in LEP is now in an active phase. The two aims are: 1/ Energy calibration at the Z^0 energy to an accuracy better than the presently achieved 12 MeV per beam [1], 2/ Physics with longitudinally polarized beam at the Z^0 peak, for an accurate check of the Standard Model [2]. The LEP polarization project is described in detail in [3]. It incorporates strongly asymmetrical wigglers which reduce the polarization rise-time from 300 to 36 minutes, a Compton polarimeter [4] and a Richter-Schwytter spin rotator [3].

2 EXPECTED POLARIZATION LEVEL

As discussed in [3], the depolarization in LEP is significantly enhanced compared to SPEAR and PETRA, due to the higher energy and to the higher-order spin resonances driven by the large beam energy spread. An estimate of the tolerances on the alignment and the field imperfections allowed the computation of the disturbed optics and asymptotic polarization level. Even with the enhanced beam energy spread due to the wigglers, a detectable polarization level should be observed.

In practice, however, the optics imperfections have been measured to be larger than expected and the predicted polarization much lower:

Optics parameters	$<x>$ mm at $\bar{\beta}$	$<y>$ mm at $\bar{\beta}$	$<D_y>$ m at $\bar{\beta}$	ϵ_y/ϵ_x %	P_∞ (calculated in linear approx) %
Expected	0.75	0.75	6	≤1	20 to 40
Measured	1.40	1.00	17 to 21	5 to 10	5

Simulation of the spherical spin motion showed that, for low polarization levels, the higher-order spin resonances further decrease P_∞ by a factor of 2. Therefore no measurable polarization can develop on the standard LEP optics (figure 1).

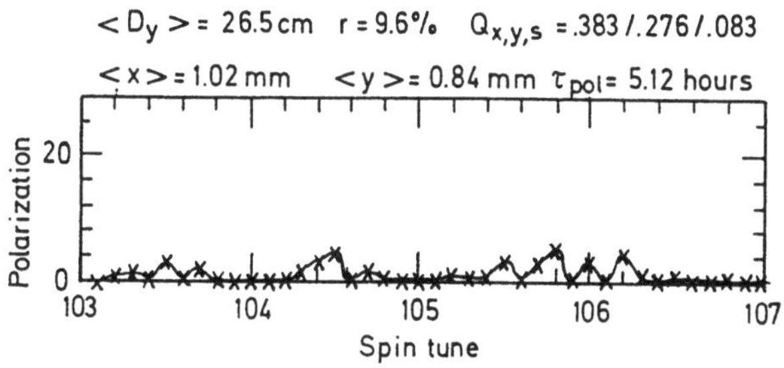

Figure 1: Expected polarization on the LEP physics optics

Another difficulty arose from the presence of a parasitic skew gradient in the vacuum chamber [5] which limits the polarization to some 50 %. In order to minimize the resulting linear coupling, the optics was retuned from $Q_x = 70.38/Q_y = 78.28$ to $Q_x = 71.38/Q_y = 77.28$. A consequence is the change of the selection of the systematic spin resonances following [6]: $\nu = Q_{int} + S_b k$, where Q_{int} is the integer part of the betatron tunes, S_b the superperiodicity of the bending and k any integer.

Figure 2 shows indeed that there is no polarization at the Z^0 peak, close to the systematic resonance $\nu = 103 = 71 + 8 \times 4$, even if the machine was otherwise perfect.

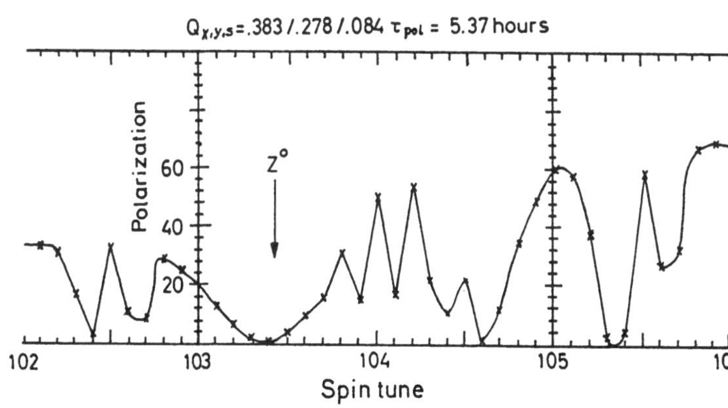

Figure 2: Depolarization due to the parasitic skew gradient in an otherwise perfect machine

3 MACHINE STUDIES & POLARIZATION PARAMETERS

The favourable energy domain is between systematic spin resonances: $\nu \in [104.5, 107.5]$. Due to the limited amount of time, only the spin tunes $\nu = 105$ (zero polarization) and $\nu = 105.6$ were explored.

3.1 BETATRON TUNES

The working point should be as close as possible to the integer to leave a maximum resonance-free area around the half-integer spin tune. There is a more specific requirement to lower Q_x so as to avoid the horizontal betatron spin resonance excited by the parasitic skew gradient in the vacuum chamber.

Starting from the tunes initially used for physics ($Q_x = 71.38, \quad Q_y = 77.28$), a new working point meeting the polarization and beam life and size criteria was found at $Q_x = .12, \quad Q_y = .20$. In spite of the closeness of strong synchro-betatron resonances, the beam was very stable and the emittance ratio lower.

3.2 CLOSED ORBIT AND DISPERSION

During normal operation, the residual vertical closed orbit is corrected down to about 1 mm rms. The vertical dispersion measured is then 17 to 21 cm rms and the emittance ratio estimated to be some 5 %.

The depolarization goes with D_y^2 and higher powers. The gain to be expected from lowering the vertical dispersion is large, once the betatron tunes have been lowered. To achieve it, the closed orbit correction procedure was pushed to yield 0.5 mm rms residuals for the last polarization runs. To obtain these results, it is necessary to eliminate the 'faulty' PU's. Visual inspection, distribution cuts and harmonic analysis were used. Not more than 15 % of the PU's were eliminated. The MICADO algorithm was used with a large number of orbit correctors. Harmonic analysis of the residuals showed that 0.5 mm rms was not the ultimate. More refined tools will need to be prepared to go beyond.

Following orbit correction, the vertical dispersion was measured to be only $< D_y > = 11$ cm. This fortunate result is beyond the prevision and will require

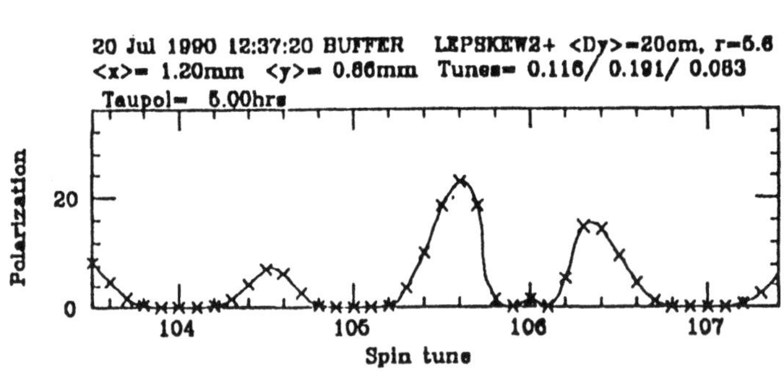

Figure 3: Expected polarization on the improved polarization optics

further analysis. The emittance ratio decreased to less than 2 %. Altogether the optics improvements are calculated to allow the polarization to build up to around 20 % at the chosen spin tune of 105.5 (figure 3). The higher-order spin resonances should reduce this value by up to a factor of 2.

3.3 PARAMETER LIST OF THE OPTICS FOR POLARIZATION

Optics	Injection (Detuned low-β)
E	46.51 GeV
ν	105.55
Q_x	71.12
Q_y	77.20
Q_s	0.085
$< x >$	1.4 mm
$< y >$	0.5 mm
$< D_y >$	11 cm
ϵ_y/ϵ_x	$\leq 2\%$
I_b (4 bunches)	200 μA per bunch

The experimental solenoids, skew quadrupoles, wiggler magnets and electrostatic separators were switched off.

It took 3 to 4 hours to establish the optics out of a 16 hour run. The polished optics for polarization was found to be neither reproducible nor completely stable.

4 RESULTS

The measurement using the Compton polarimeter is discussed in [4] (figure 4). The data, all collected at $\nu = 105.55$, are divided in 10 periods of time where different manipulations took place: natural rise after depolarization, depolarization by the excitation of the spin resonance 106 by means of vertical orbit bumps, depolarization by bumps simulating the solenoids. In all cases, the evolution of the polarization is governed by:

$$P(t) = P_\infty + (P_{init} - P_\infty)e^{-(1/\tau_p+1/\tau_d)t}$$

where P_{init} is the initial polarization level. P_∞ is provided by the polarimeter. The rise time may either be calculated from P_∞ or from the known excitation of the spin resonance. The results are presented in a table:

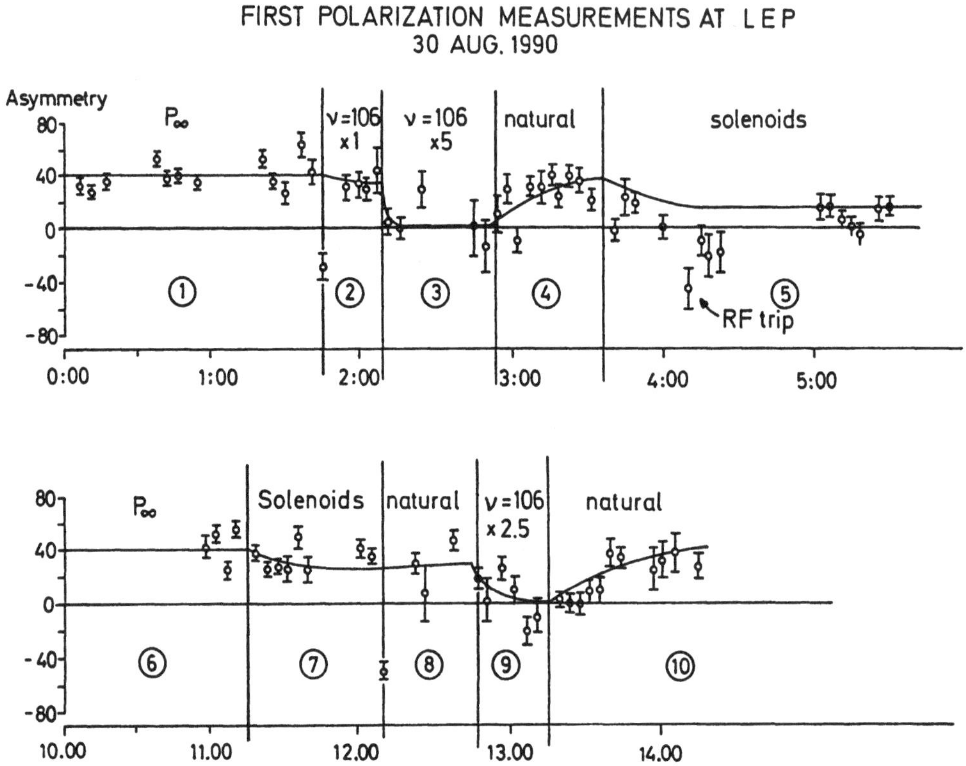

Figure 4: Polarization measurement

Period	P_∞ %	τ_{calc} minutes	τ_{fit} minutes	Comment
1	9	29		asymptotic polarization
2		24	18	excitation of ν=106 × 1
3		1	8	excitation of ν=106 × 5
4	9	29	47	natural rise
5	2.1	9	7	solenoids
6	9	29		asymptotic level
7	2.1	9	6	solenoids
8	9	29		asymptotic level
9	1	3.3	7	excitation of ν=106 × 2.5
10	9	29	44	natural rise

5 CONCLUSION AND FUTURE PLANS

The measured asymmetry of the back-scattered photons is consistent with a polarization signal. This first measurement is important as there has been much debate as to wether polarization could ever be observed at LEP. The ability to polarize the LEP beam even at this low level relied on the predictivity of the theory of the polarization which led us to change the LEP optics accordingly. It is satisfactory and very promising that the experiment confirmed the previsions. Next year's programme will focus on the LEP energy calibration and the improvement of the polarization level.

References

[1] A. Hofmann *et al.*, 2nd European Particle Acc. Conf., Nice 1990.

[2] 'Polarization at LEP', CERN 88-06, 1988.

[3] A. Blondel *et al.*, 8th Int. Symp. on H.-E. Spin Physics, Minneapolis 1988 and this conference.

[4] M. Placidi *et al.*, this conference.

[5] J. Billan *et al.*, 2nd EPAC, Nice 1990.

[6] J.P. Koutchouk, 8th Int. Symp. on H.-E. Spin Physics, Minneapolis 1988.

Compensation of Depolarizing Effects at HERA

E. Gianfelice-Wendt

Deutsches Elektronen-Synchrotron, Notkestraße 85,
W-2000 Hamburg, Fed. Rep. of Germany

A method is described for correcting the tilt of the equilibrium polarization axis due to machine misalignements. Results for HERA are presented.

Introduction

The expectation value, $\vec{S}$, of the spin vector operator for a non-radiating particle moves according to the Thomas-BMT equation

$$\frac{d\vec{S}}{dt} = \vec{\Omega} \times \vec{S}, \tag{1}$$

where in MKS units

$$\vec{\Omega} = -\frac{e}{m_0}\left[\left(a+\frac{1}{\gamma}\right)\vec{B} - \frac{a\gamma}{\gamma+1}\vec{\beta}\cdot\vec{B}\vec{\beta} - \left(a+\frac{1}{\gamma+1}\right)\vec{\beta}\times\vec{E}\right]$$

with $a = (g-2)/2$. Eq.(1) represents a precession of $\vec{S}$ around the direction of $\vec{\Omega}$ (in a ring generally the direction of the main guiding field) with frequency $\Omega/2\pi$. In the general case, $\vec{\Omega}$ depends on the particle trajectory, so that each closed trajectory corresponds a different periodic solution (unique, unless particular conditions are fulfilled). The number of precessions per turn ν is called spin tune; on the design orbit $\nu \simeq a\gamma$. Spin Rotators are used in order to modify such periodic solutions in a suitable way at some points of a machine. When the spin tune is close to an integer, any perturbing field, non parallel to the direction of $\hat{n}_0$, results in a large deviation of the polarization axis from the nominal one.

In a storage ring electrons have the tendency to selfpolarize along the $\hat{n}_0$ direction. In the presence of betatron motion, the equilibrium polarization is given by[1],[2]

$$P_{eq} = -\frac{8}{5\sqrt{3}}\frac{<\frac{1}{|\rho|^3}\hat{b}\cdot(\hat{n}-\gamma\frac{\partial\hat{n}}{\partial\gamma})>}{<\frac{1}{|\rho|^3}[1-\frac{2}{9}(\hat{n}\cdot\hat{v})^2+\frac{11}{18}|\gamma\frac{\partial\hat{n}}{\partial\gamma}|^2]>} \tag{2}$$

where $b \equiv \vec{v}\times\dot{\vec{v}}/|\vec{v}\times\dot{\vec{v}}|$ and $\hat{n}$ are quasi-periodic solutions[1] of (1). The term $\gamma\partial\hat{n}/\partial\gamma$ quantifies depolarizing effects resulting from trajectory perturbations due to radiation. It appears squared in the denominator of the rhs of (2) leading always to a decreasing of polarization (quantum diffusion). In linear approximation, quantum diffusion is strongest when the resonance conditions $\nu = m \pm Q_{x,z,s}$ are approached. For a ribbon beam circulating in a perfect flat machine, $\gamma\partial\hat{n}/\partial\gamma$ is zero, the special solutions $\hat{n}$ of (1) are not energy dependent and the machine is spin transparent. In a machine including rotators the transparency has to be restored via "spin matching"[3]. However, in a real machine, in the presence of errors such as element misalignment etc., the closed orbit and then the $\hat{n}_0$ axis are distorted, so that the spin transparency is randomly broken with consequent loss of polarization. Thus, methods to improve the polarization level in presence of such errors have to be found.

Description of the correcting method

The basic idea of the method described here is to return $\hat{n}_0$ to the design orientation. This has been done with success at PETRA[4], but some assumptions were made about the regularity of the lattice. So the formalism has been reviewed and extended[5] in order to overcome such difficulties and to handle a non flat machine containing rotators as at HERA.

With small perturbations, one can write the periodic solution of (1) along the c.o. in the form $\vec{N}_0 = \hat{n}_0 + \delta\hat{n}_0 = \hat{n}_0 + \alpha\hat{m} + \beta\hat{l}$, where $\hat{m}$ and $\hat{l}$ are constructed from the non-periodic eigenvector of (1) on the nominal c.o. in such way as to form a periodic orthogonal right handed reference system along this orbit. In linear approximation, α and β are related to the errors by[5]

$$(\alpha - i\beta)(s) = -i\frac{L}{2\pi}\sum_k \frac{f_k}{k-\nu}e^{i2\pi ks/L} \tag{3}$$

where f_k are the Fourier components of $f \equiv d_1 - id_2$, with

$$\vec{d} \equiv \begin{pmatrix} d_1 \\ d_2 \end{pmatrix} = \mathbf{L}\left[\mathbf{F}\vec{y} - \frac{1}{B\rho}\begin{pmatrix} \Delta B_\tau(1+a) \\ \Delta B_x(1+a\gamma) \\ \Delta B_z(1+a\gamma) \end{pmatrix}\right] \tag{4}$$

$\Delta\vec{B}$ is the extra field along the design c.o., i.e. the field causing the c.o. distortion, $\mathbf{L}$ is a 2×3 matrix containing the components of $\hat{m}$ and $\hat{l}$ in the orbital reference system, $\mathbf{F}$ is a 3×6 (energy dependent) matrix containing the nominal fields and $\vec{y}$ is the 6-dimensional c.o. From eq.(3) one can see that $\delta\hat{n}_0$ is sensitive in particular to those harmonics of f closest to the spin tune. One can reduce α and β by introducing dipolar fields which compensate such harmonics. This leads to a matrix equation of the form (see[5] for explicit expressions) $\mathbf{M}\vec{\theta} = -\vec{a}$, where $\vec{\theta}$ and $\vec{a}$ are correcting kicks and Fourier components of f respectively. From (4), one could expect that c.o. correction leads automatically to a strong reduction of the polarization axis tilt. In order to understand the relationship between c.o. and polarization axis tilt, let me consider as the source of perturbations only the vertical random displacement of the quadrupoles; the Fourier components of $\vec{d}$ can be written in terms of those of the c.o., z_k, and of the perturbing fields, b_k,

$$\vec{d}_k = \sum_m\sum_n \mathbf{L}_m\mathbf{F}_{n-m}z_{k-n} + b_k. \tag{5}$$

In this formalism the strongest component of $\mathbf{L}$ is the zero component, so that the main contribution to the harmonic $\bar{\nu}$ (closest integer to ν) of f comes from the orbit harmonics $q = \bar{\nu} \pm mP$, with P periodicity of the focusing structure and m integer. The most unlucky case is when one of these conditions corresponds just to the vertical betatron tune. In HERA, with a periodicity almost of 4, a spin tune between 66 and 67 (in the energy range here considered) and a betatron tune of 48.28, one can say that

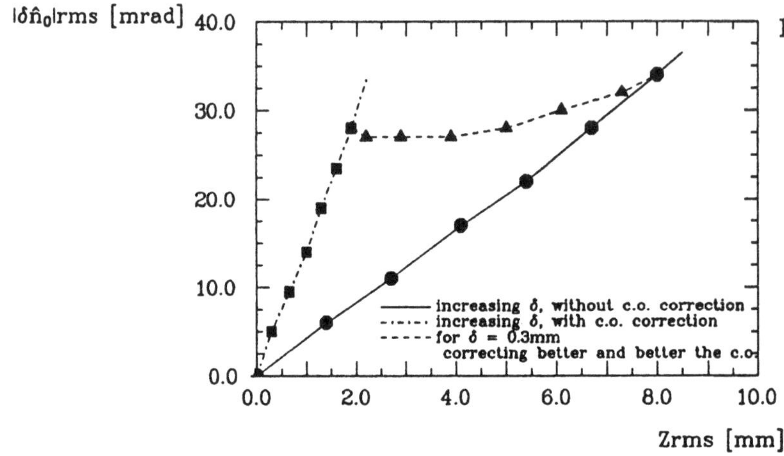

Figure 1.

$\delta\hat{n}_0$ and the closed orbit are sensitive to different field harmonics. This also means that the contribution of b_k in (5) cannot be always neglected. Thus in general correction of the c.o. is not sufficient to guarantee a reduction of the $\hat{n}_0$ axis tilt and one should not compare results concerning different rms values δ of the displacements, although the rms c.o. can be, via c.o. correction, the same (see Fig.1).

The $\hat{n}_0$ axis tilt correction will instead in general increase the c.o. distortion. So it is necessary to correct the $\hat{n}_0$ axis tilt by using closed bumps (method B) as done for LEP[6] or adding to the original system of equations the condition that the orbit fundamental harmonics should remain unchanged (method H).

Application to HERA and numerical results

I will consider here as source of distortions only the quadrupole vertical misalignment which is the most dangerous for the beam polarization. This kind of error produces mainly a distorted vertical c.o. (and thus a tilt of $\hat{n}_0$), as well as a spurious vertical dispersion. The latter leads to depolarization even in the case of an unperturbed $\hat{n}_0$. This effect, which cannot be corrected with the method described here, can be corrected by "harmonic spin matching"[3].

The rms value of the vertical displacement of the quadrupoles assumed is 0.3mm, according to the HERA specifications and in agreement with the experimental observations of last year. The rms c.o. distortion is in general large and in some cases the machine is even unstable, so the c.o. must be previously corrected and c.o. correction has to be included in the simulation. Two methods have been used, the well known method of the best corrector, as well as the harmonic method. Application of the latter is particularly simple and convenient for phase equidistant correctors[7], a condition which is not completely fulfilled in HERA. So the nice features of this method apply only partially here. After c.o. correction, the code FIDO[8] is used in order to correct the polarization axis tilt as described above. The FIDO routines are basically those of the well known code SLIM.

I observed that for HERA 4 harmonics of f around the spin tune must be controlled together with the harmonics 48 and 49 of the c.o. This means that 12 correctors are needed. To control the same 4 harmonics of f using closed bumps one needs 8 bumps.

Since in practice the Fourier components of f are unknown and have to be corrected empirically, it is important that the simulation demonstrates that only a few harmonics must be corrected and that this can be done by using the same family of coils for any c.o. In HERA there are 267 vertical correcting coils. Among these one has to choose between 12 and 16 coils. The criterion used here is to select schemes with large determinants of the matrices $\mathbf{M}$.

Numerical results are summarized in the table below. The values are obtained at fixed energy (29.3 GeV) averaging over 10 orbits.

It is worth nothing that the same family of 12 coils (method H) and the same 8 bumps (method B) have been used.

In Figs.2 and 3 the expected polarization in HERA before and after correction is shown vs energy for two particular orbits.

Finally I remark that the numerical studies performed indicate that the c.o. correction reduces significantly the spurious vertical dispersion, so that harmonic spin matching is probably not needed, and that the main limit to the polarization is given by the tilt of $\hat{n}_0$ in the quadrupoles at locations of high horizontal dispersion. So one can expect that a preferential correction of the $\hat{n}_0$ axis tilt as described in[9] might be more effective.

Table 1.

		with c.o. correction			
	without corrections	best corrector method		harmonic method	
$z_{rms}(mm)$	9.1 ± 3.0	1.0 ± 0.1		2.4 ± 0.5	
$P(\%)$	7.2 ± 7.0	6.4 ± 4.9		9.0 ± 7.9	
$z_{rms}(mm)$		1.2 ± 0.2	3.0 ± 1.2	2.4 ± 0.5	2.9 ± 0.5
$P(\%)$		62.0 ± 10.5	57.8 ± 9.9	57.7 ± 9.3	56.3 ± 9.1
		+B	+H	+B	+H

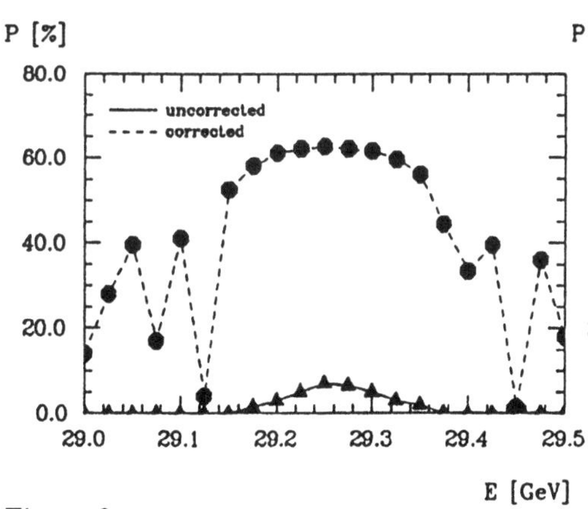

Figure 2.

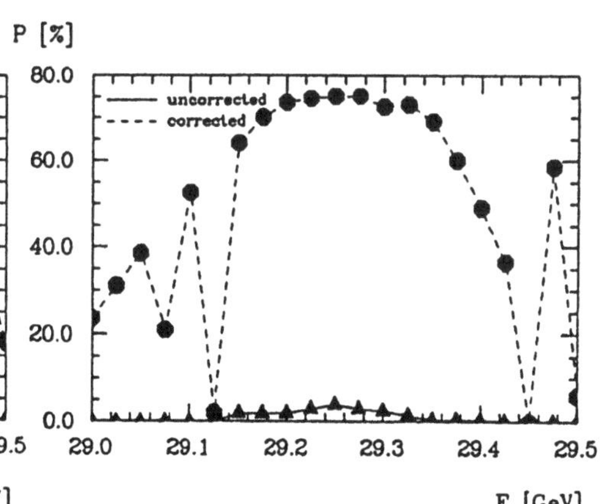

Figure 3.

Conclusions

The polarization level of the electron beam in HERA, dramatically low when errors in the machine are considered, can be greatly improved as shown here. The question is how much of this polarization will survive higher order polarization calculations[10].

Acknowledgements

I would like to thank D. P. Barber for his guidance, J. P. Koutchouk for stimulating discussions and G. A. Voss for encouragement.

References

[1] Ya. S. Derbenev and A. M. Kondratenko, Sov. Phys. JETP **37**, 968 (1973).
[2] S. R. Mane, Phys. Rev. A **36**, 105 (1987).
[3] K. Steffen, DESY HERA 82/02 (1982).
[4] R. Rossmanith and R. Schmidt, DESY 84-105 (1984).
[5] D. P. Barber et al., DESY 85-044 (1985).
[6] J. P. Koutchouk and T. Limberg, CERN 88-06 **2**, 204 (1988).
[7] E. Gianfelice and R. Giannini, CERN/PS 85-42 (LEA) (1985).
[8] S. R. Mane, DESY HERA 85-23 (1985).
[9] D. P. Barber and G. Ripken, DESY HERA 87-15 (1987).
[10] D. P. Barber, Proceedings of the Conference accompanying this Workshop.

Spin Matching Conditions in Large Electron Storage Rings with Purely Horizontal Beam Polarization*

R. Rossmanith

CEBAF, 12000 Jefferson Avenue, Newport News, VA 23606, USA

Introduction

In the last few years the self-polarization of electrons and positrons in storage rings (Sokolov-Ternov effect[1]) was observed in many storage rings: ACO,[2] VEPP2,[3] VEPP4,[4] SPEAR,[5] DORIS,[6] PETRA,[7] and CESR.[8] In all these machines the beam was polarized by synchrotron radiation emission in the bending magnets of the arcs. As a consequence the spin axis in the arc must be parallel or antiparallel to the bending field, depending on the particle species.

At LEP (at least in phase I) the situation is different. The beam is polarized by so-called asymmetric wigglers at a point or points of the machine.[9] In the very beginning these wigglers will produce transversal polarization and the arcs act only as a spin transport system. Many years ago it was discussed whether it is possible to rotate the wigglers by 90 degrees and produce horizontally polarized beams.[10] It was shown experimentally at low energies[11] that such a system could work. It is also well known that a horizontal polarization can be maintained by so-called Siberian Snakes[12] in proton machines, as has been experimentally shown recently.[13] A Siberian Snake rotates the spin around the momentum axis by 180 degrees. As a result most of the perturbations which would otherwise add up over many revolutions compensate each other after two revolutions.

It is interesting at least from an intellectual point of view whether the Siberian Snake can be applied to electron machines with an asymmetric wiggler as a polarizer. It is assumed that the wiggler polarizes the beam in the horizontal direction and that the Siberian Snake is opposite to the wiggler (fig. 1). The only difference from a proton machine is the emission of synchrotron radiation in the arcs. The calculations are interesting for two reasons:

a.) the configuration shown in fig. 1 is less perturbing for the whole machine compared to spin rotators at each experiment.

b.) it is the present understanding that the degree of polarization is limited in big storage rings by the so-called nonlinear spin acceptance[14]: at higher energies the nonlinear effects contribute significantly to depolarizing effects. A horizontal spin could push this fundamental limit to higher values.

It was first shown by A. Chao and later by C. Prescott in unpublished contributions[15] that for LEP a purely horizontal polarization would have a depolarization time

* This work was supported by the U.S. Department of Energy under contract DE-AC05-84ER40150.

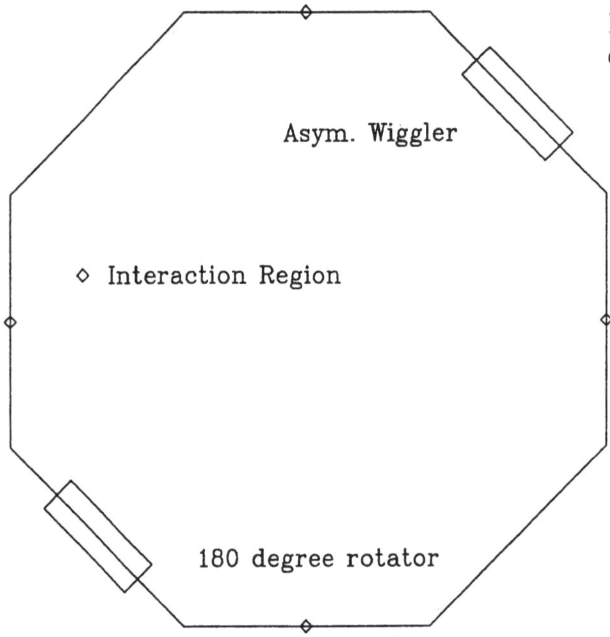

Figure 1. Schematic diagram of storage ring

of only a few seconds. The reason is that the particles emit photons along the arcs. As a result the particle spins precess with different angular velocities around the field axis. Since the emission point can be anywhere in the ring the perturbation does not repeat after one revolution and the stability condition of a Siberian Snake is not fulfilled.

The aim of this paper is to define spin matching conditions that compensate this effect.

Depolarization of a purely horizontal polarization without Snakes

A purely horizontal spin precesses around the magnetic bending field. When the particle is deflected by the angle α the spin is deflected by an angle ϕ. The angles are related to each other by

$$\phi = a\gamma\alpha, \tag{1}$$

where $a = (g-2)/2$ and γ is the Lorentz factor. After a full circulation α is 2π. Assume two particles: one at the nominal energy and one with an energy deviation $\Delta\gamma_0$ caused by the emission of a photon. The spins rotate with different angular velocities. Particles emitting synchroton photons perform synchrotron oscillations. As a result $\Delta\gamma$ varies with time.

$$\Delta\gamma = (\Delta\gamma_0)cos(\omega_s t)e^{-t/\tau_s}, \tag{2}$$

where ω_s is synchrotron frequency and τ_s is the damping time for synchrotron oscillations. After damping, the spins of the particles deviate slightly from each other

$$\Delta\phi = 2\pi \int_0^\infty a(\Delta\gamma)dt = 2\pi a(\Delta\gamma_0)\left(\frac{b}{b^2+\omega_s^2}\right), \tag{3}$$

where b is $1/\tau_s$. Note that after the damping period the deviation of the two spin directions is small but not zero. A spin tune of 0.08, a revolution frequency of 10 MHz and a damping time of 34 msec (LEP at 45 GeV) leads to a deviation of 10^{-9} radian when the emitted photon changes the energy of the electron by 10^{-3}.

In order to compensate this effect a spin shift proportional to $-\Delta\gamma$ with the proportional constants defined in equation (3) has to be applied. This can be done by beam bumps. In the following the principal concept is explained and in the appendix the beam bump concept is discussed in more detail.

Starting point for the proof of principle is equation (1). Assume a section of the machine where the particle is deflected by an angle α. Assume further that before and after this section the spin is rotated into the vertical direction. As a result the spin motion of a particle with the energy γ is retarded by $a\gamma\alpha$. The spin motion of a particle with the energy $\gamma + \Delta\gamma$ is retarded by $a\gamma\alpha + a(\Delta\gamma)\alpha$. The difference in the spin motion is $a(\Delta\gamma)\alpha$. By choosing the angle α the spin motion can be influenced proportional to $\Delta\gamma$ with an adjustable proportionality constant. Note that a negative angle α leads to a negative spin shift.

According to equation (3) the required spin shift is very small. Therefore in a real machine is required not an impractical 90 degree spin rotation but simply a deviation of the spin from the horizontal plane in one or several bending magnets. It will be shown later that this is sufficient.

It has to be taken into account that the spin shift is performed every revolution. Using equation (2) the spin shift has to be summed over many revolutions:

$$\Delta\phi_f = 2\pi a(\Delta\gamma_0) \sum_{n=0}^{\infty} \left(e^{-nt_0/\tau_s} cos(nt_0\omega_s) \right)$$

$$= 2\pi a(\Delta\gamma_0) \frac{1 - e^{-t_0/\tau_s} cos(\omega_s t_0)}{1 - 2e^{-t_0/\tau_s} cos(\omega_s t_0) + e^{-2t_0/\tau_s}}. \qquad (4)$$

The spin matching condition for this case is $\Delta\phi = -\Delta\phi_f$.

Spin matching condition with concentrated cavities

Equation (2) is an idealized description of particles in a storage ring. In a real storage ring the cavities are concentrated in one or two sections. The particle energy is changed by emission of a photon in the arc. After the emission the energy is not changed until the particle passes through the cavities. Assume now that a spin shift proportional to the energy is performed in front of and after each set of cavities. The spin is changed in front of the first set of cavities proportionally to $-\Delta\gamma_0$ by an angle $\Delta\phi_0$. The cavities change the energy to $\Delta\gamma_1$ and the spin is shifted after the cavities proportionally to $\Delta\gamma_1$ by an angle of $-\Delta\phi_1$. The energy is not changed until the spin comes to the next set of cavities. Before the particle enters the next set of cavities the shift is proportional to $-\Delta\gamma_1$, namely $\Delta\phi_1$. As a result the effect of the second shift is canceled. Finally all shifts are cancelled except the first one:

$$\Delta\phi_f = \Delta\phi_0 - \Delta\phi_1 + \Delta\phi_1 - \Delta\phi_2 + \Delta\phi_2 - ..... = \Delta\phi_0. \qquad (5)$$

The spin matching condition is $\Delta\phi = -\Delta\phi_0$. The shift is proportional to $-\Delta\gamma_0$. A comparison with formula (5) shows that the spin shift no longer depends on storage ring parameters.

The spin shift proportional to the energy before and after the cavity is equivalent to a spin shift proportional to Δs, where Δs is the deviation of the particle from the center of the bunch. This can be shown in the following way: Δs is expressed as

$$\Delta s \approx (\Delta\gamma) \int_{s_1}^{s_2} \alpha(s) ds, \tag{6}$$

where $\alpha(s)$ is the momentum compaction factor. On the other hand, the energy gain in a cavity is proportional to Δs. When the spin is shifted before and after a cavity proportionally to $\Delta\gamma$ with different signs the shift will be therefore proportional to Δs.

Spin matching with concentrated cavities and rotator

The principal layout is shown in fig. 1. Note that there is only one stable condition possible in such a configuration ("stable" means that the spin returns after one revolution into the same position). This stable position or n-axis in the polarizing element points in the radial direction. As a result the spin is aiming in the direction of the n-axis when the particle is polarized in the polarizing element or when a photon is emitted in the polarizing wiggler. Assume now that a photon is emitted somewhere in the arc. The energy is reduced and the spin comes to the rotator with a deviation of

$$\Delta\phi_n = a(\Delta\gamma_0)\left(\pi - \frac{s_2 - s_1}{R}\right) \tag{7}$$

from the n-axis, where s_1 is the position of the rotator and s_2 is the position where the photon is emitted. s is only counted in the bending magnets. R is the bending radius in the magnets. $\Delta\phi_n$ is a deviation from the n-axis and remains a deviation after damping and finally leads to depolarization.

The scheme for compensation for this effect is similar to the scheme shown in the previous chapter. If it is assumed that Δs in equation (6) and $s_2 - s_1$ in equation (7) are proportional to each other (which is true for most storage rings), the effect can be compensated by a spin rotation before and after the cavity in the previously described manner.

This concept is explained in more detail in the following. The deviation of the spin from the n-axis is, according to equation (7),

$$\Delta\phi_n = a(\Delta\gamma_0)\left(\pi - \frac{s_{cavity} - s_{emission}}{R}\right). \tag{8}$$

The change in energy in a cavity according to equation (6) is

$$\Delta\gamma_{cavity} = k(\Delta\gamma_0) \int_{s_{emission}}^{s_{cavity}} \alpha(s) ds,$$

$$\approx k_1(\Delta\gamma_0)\left(\frac{s_{cavity} - s_{emission}}{R}\right), \tag{9}$$

k and k_1 are constants. As mentioned, the spin is shifted proportionally to Δs or $\Delta\gamma_{cavity}$,

$$\begin{aligned}\Delta\phi_{total} &= k_2\Delta\phi_n + k_3\Delta\gamma_{cavity}, \\ &= k_2(a\Delta\gamma_0)\left(\pi - \frac{s_{cavity} - s_{emission}}{R}\right) \\ &\quad + k_3\Delta\gamma_0\left(\frac{s_{cavity} - s_{emission}}{R}\right)\end{aligned} \tag{10}$$

If k_2 and k_3 are chosen in an appropriate way,

$$\Delta\phi_{total} \approx a(\Delta\gamma_0)\pi \tag{11}$$

is independent of the point of emission.

Equation (11) is valid only when cavities and spin rotator are not separated by elements with strong synchrotron emission. If there is a significant synchrotron radiation emission between cavity and spin rotator the polarization will be destroyed. In order to avoid this an additional spin shift proportional to Δs has to be installed. This can be done by using a cavity in front and after the spin rotator. The cavities have opposite phase and therefore do not act on the beam.

Appendix: spin matching by closed orbit bumps

As shown in previous papers, transverse polarization in high energy electron-positron storage rings can only be obtained by applying closed orbit deviations in an intelligent manner to the existing orbit. The method is based more or less on trial and error; no beam position monitors are able to detect the small orbit variations between a low and a high degree of polarization. In the following it is shown that a similar technique can be applied for pure horizontal polarization.

Assume a closed orbit kick at a certain position in the machine. Let us assume that the spin is rotated as a result of this kick around the momentum axis by an angle β_1. Afterwards the spin is rotated in a bending magnet by Δ around the vertical axis and afterwards by another kick or set of kicks around the momentum axis by β_2. The spin transfer matrix for this case is

$$\begin{pmatrix} cos\Delta cos\beta_1 cos\beta_2 - sin\beta_1 sin\beta_2 & -sin\Delta cos\beta_2 & -\cos\Delta sin\beta_1 cos\beta_2 - cos\beta_1 sin\beta_2 \\ sin\Delta cos\beta_1 & cos\Delta & -sin\Delta sin\beta_1 \\ cos\Delta cos\beta_1 cos\beta_2 + sin\beta_1 cos\beta_2 & -\sin\Delta sin\beta_2 & -cos\Delta sin\beta_1 sin\beta_2 + cos\beta_1 cos\beta_2 \end{pmatrix} \tag{12}$$

If the starting point (where β_1 is applied) is (x_0, y_0, z_0), and the final point after three deflections is (x_3, y_3, z_3), and z_0 and z_3 are close to zero (purely horizontal spin), the total spin deflection angle can be written in the form

$$\Psi_m = f(\frac{y_0}{x_0}, \gamma) = \sum_{n=0}^{\infty} a_n \left(\frac{y_0}{x_0}\right)\gamma^n. \tag{13}$$

Applying several bumps in a proper way leads to

$$\Psi = \sum_m \Psi_m = \sum_m \sum_{n=0}^{\infty} a_{n,m}\left(\frac{y_0}{x_0}\right).\gamma^n \approx k\gamma + \text{higher orders.} \tag{14}$$

k can be chosen by selecting a series of suitable bumps. As a result the spin matching in the horizontal case is similar to the correction scheme which has to be applied for the vertical spin direction.[16]

Summary

In a storage ring with a purely horizontal spin and a Siberian Snake, the spin matching conditions are similar to the spin matching conditions for vertical polarization; a combination of beam bumps has to be found which compensate the depolarizing effects. These bumps compensate the random emission of synchrotron emission on the spin.

Literature

[1] A. A. Sokolov and I. M. Ternov, Dokl. Akad. Nauk SSSR [Sov. Phys.-Doklady] 8, 1203 (1964).
[2] R. Belbeoch et al., Proc. USSR Natl. Conf. Part. Acc. 1968, p. 129.
[3] S. I. Seredyaokov et al., Zh. Eksp. Teor. Fiz. [Sov. Phys.-JETP] 44, 1063 (1976).
[4] A. S. Artamanov et al., Phys. Lett. B118, 225 (1982).
[5] R. F. Schwitters et al., Phys. Rev. Lett. 35, 1320 (1975).
[6] D. P. Barber et al., Phys. Lett. B135, 498 (1984).
[7] D. P. Barber et al., IEEE Trans. Nucl. Sci., NS30, 2710 (1983).
[8] W. W. Mackay et al., Phys. Rev. D29, 2483 (1984).
[9] A. Blondel and J. M. Jowett, LEP note 606.
[10] M. Placidi and R. Rossmanith, LEP note 539 (1985).
[11] Y. Shatunov et al., AIP Conf. Proc. No. 187, p. 305.
[12] Ya. S. Derbenev and A. M. Kondratenko, Part. Acc. 8, 115 (1978).
[13] A. D. Krisch et al., Phys. Rev. Lett. 63, 1137 (1989).
[14] J. Kewisch et al., Phys. Rev. Lett. 62, 419 (1989).
[15] A. Chao, private contribution, C. Prescott, private contribution, CERN 1985.
[16] R. Rossmanith and R. Schmidt, Nucl. Instrum. Methods A236, 231 (1985).

The New Conceptual Design of Snakes and Spin Rotators in RHIC*

S.Y. Lee and E.D. Courant

Accelerator Development Department, Brookhaven National Laboratory, Upton, NY 11973, USA

We discuss the generalized snake configurations, which offers either the advantages of shorter total snake length and smaller horizontal orbit displacement in the compact configuration or the dual functions of a snake and a 90° spin rotation for the helicity state. The generalized snake is then applied to the polarized proton collision in RHIC. The possible schemes of obtaining high luminosity are discussed.

1. Introduction

Recently, K. Steffen[1] has discovered a family of snakes with the magnet sequence as

$$S = (-H,\ -V,\ 2H,\ 2V,\ -2H,\ -V,\ H) \tag{1}$$

where H and V are respectively the horizontal and vertical bending magnets. To satisfy the snake criteria, the sequence of magnets must not alter the particle orbit outside the snake and the spinor of the particle is transformed according to $e^{i\frac{\varphi}{2}\hat{n}_s\cdot\vec{\sigma}}$, where φ is the spin rotation angle and $\hat{n}_s = (\cos\varphi_s, \sin\varphi_s, 0)$ is the snake axis. A 100% snake rotates the spin with angle $\varphi = \pi$. A 90° spin rotation corresponds to $\varphi = \pi/2$. A partial snake corresponds to $|\varphi| < \pi$.

An interesting feature of the Steffen's snake is that the snake axis φ_s and the spin rotating angle φ can be varied by varying the excitation of H and V magnets, i.e. for $\varphi = \pi$, we obtain

$$\cos^2\psi_y + \cos 2\psi_x \sin^2\psi_y = 0 \tag{2}$$

$$\sin\varphi_s = \sqrt{2}\cos\psi_x \tag{3}$$

where ψ_x and ψ_y are respectively the spin rotation angle of H and V magnets. The relation between ψ_x and ψ_y in Eq. (2) ensures the snake condition of the magnet sequence and Eq. (3) determines the snake axis. Figure 1 shows ψ_x, ψ_y relationship of Eq. (2) and φ_s vs. ψ_y. When $\varphi_s = 0$, or π, the snake axis is along the radial $\hat{x}$ axis. The total integrated magnet strength is given by

$$\int B d\ell = 1.746(6\psi_x + 4\psi_y)\ [\mathrm{T-m}] \tag{4}$$

and the corresponding orbit displacements are given by

$$D_x = (\ell_x + \ell_y + 2\,\ell_g)\frac{\psi_x}{G\gamma}\ ;\ D_y = (2\,\ell_x + \ell_y + 2\,\ell_g)\frac{\psi_y}{G\gamma} \tag{5}$$

where ℓ_x, ℓ_y and ℓ_g are respectively the length of H and V magnets and the distance

* Work performed under the auspices of the U.S. Department of Energy

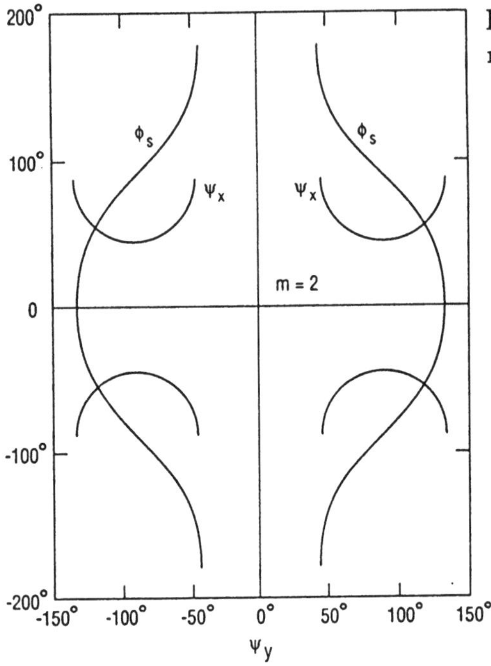

Figure 1: Spin rotation angle of H, V magnets, ψ_x, ψ_y is shown for 100% snake.

between the adjacent magnets and where $G = (g-2)/2$ is the Pauli anomalous g factor. Since the lengths of the magnets ℓ_x, ℓ_y are given by

$$\ell_{x,y} = 1.746 \frac{\psi_{x,y}}{B[T]} \tag{6}$$

where the magnetic flux density B is in Tesla, therefore the orbit displacement is inversely proportional to the magnetic field strength of the dipole.

The advantage of the Steffen snake is that the snake axis $\hat{n}_s$ can be changed continuously by proper ψ_x, ψ_y excitations. However, the snake configuration suffers the rigid structure of magnet position, i.e. the total length of the snake is given by

$$L = 6\ \ell_x + 4\ \ell_y + 6\ \ell_g + [\ell_x + 2\ \ell_g]\ , \tag{7}$$

where extra free space in the bracket of Eq. (7) is wasted. On the other hand, if the snake can be divided into two parts, then the snake can be fitted into two adjacent straight sections. In this case, the helicity state of the particles at the mid–section of the snake can be achieved. Such a modified snake configuration has been worked out recently.[2] In this paper, we apply the idea to study the possible polarized proton collision mode in RHIC (the Relativistic Heavy Ion Collider). Section 2 reviews the modified snake configuration. Section 3 discusses the AGS–RHIC requirement. The conclusion is in Section 4.

2. Review of the Modified Snake Configuration

The essential feature of the Steffen snake is the symmetric arrangement of vertical bending magnets and the anti–symmetric horizontal bending magnets. These features can be preserved in the following modified snake configuration

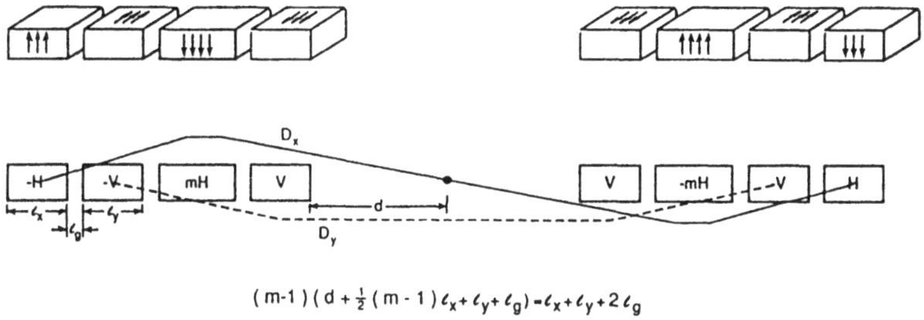

Figure 2: Schematic drawing of the modified snake configuration.

$$S_m = (-H,\ -V,\ mH,\ 2V,\ -mH,\ -V,\ H)$$

where the number m is determined by the geometry. Figure 2 shows the schematic drawing of the magnet arrangement, where m is given by the geometric condition

$$(m-1)\left(d + \frac{1}{2}(m-1)\ell_x + \ell_y + \ell_g\right) = \ell_x + \ell_y + 2\,\ell_g\,. \tag{8}$$

The spin rotation angle φ and the snake axis angle φ_s are given by

$$\cos\frac{\varphi}{2} = \cos^2\psi_y + \cos m\psi_x \sin^2\psi_y \tag{9}$$

$$\cos\varphi_s = \frac{-\sin\frac{m\psi_x}{2}\cos\psi_y}{\sqrt{\cos^2\frac{m\psi_x}{2} + \sin^2\frac{m\psi_x}{2}\cos^2\psi_y}} \tag{10}$$

Note that $m\psi_x$ and ψ_y are the relevant variable in determining φ and φ_s.

2.1 The Compact Snake Configuration

The compact snake configuration corresponds to the parameter $d = 0$. Depending on ℓ_x, ℓ_y and ℓ_g, m can be determined from the geometric consideration. For example, assuming $B = 2$ Tesla, $\ell_g = 0.15$ m and $\varphi_s = 180^o$, we obtain m=2.334 with a total length of 11.48 meters for the compact snake configuration in comparison with a total length of 13.54 meters for m=2 configuration. The compact snake configuration saves about 15% of the total length. Besides, the total $\int Bd\ell$ and D_x are also reduced.

2.2 Split Snake Configuration

By adjusting the parameter m, we can obtain proper distance $2d$ between the two halves of the snake. The orbit displacements at the middle of two halves of the snake can be corrected by a orbit shifter of $(-V',\ V')$ at both ends of the snakes. However when the distance $2d$ becomes large, we will have m$\simeq$1. Obviously, the total length of the snake and the radial orbital displacement increase as well. Such a split snake configuration would not be practical for the insertion detector area. However the split snake can be fitted into two adjacent straight sections separated by a quadrupole. Such a snake configuration ease the design criteria of the low energy ($\leq$ 30 GeV) accelerators.

To fit a collider IR onto the space between the split snake, the snake configuration shown in Fig. 3 is most appropriate.

The advantage of the split snake configuration is that the spin in the mid section of the snake will be on the horizontal plane. Thus such a snake serves a dual purpose

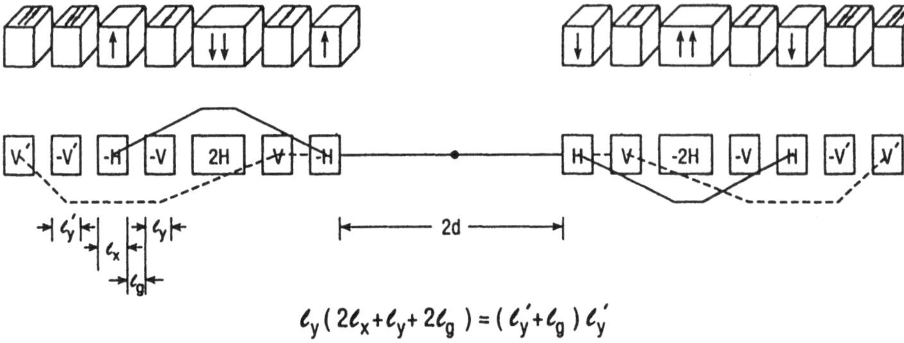

Figure 3: Schematic drawing of the split snake configuration.

of being a snake and a spin rotator for the helicity state experiments. For a spin up particle passing through the half snake, the spin orientation becomes

$$S_x = -\sin m\psi_x \sin\psi_y \ ; \ S_y = \sin^2\frac{m\psi_x}{2}\sin 2\psi_y \ ; \ S_z = 0 \tag{11}$$

where $m = 2$ for the snake configuration shown in Fig. 3. Because of the horizontal orbit compensation magnet, the spin is further rotated ψ_x angle around the vertical axis. Thus by changing ψ_x, ψ_y excitation, one can study helicity experiments or transverse spin experiments in the same IR. Such a scheme can save the need of four spin rotators in the polarized proton experiment.

3. Polarized Proton Collision in RHIC

To achieve a good luminosity in the polarized proton collision in RHIC, we must try to achieve the following parameters:

$$N_B \geq 10^{11}; \epsilon_N \leq 20\ \pi\text{mm} - \text{mrad}$$

where N_B is the number of particles per bunch and ϵ_N is the 95% normalized emittance. To achieve these goals, 20 linac pulses accumulation in the Booster is needed. These polarized protons are then accelerated in AGS to reach $G\gamma \simeq 48$. Partial snake in AGS was studied recently[3] to correct the imperfection resonances. We foresee no difficulty in reaching $G\gamma = 48$ with about 50% polarization.

Once the polarized proton reach $G\gamma \simeq 48$, the beam can be transferred to RHIC without depolarization.[4] The depolarization resonance strength in RHIC is shown on Fig. 4, i.e. the important resonance location will be

$$K = 81n \pm \nu_B; \nu_B = \nu_y - 6 \tag{12}$$

where ν_B is the phase advance accumulated across the dipole cells. The maximum resonance strength will be of the order $\epsilon_K \leq 0.5$ for $\epsilon_N \leq 10\ \pi$ mm-mrad. The imperfection resonance will be of the order 0.06, when the closed orbit is corrected to within 0.3 mm rms. Careful numerical simulation and analytic study[5] indicates that two snakes will be sufficient to maintain the spin polarization in RHIC. Using the concept of Section 2, one can employ two snakes in each ring. The corresponding luminosity at the top energy of 250 GeV is given by

$$\mathcal{L} = 5.7\times 10^{31}\frac{\left(\frac{N_B}{10^{11}}\right)^2\left(\frac{B}{57}\right)^2}{\left(\frac{\epsilon_N}{20\pi}\right)^2}\ \text{cm}^{-2}\text{sec}^{-1} \tag{13}$$

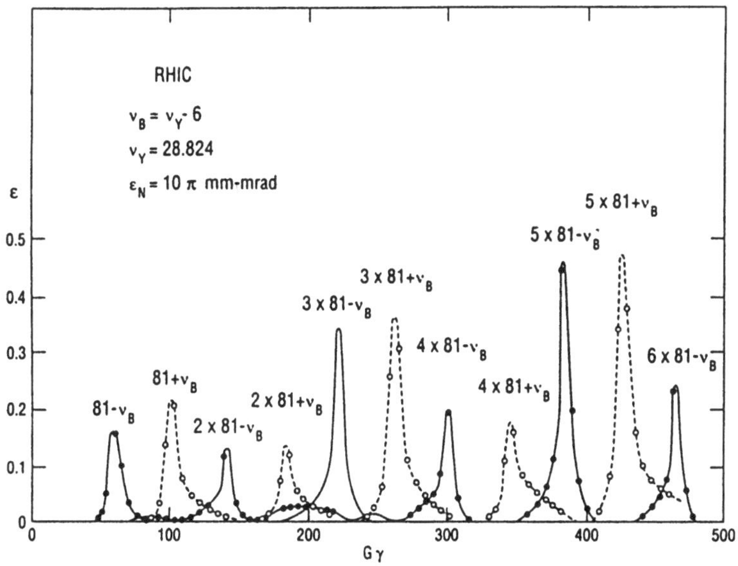

Figure 4: Intrinsic resonance strength for RHIC for $\epsilon_N = 10\ \pi$ mm-mrad.

where B is the number of bunches in each ring. Note here that the importance of smaller emittance in obtaining higher luminosity. The luminosity will gain a factor of 4 when the emittance is smaller by a factor of 2.

4. Conclusion

We have made feasibility study for the possible polarized proton operation in RHIC. The important tasks are:

1. multi–pulses accumulation in the AGS–Booster with small emittance,
2. accelerator through AGS to reach $G\gamma \simeq 48$, where the transfer line between the AGS and RHIC is spin transparent, and
3. construction of four snakes in RHIC.

When these tasks are accomplished, the polarized proton luminosity will be larger than 5×10^{31} cm^{-2}sec^{-1} at $\epsilon_N = 20\pi$ mm-mrad. By carefully maintaining the emittance at 10π mm-mrad, the luminosity can achieve 2×10^{32} cm^{-2}sec^{-1}, which corresponds to 2 interactions per crossing.

References

1. K. Steffen, Particle Accelerator 24, 45 (1989).
2. S.Y. Lee, "Snakes and Spin Rotators", BNL 52248/UC–414 (1990).
3. T. Roser, "Properties of Partially Excited Siberian Snakes", AIP Conf. Proc. No. 187, p. 1442 (1988).
   L. Ratner, ed., Workshop on Partial Snake in AGS. June 1990.
4. S.Y. Lee and E.D. Courant, "Effect of the Vertical Bends in ATR Transfer Line on the Polarized Proton Operation", AD/RHIC–63 (1990).
5. S.Y. Lee and E.D. Courant, Phys. Rev. D41, 292 (1990).

The Capability of Polarized Beam Acceleration at the Moscow KAON Factory

N. Golubeva and Yu. Senichev

Institute for Nuclear Research of the Academy of Sciences of the USSR, SU-117312 Moscow, USSR

In the design of the Moscow Kaon Factory great emphasis bas been put not only on the production of a high intensity beam (250 μA in the Booster and 125 μA in the Main Ring), but also on the capability of the acceleration of polarized proton beam up to the final energy (45 GeV) and with lesser intensity. The expected depolarizing effects and the conservation methods of the polarization have been investigated. In the Booster at present we propose to cross the depolarizing resonances either with a fast tune jump method or a harmonic correction method.

Depolarization in the Booster

The Booster lattice for the Moscow Kaon Factory is the racetrack with two arcs and two long straight sections. The arcs have a FODO structure with 32 cells. To increase the transition energy the eight periodicity is introduced on the arcs. A regular dipole arrangement is destroyed so that two central halves the cells on the superperiod are empty. Besides there are the small modulation of quadrupole strengths on the arc superperiods. To produce zero dispersion in both straight sections the arcs are tuned to the integer value of the tune. The superperiodicity of the Booster ring equals two [1].

During the acceleration from 600 MeV to 7.5 GeV 14 imperfection and 13 intrinsic resonances will be encountered. The strengths of the resonances have been evaluated with the help of the programme DEPOL written by E. Courant [2].

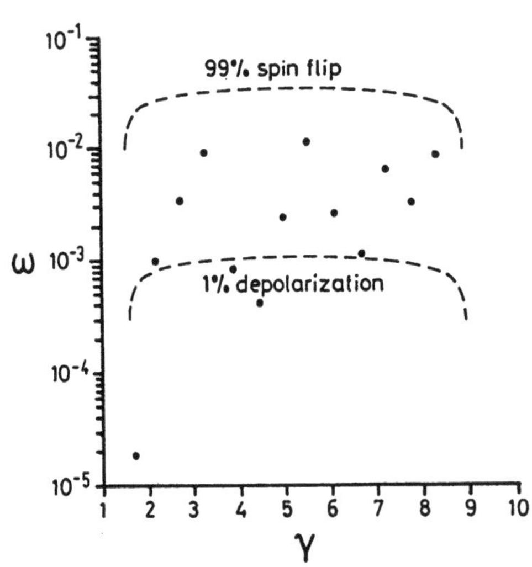

Fig.1 Imperfection resonances in the Booster

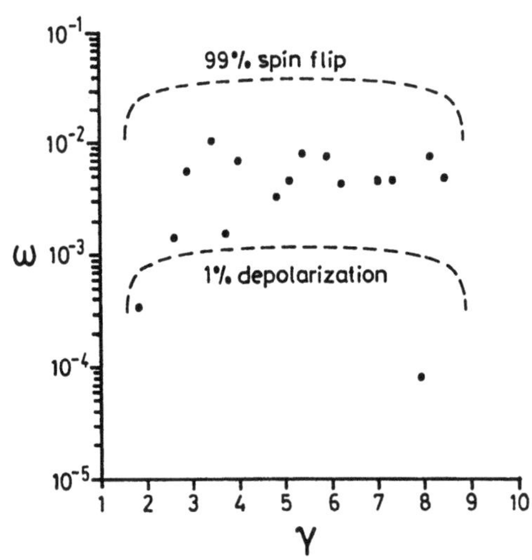

Fig.2 Intrinsic resonances in the Booster

For the imperfection resonances it is assumed that all magnets are misaligned in all dimensions according to a Gaussian distribution. Figure 1 shows the resonance strengths of the Booster for the closed orbit deviations of 5.14 mm and 0.4 mm, that corresponds to the uncorrected and corrected orbit.

The effect of the intrinsic resonances have been calculated with normalized emittance of 10 π mm mrad. Figure 2 shows a diagram of the intrinsic resonances (small stars). Most of the resonances have the intermediate strengths and arise due to the fact that the straight sections have a noninteger tune und therefore disturb the high periodicity of the arcs.

The lattice structure of the MKF Booster allows to reduce the number of depolarizing intrinsic resonances, that was investigated for the first time by U. Wienands for the racetrack lattice for the TRIUMF Kaon Factory Driver ring [3]. Since the racetrack lattice of the MKF Booster has twofold symmetry, 14 first-order intrinsic resonances within the energy range of the ring are given by the conditions

$$\gamma G = 2k \pm \nu_y, \quad k = 0, 1, 2, \cdots, \tag{1}$$

where γ is the relativistic energy, G is the gyromagnetic anomaly of the proton and G = 1.7928. The periodicity of the arcs is eight. In the inital racetrack Booster the vertical phase advance of the straight sections was 1.25 ($\nu_y = 6 + 1.25 = 7.25$). We can tune the straight sections to have an integer multiple of 2π phase advance, for example, $2 \times 2\pi$, and can expect that this straight sections will be transparent for the beams as far as spin motion is concerned. Then the conditions of depolarizing resonances are

$$\gamma G = 2k \pm (\nu_{yarc} + 2) = 8k \pm \nu_{yarc}, \quad k = 0, 1, 2, \cdots, \tag{2}$$

where ν_{yarc} is the tune of the arcs. The tune of the arcs becomes noninteger $\nu_{yarc} = 6.25$. This may be done in both planes, then the straight sections are completely invisible to the spin motion. But since the first-order depolarizing resonances are excited by the vertical motion of the beam, it should be sufficient to tune only the vertical phase advance of the straight sections. This maintains zero dispersion in the straight section.

If we retune the Booster lattice in the vertical plane according this rule, there are three intrinsic resonances within the energy range. Figure 2 shows the strengths of this resonances by big stars. One of them $\gamma G = 6 + \nu_y = 6 + 2 + \nu_{yarc} = 8 + \nu_{yarc}$ is very week, that it is possible due to the missing magnet arrangement of the arcs. The remaining resonances are due to the eightfold symmetry of the arcs created by the modulation of the ρ-function and beta function. Table 1

Table 1. Depolarizing resonances in the Booster racetrack with invisible straight sections

kS $\pm\nu_y$	γ	$\mid\epsilon\mid$	P_i/P_O $\delta = 0$	P_i/P_O $\delta = 0.1$
2 - ν_y	3.486	9.440 10^{-3}	0.2444	0.9816
18 - ν_y	5.438	7.280 10^{-3}	0.2801	0.9805
6 - ν_y	7.949	0.074 10^{-3}	1	-
Total				0.9624

summarizes the strengths of the resonances and the expected depolarization. The depolarization is calculated for each resonance either without fast tune jump for the crossing speed for the sinusoidal magnet angle or assuming fast crossing with a tune jump of 0.2 and a crossing speed of 0.035, which has been established within one turn. Formulae of Froissart and Stora and Courant and Ruth have been used to evaluate the depolarizaton [2]. All resonances can be crossed very efficiently using the tune jump. The total depolarization is approximately 5 %. If we take into account 5 % residual depolarization from imperfection resonances, more than 90 % of the polarization can be preserved up to maximum energy.

It appears that a racetrack lattice of the MKF Booster can be very well suited for acceleration of polarized protons, if the arcs have high symmetry and the straight sections are invisible to the spin. But since the vertical phase advance of the arcs is noninteger, the dynamic aperture of the ring is decreased considerably. The intensity of the polarized beam is about 10 μA (rather than the design curent of 250 μA). We suppose that such dynamic aperture will be enough for the polarized proton. There are not the restrictions on the horizontal tune, that permits the control of the dispersion function in the straight sections.

For the tune jump it is necessary to accomodate the pulsed quadrupoles in the suitable places on the ring. There are enough free space on the straight sections.

Since we have long straight sections (every about 45 m) in the Booster lattice, it is possible to consider the problem of Siberian Snakes adoption for the Booster lattice.

Depolarization in the Main Ring

The synchrotron lattice uses a FODO focusing structure with two arcs joined by dispersion suppressors and straight sections. The transition energy is determined by the horizontal tune, that equals $\nu_x = 6$. Every arc consists of 18 regular cells with a phase advance per cell of 60° in both planes. The dispersion suppressors consists of two cells, which have the same focussing structure as the regular cells of the arc but half the bending strength. All the bending power is concentrated in one cell and the second cell is free of any bending magnet.

Due to the low periodicity and large acceleration range from 7.5 GeV to 45 GeV this lattice have a high number of resonances: 72 intrinsic and 71 imperfection. The strengths of these resonances have been evaluated with the help of DEPOL similarly as in the Booster. The result is shown in Figure 3. It is obvious that crossing all these resonances will be very difficult. We therefore plan to adopt the Siberain Snake scheme to avoid the passage through a depolarizing resonances. The length of the straight sections is about 170 m and provides space long enough for Siberian Snakes. At present we work at the problem of the Siberian Snakes adoption.

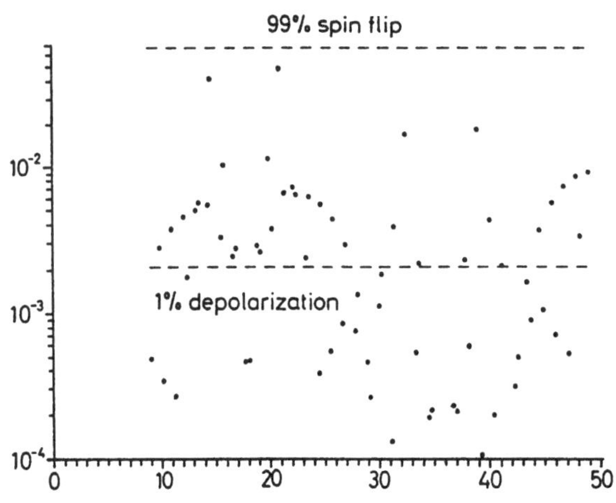

Fig. 3 Intrinsic resonances in Main Ring.

References

1. Golubeva N.I., Iliev A.I., Senichev Yu.V. The New Lattices for the Booster of Moscow Kaon Factory, Proc. of INES, Moscow, 1989.

2. Courant E.D. and Ruth R.D. The Acceleration of Polarized Protons in Circular Accelerators, BNL 51270, UC-28, 1980.

3. Wienands U. Polarization in the New Racetrack Driver Ring of the TRIUMF Kaon Factory, TRI-PP-88-14.

Polarized Protons in the TRIUMF KAON Factory

U. Wienands

TRIUMF, 4004 Wesbrook Mall, Vancouver, B.C., Canada V6T 2A3

The TRIUMF KAON Factory has the potential to deliver polarized beams at an intensity up to two orders of magnitude higher than available at other facilities. The provisions made for avoiding depolarization in the accelerators are discussed. The 3 GeV Booster lattice has by virtue of its lattice design only one intrinsic and five imperfection resonances; all are weak and produce little depolarization. For the 30 GeV Driver and also the Extender helical Siberian snakes are proposed, designed using 2 T room temperature dipole magnets. Care has been taken in order to ensure proper matching of the spin on the transfer between rings with and without snakes.

1 Introduction

At TRIUMF a 500 MeV polarized proton beam of 500 μA intensity and 80% polarization has been in use for many years. The new optically pumped ion source being commissioned at this time has recently produced more than 20 μA of H^- beam and allowed more than 5 μA protons with 50% polarization to be extracted from the cyclotron. As commissioning continues, the intensity and degree of polarization are expected to increase substantially. Accelerated to 30 GeV in the proposed KAON Factory[1] this beam would surpass in intensity any other high-energy polarized proton beam by about 2 orders of magnitude.

Acceleration of polarized proton beams in a synchrotron without depolarization is possible if the polarization vector is aligned at injection with the local precession axis, the axis about which the polarization vector appears to precess in subsequent turns at a fixed azimuth. This axis defines a stable spin direction. The number of revolutions per turn of the spin around this axis is called the spin tune and is given by $\gamma G = \gamma(g-2)/2$, where γ is the total energy in units of the rest mass and G is the anomalous magnetic moment; for protons $G = 1.7928$. Due to the dominating vertical guiding field this direction is usually vertical. However, if the effect of horizontal field components add up coherently over many turns a resonance condition occurs and the local precession axis will be rotated away from the vertical direction. The speed at which this occurs, in other words the crossing speed of the resonance, determines whether or not the polarization will be preserved crossing a resonance. The horizontal field components occur due to either the betatron oscillations of the particles, leading to the intrinsic resonances, or due to the unavoidable closed-orbit excursions in any real machine, leading to the imperfection resonances. First-order intrinsic resonances occur at

$$\gamma G = kP \pm \nu_y, \tag{1}$$

where k is any integer and P is the periodicity of the machine. Imperfection resonances occur at any integer value of the spin tune, *i.e.*

$$\gamma G = n. \tag{2}$$

The resonance strength ϵ indicates by what fraction of 2π the polarization vector is rotated away from the vertical axis during one turn.

While crossing a resonance the local precession axis rotates from the vertical into the horizontal direction right on the resonance and into the vertical direction again with opposite direction after leaving the resonance. Froissart and Stora have treated resonance crossing analytically and derived a formula for the depolarization due to asymptotic crossing of an isolated resonance[2],

$$\frac{P}{P_0} = 2\exp\left(-\pi^2\epsilon^2/2\alpha\right) - 1\,, \tag{3}$$

where α is the crossing speed, $\mathrm{d}\gamma G/\gamma G$. It is thus the expression ϵ^2/α that determines the degree of depolarization. Polarization will be unchanged in magnitude if this expression is either very large (slow crossing, very strong resonance) or very small (fast crossing, very weak resonance). Depolarization will occur if the magnitude of this expression is in between these cases (moderate resonance, intermediate crossing speed).

For intrinsic resonances the crossing speed can be increased by pulsing the machine tune, thereby adding a term $\mathrm{d}\nu_y/\nu_y$ to α. Hereupon based are the resonance-jumping schemes employed first at the ZGS and now at the AGS.[3,4] For imperfection resonances no such tune dependence exists; the strength of these resonances is reduced during crossing by orbit correction cancelling the offending harmonic of the horizontal fields.

2 The KAON Factory Accelerators

In order to be able to accelerate the full intensity of 100 μA delivered from the TRIUMF cyclotron up to 30 GeV the TRIUMF KAON Factory consists of two fast-cycling synchrotrons and three storage rings, as shown in Fig. 1. The first synchrotron, "B" (Booster), accelerates the beam from 0.45 GeV to 3 GeV in 10 ms, while the second synchrotron, "D"

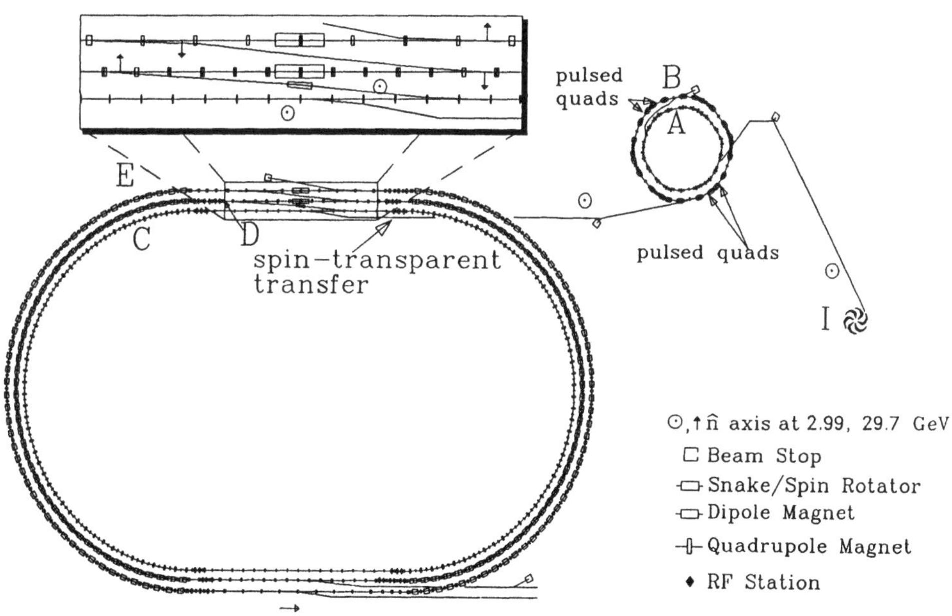

Figure 1. Layout of the accelerators and storage rings of the KAON Factory

(Driver) accelerates to the final energy in 75 ms. The storage rings are (A) to accumulate the beam from the cyclotron, (C) to "park" five fillings of the Booster before transfer into the Driver, and (E) for slow extraction to regain a high duty factor. Only in the accelerators are depolarizing resonances to be crossed as the energy of the storage rings is in between the resonances.

The depolarizing-resonance spectrum for both synchrotrons of the TRIUMF KAON Factory were calculated using the program DEPOL[5] and are shown in Fig. 2. The Booster has six-fold symmetry; in addition its vertical tune was lowered from the regular high-intensity tune of 7.2 to 4.2 leaving only one intrinsic resonance in the acceleration range. This resonance is quite weak as are the 5 imperfection resonances. As the acceleration and therefore the resonance crossing is very fast depolarization is expected to be quite small even if no corrective measures are taken. However, there is room in the lattice for 4 pulsed quadrupoles in case they turn out to be necessary and correction of the imperfection resonances can be done by the 24 vertical orbit correctors already planned for in the machine.

The Driver lattice has only 3 strong intrinsic resonances if the lattice is tuned in a special way, enhancing its apparent symmetry. But there are 52 imperfection resonances which are sufficiently strong to cause severe depolarization, especially towards the upper energy. Correcting these resonances would be very difficult, especially for a fast cycling machine. It is for this reason, and to avoid possible problems with the spin flip crossing the strong intrinsic resonances, that we propose to use Siberian snakes as a means to avoid depolarization in the Driver.

Siberian snakes are 180° spin rotators with the spin-precession axis in the horizontal plane. By flipping the vertical component of the spin on each turn in a circular machine

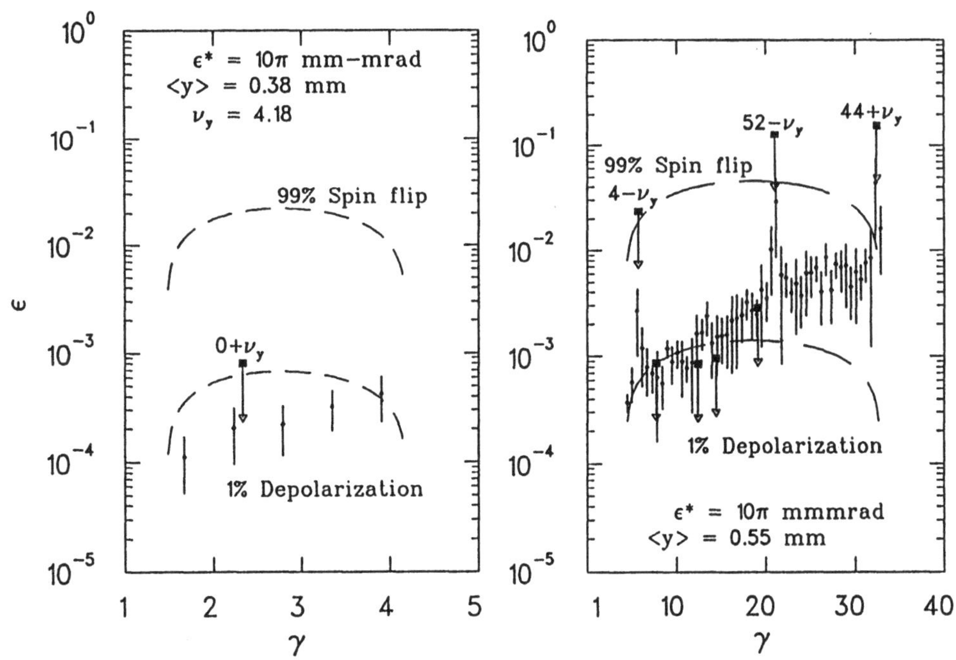

Figure 2. Spectrum of depolarizing resonances in the Booster (left) and Driver (right). The intrinsic resonances (↓) are labelled according to their harmonics; the arrows point to the strength at 1π mm-mrad emittance. The bars on the imperfection resonances indicate the $\pm 1\sigma$ uncertainty of their strength.

the effect of horizontal depolarizing fields can be made to cancel over two turns, thus eliminating all depolarizing resonances.[6] In a ring equipped with a snake the spin tune is fixed at 1/2, which is a non-resonant value. Snakes can be classified according to the orientation of their spin-precession axis; snakes of the 1st kind have a longitudinal precession axis while snakes of the 2nd kind have a radial axis; however, any other orientation of the precession axis is possible as long as it is in the horizontal plane. For use at high energies snakes have to be designed using transverse dipole fields as these are more efficient than longitudinal fields by a factor $\gamma G/(1+G)$ in rotating the spin. The transverse fields lead to excursions of the orbit within the device, hence the name "snake". Recently experimental proof has been given of the snake principle.[7]

For the TRIUMF Driver ring the challenge was to design a snake with sufficiently small orbit excursions at 3 GeV injection energy. Helical snakes, using a cork-screw like field, have been shown to produce very small orbit excursions and at the same time to be very efficient in the total field integral needed to achieve the 180° spin rotation.[8] At TRIUMF we have shown that helical snakes can also be designed with straight dipole magnets at little increase in orbit excursions and field integral.[9] For the Driver we designed a helical snake with $2\frac{1}{4}$ twists having 9 straight dipole magnets plus 4 orbit restoring dipoles, all room-temperature dc-magnets at 2 T field. At 3 GeV orbit excursions are about 10 cm. The total length of the snake is 13 m and the field integral is 23.6 Tm. A diagram of the orbit and the magnet array of the snake is shown in Fig 3. Such a snake has been successfully matched into the straight section of the Driver lattice. Since the snake has a constant magnetic field the matching conditions change with energy, therefore four quadrupoles at each end of the snake have to be programmed in order to maintain matching at all energies.

In addition to its fixing the spin tune at 1/2 the snake also turns the stable spin direction into the horizontal plane. This implies that the polarization, which is vertical in the C ring, has to be rotated to match the local precession axis at the injection point in the Driver. In order to facilitate this injection will be in the same straight section where the snake is located. At a transfer energy of 2.99 GeV, *i.e.* a spin tune of 7.5, the local precession axis of the Driver is radial at the snake azimuth and matching of the spin can be done with a solenoid rotating the spin by 90°. This avoids a spin rotator with a non-longitudinal axis, which would be a much more complicated device, in the transfer line. The solenoid has a field integral of 7.2 Tm and may be superconducting.

At extraction the polarization has to be matched into the Extender ring. Here matching with a spin rotator in the transfer line is impossible because of the high energy. Therefore a snake is also foreseen in the Extender ring. At a transfer energy of 29.7 GeV γG is 58.5 and the local precession axis in the snake straight is radial in both rings, so matching of the polarization into the Extender is achieved as the transfer line is designed to be spin transparent. Although the direction of the polarization at the snake azimuth varies with energy polarization is always longitudinal at the straight section opposite to the snake,

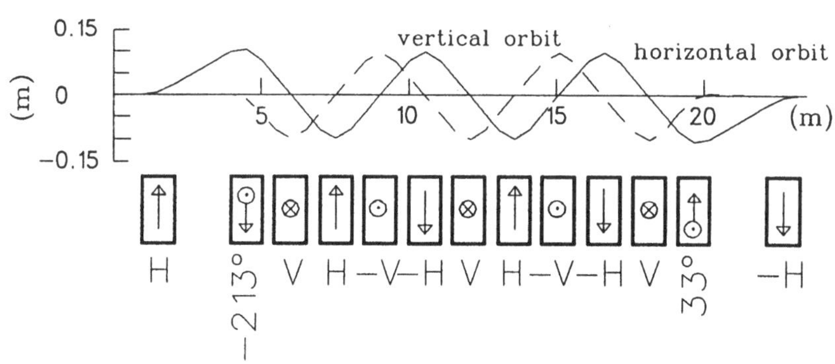

Figure 3. Diagram of the $2\frac{1}{4}$-twist helical Siberian snake.

where the extraction system is located. To accommodate this the transfer line from the E ring to the experimental area for polarization experiments has been designed to be spin transparent as well. The stable spin direction is indicated in Fig. 1.

Vertical polarization can be maintained in a ring with two snakes with orthogonal precession axes. However, no design of a snake other than of the 1st kind is known with sufficiently small orbit excursions at 3 GeV beam energy to be practical.

3 Summary

The TRIUMF KAON Factory has the potential to deliver polarized beams up to 30 GeV at an intensity about 2 orders of magnitude above what will be feasible by the time it is built. Care has been taken in the design to ensure spin transparency of the transfer lines between the large rings and to the experimental area, where the polarization will be non-vertical. The lattice of the Booster synchrotron, with a periodicity of six, is inherently well suited for acceleration of polarized proton beams and should cause little or no depolarization. For the Driver ring, Siberian snakes are the most efficient means of preserving polarization; the long straight sections of the racetrack lattices provide ample space for the placement and optical matching of the snake. As no design for an orthogonal pair of snakes with sufficiently small orbit excursions at 3 GeV is known, only a single snake in the Driver is possible. This leads to horizontal polarization in the Driver and necessitates a spin-matching solenoid in the injection line into the Driver as well as a snake in the Extender ring.

REFERENCES

1. *KAON Factory Study Accelerator Design Report* (TRIUMF, Vancouver, 1990) (unpublished).
2. R. Froissart and M. Stora, Nucl. Instrum. Meth. **1**, 297 (1960).
3. T. Khoe *et al.*, Part. Accel. **6**, 213 (1975).
4. L.G. Ratner *et al.*, IEEE Trans. Nucl. Sci. **NS-32**, 1656 (1985).
5. E.D. Courant and R.D. Ruth, *Report 51270* Brookhaven National Laboratory (Upton, NY, 1980).
6. Ya.S. Derbenev and A.N. Kondratenko, *Proc. X. Conf. on High Energy Accelerators*, Protvino, USSR, (1977), and
   Ya.S. Derbenev and A.N. Kondratenko, *Part. Accel.* **8**, 115 (1978).
7. A.D. Krisch *et al.*, *Phys. Rev. Lett.* **63**, 1137 (1989).
8. E.D. Courant, *Proc. 8th Int. Symposium on High Energy Spin Physics,* Minneapolis, MN, (1988), p. 1085*ff*.
9. U. Wienands, *Helical Siberian Snakes using Dipole Magnets*, submitted to *Particle Accelerators.*

Status of the Spin Splitter Experiment at IUCF

R. Rossmanith

CEBAF, 12000 Jefferson Avenue, Newport News, VA 23606, USA

Representing the Spin-Splitter Collaboration: N. Achurin (Iowa), Laura Badano (Genova) D. Cline (UCLA), M. Conte (Genova), R. Giacomich (Trieste), M. Giorgi (Trieste) J. Hall (New Mexico), S. Hsueh (Fermilab), H. Kreiser (Hamburg), H. Meyer (IUCF) Jane Nachtman (Iowa), T. Niinikoski (CERN), Y. Onel (Iowa), G. Pauletta (Udine) P. Pancella (IUCF), A.Penzo (Trieste), A. Pisent (CERN), M. Pusterla (Padova) R. Rossmanith (CEBAF), U. Strohbusch (Hamburg), A. Yegneswaran (CEBAF)

Introduction

During the last years a new polarization effect for protons or antiprotons was investigated theoretically[1,2]. The idea is very simple: a storage ring consists of strong nonlinear fields, the quadrupoles. The nonlinear fields act on the spin of the particle by the Stern-Gerlach effect: the particles are kicked by a very small amount. This small kick produces betatron oscillations. During the next revolution the gradient of the quadrupole is (naturally) the same but the spin direction can be altered. As a result the strength and the direction of the kick can be different from revolution to revolution. When this time dependent kick follows a betatron oscillation the kicks can add up over many revolutions and a macroscopic oscillation can be excited which can be detected by an outside detector. The principle is the same as in the one used for measuring the Q-value: small kicks add up over many revolutions and produce an oscillation with a significant amplitude.

Aim of the ongoing measurements is to establish this effect. Although the above mentioned concept is very simple and staightforward and somewhat an analogy of well established techniques it differs in two points from these techniques. First of all the excitation strength is small and therefore a long time is needed to see an effect. Long term drifts may smear out the effect. And, secondly, the driving force is related to the spin direction. This direction cannot be measured from outside in a simple way.

Concept of the Measurements

When a proton passes a quadrupole the acquired kick depends on the mometum of the particle:

$$\eta' = p_t/p$$

where $p = \beta\gamma m_p c$, $\gamma = (1-\beta^2)^{-1/2}$ and m_p is the (anti)proton mass. p_t depends on the force F and the length of the interaction τ. When L is the length of the interaction and βc is the velocity, η' becomes:

$$\eta' = \frac{FL}{\beta^2 \gamma m_p c^2}$$

The Stern-Gerlach force is $F = \nabla(\mu.B)$ with $\mu = 1,41.10^{-10} JT^{-1}$. For low energies $\gamma \approx 1, \beta \approx 0.1$, a field of 20 Tm^{-1} the kick is 9.4x10^{-14} rad. In order to calculate a displacement in the machine the kicks have to be multiplied with the square roots of the beta function. Assuming that the betafunction both at the point of observation and the point where the kick is applied is 100 m the betatron amplitude is $\approx 10^{-8}$mm. 10^8 revolutions are required to accumulate a betatron amplitude of 1 mm.

Since most of the low energy storage rings have revolution times in between 10^{-6} and 10^{-7} secs a measurable amplitude should be accumulated within 10 to 100 seconds. During this time the spin motion and the betatron motion must be in phase. The resonance condition is

$$G\gamma = Q \pm n \tag{1}$$

where Q is the Q-value of the machine and n is any integer.

It should be mentioned that the addition of Stern-Gerlach kicks on single particles is a well established effect[3]. The difference to the single particle technique is only that an ensemble of particles and an esemble of spins have to be considered. This means the spin motion must be coherent over the revolutions and the betatron motion must be coherent too.

Coherent spin motions

The simplest way of obtaining a coherent spin motion is to inject the beam into the storage ring with a polarization along the n-axis. Since the spin motion at the n-axis is the same from revolution to revolution the Q-value must be an integer in order to fulfill the resonance condition equation (1). This is not a stable condition for running the machine.

In order to overcome this problem solenoids can be used. When we considered this experiment first an Ann Arbor-IUCF group[4] discussed a Siberian-Snake experiment with a strong solenoid at IUCF, Indiana, USA. IUCF is a proton storage ring with an electron cooler operating in the 100 MeV range. After the Siberian Snake group have finished parts of their shifts we asked for a few shifts to test some of our ideas.

The tests are incomplete. Nevertheless, they produced already interesting results.

First we wanted to accumulate a relatively high polarized current in the machine. The aim was to measure with only one filling the polarization with high accuracy. We injected the beam with a polarization along the n-axis. The n-axis is parallel to the field axis of the solenoid. All other spin directions would rotate around the n-axis. Currents of up to $2\mu A$ polarized beam at 107 MeV ($G\gamma$=2) can be accumulated. The solenoid was not compensated during these shifts by the skew quadrupoles. The polarimeter shows a polarization of typically

$$P_H = 86 \pm 10 percent, P_V = 4 \pm 10 percent$$

The next measurement was performed in order to test if a polarization perpendicular to the the n-axis oscillates in a stable way around the n-axis. In order to proof this we proceeded first as before. After accumulation we switched slowly the solenoid off, waited a few minutes and measured the polarization. With the switched off solenoid the n-axis is vertical and the horizontal polarization has to oscillate around the vertical n-axis. The measurement shows that this is the case: a clear horizontal and vertical component can be detected.

$$P_H = 84 \pm 13 percent, P_V = 53 \pm 13 percent$$

We were able to store up to $20\mu A$ polarized beam when, in the next step, the solenoid was completely compensated by skew quadrupoles. The compensation was verified by measuring the coupling between the horizontal and vertical plane. We tried to inject a horizontal polarization 90 degree relative to the n-axis. The polarization should flip from revolution to revolution by 180 degree. By timing the injection in a proper way we make sure that the newly injected beam is only injected when the stored beam has the right polarization. In the used accumulation scheme the beam has to be debunched. Due to the used electronic scheme we were not able to keep the debunching time short enough. The beams are debunched during a period of 10^6 revolutions. A particle with an energy deviation of 10^{-3} performs 1000 revolutions more or less than a particle with nominal energy and the spin flips thousand times more or less often leading to a depolarization of the stacked beam. A remedy could be an additional cavity keeping the bunches together or a shorter time in which the beam is debunched.

Summarizing these measurements: non equilibrium spin motions were investigated and everything seen so far does not seem to be in any contradiction to the assumptions made for the spin splitter concept. Nevertheless, more investigations are necessary.

Coherent Betatron Motions

A rather short time was spent to get more information on Landau effects. Equation (1) assumes that the Q-value is the same for all particles. In reality the Q-value depends on γ, the phase and the amplitude of the betatron motion. The chromaticity (dependance of Q on γ) is not compensated at IUCF. One would assume that a single kick applied to the beam would lead to a incoherent oscillation. This was verified experimentally[5]. In a next step presumably the chromaticity has to be compensated by sextupoles.

Literature

[1] T. Niinikoski and R. Rossmanith, Nucl. Instr. Meth. A 255,460 (1987)
[2] M. Conte and M. Pusterla, Il Nuovo Cimento, 103A, 1087 (1990)
[3] R. van Dyck, P. Schwinberg and H.Dehmelt, Phys. Rev. D, 722 (1986)
[4] A. D. Krisch et al., Phys. Rev. Lett., 63, 1137 (1989)
[5] T. Ellison, private communication, IUCF 1990

An Intense Polarized Beam by a Laser Ionization Injection

C. Ohmori[1], *S. Hiramatsu*[2], *and T. Nakamura*[3]

[1]Institute for Nuclear Study, University of Tokyo,
Tanashi-shi, Tokyo 188, Japan
[2]National Laboratory for High Energy Physics,
Tsukuba-shi, Ibaraki 305, Japan
[3]Electrotechnical Laboratory, Tsukuba-shi, Ibaraki 305, Japan

Accumulation of protons and polarized protons by photo-ionization injection are described. This method consists of (1)producing the neutral hydrogen beam by Lorentz stripping, (2)excitation of the neutral hydrogen beam with a laser, and (3)ionization of the hydrogen beam in the 2P excited state with another laser. When the laser for the excitation is circularly polarized, we can get a polarized proton beam. An ionization efficiency of 98% and a polarization of 80% can be expected by an intense laser beam from a FEL(Free Electron Laser).

INTRODUCTION

To accumulate an intense proton beam in a synchrotron, multi-turn injection using the charge exchange reaction is one of the most available methods, and is used at many accelerators. In this charge exchange injection, generally, beam loss and emittance growth from multiple scattering on a thin foil are severe problems for accumulating the very intense beam. For example, the beam loss becomes to be few μA for a beam current of a few hundred micro amperes even if the loss rate is less than 1%.

To store such an intense beam, a new accumulation method by photo-ionization injection has been proposed to eliminate the charge exchange foil[1]. The scheme is shown in Fig. 1. The first step of this ionization process is neutralization of the negative hydrogen beam with Lorentz stripping. The second step is excitation of neutral hydrogen with a intense laser flux. The final step is ionization of the 2P-excited hydro-

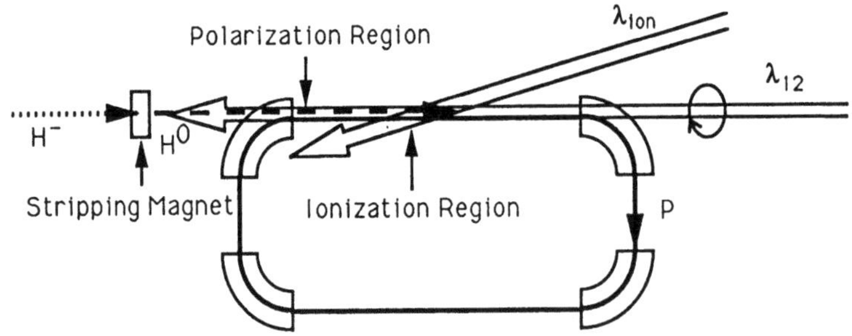

Fig. 1. The H^0 beam(dashed line) is excited and polarized by the laser with a wavelength of λ_{12} in the polarization region. Then, it is ionized with simultaneous irradiation by another laser (λ_{ion}) in the ionization region.

gen with another intense laser flux. The reason why we don't use the simple one-photon process is that the cross section of this two photon process is much larger than that of one photon($13.6eV$) absorption.

At the same time, we can obtain an intense polarized proton beam if we use an intense and circularly polarized photon flux for the excitation(optical pumping)[2]. It is difficult to obtain such an intense and polarized photon flux with a commercially available laser. In a case of FEL with a helical wiggler, we can expect an intense circularly polarized photon flux. The laser amplification gain and obtainable flux density in the FEL are estimated with a one-dimension calculation.

PRINCIPLE

The required wavelengths of the photons for the optical pumping to the 2P excited state and ionization in the H^0 rest frame, are $121.6nm$ and $365nm$, respectively. These required wavelengths in the laboratory frame are lengthened with Doppler shift for the head-on collision, as shown in Fig. 1[2]. For a 1GeV H^0 beam, the required wavelengths in the laboratory frame are $470nm$ and $1.4\mu m$ for the optical pumping and ionization, respectively. The wavelength in the laboratory frame($\lambda_{Lab.}$) is given by

$$\lambda_{Lab.} = \lambda_0 \gamma/(1 - \beta cos\theta), \tag{1}$$

where λ_0 is the wavelength in the H^0 rest frame and θ is the collision angle between the laser light and H^0 beam. The wavelengths in the laboratory frame are listed in Table 1.

Table 1. The wavelengths λ_{12} and λ_{ion} in the laboratory frame for the 1S-2P excitation and ionization of the 2P excited state hydrogen, respectively.

Kinetic Energy (GeV)		0.44	0.8	1.0
λ_{12}	(nm)	309.5	414.9	471.7
$\lambda_{ion}(Max.)$	(nm)	922.6	1245	1413

The cross section of the 1S-2P transition is given as follows[2],

$$\sigma_{12} = \frac{\lambda^2}{(2\pi)^2} \frac{A_{12}}{\Delta\nu_D} \frac{g_2}{g_1} \sim 2.3 \times 10^{-15} cm^2. \tag{2}$$

In this expression, $\lambda(= 121.6nm)$ is the wavelength for the 1S-2P transition in the H^0 rest frame, $A_{12}(= 1/1.6ns)$ is the probability of a spontaneous transition of $2P \rightarrow 1S$, $\Delta\nu_D$ is the Doppler broadening of an absorption line associated with the momentum spread of the beam, and g_2/g_1 is the ratio of statistical weight of the ground and excited levels. As the band width of the absorption, $1/A_{12} = 6.6 \times 10^8 Hz$, is usually much smaller than the Doppler broadening width, the wavelength spread of a laser should be as large as the momentum spread. Such a wide line spread is difficult for a commercially available laser, but the FEL has a natural wide wavelength spread of about $10^{-3} - 10^{-2}$.

The 1S-2P transition probability w_{12} in the H^0 rest frame is given by,

$$w_{12} = \frac{\sigma_{12}}{h\nu} \frac{1+\beta}{1-\beta} q_{12}, \tag{3}$$

where $h\nu(=10.2eV)$ is the photon energy and q_{12} is the power flux in the laboratory frame. At $1GeV$, for example, the required flux density of $470nm$ light for the optical pumping is $33kW/cm^2$ by assuming $w_{12}=5A_{12}$, as listed in Table 2. Such an intense flux is necessary to obtain a polarized beam as described in the next section.

The excited hydrogen beam can be ionized with $\lambda \leq 364.8nm$ light. The ionization cross section from the 2P state is given as follows using the Born approximation,

$$\sigma_{ion} = \frac{\pi}{60} 4\sqrt{2}\alpha^6 r_0^2 (\frac{mc^2}{h\nu_{ion}})^{\frac{9}{2}} \times \frac{5}{3} \leq 1.2 \times 10^{-15} cm^2. \tag{4}$$

In this expression, $\alpha(=1/137)$ is the fine structure constant and r_0 is the classical electron radius. The required photon flux for ionization, to obtain the probability of $w_{ion}=2A_{12}$, is in Table 2. This probability means that 98% hydrogen can be ionized in the 2m ionization region. However, it should be noted that Eq.(4) gives a too large cross section by a factor of several times at the wavelength in the vicinity of the threshold. Therefore, the required flux must be larger than the listed value. In Table 2, the required power flux of the YAG laser($\lambda = 1.06\mu m$) is also listed, which will be available for ionization of the 800MeV and 1GeV H^0 beam.

Table 2. The required power flux. q_{12} means the required power flux in the laboratory frame assuming the 1S-2P excitation assuming the probability of $w_{12}=5A_{12}$. q_{ion} and q_{ion}^{YAG} mean the power flux for the ionization assuming the probability of $w_{ion}=2A_{12}$ at the threshold wavelength and at YAG laser's one, respectively.

Kinetic Energy	(GeV)	0.44	0.8	1.0
q_{12}	(kW/cm^2)	72	43	33
q_{ion}	(kW/cm^2)	84	50	39
q_{ion}^{YAG}	(kW/cm^2)		90	120

BEAM POLARIZATION

When the H^0 beam is irradiated in the long flight region with circularly polarized laser radiation for optical pumping, the hydrogen atom proton is polarized[2]. The polarization of protons depends on the power flux of irradiation photon and irradiation time as shown in Fig. 2. When the probability is A_{12}, the $10m$ polarization region is necessary to get the 80% polarized proton beam. When the transition probability is $5A_{12}$, it can be obtained in the $5m$ polarization region. However, a flux intensity several times larger than saturation($w \sim 5A_{12}$) flux is not so effective for the polarization since the polarization time is governed by the spontaneous transition.

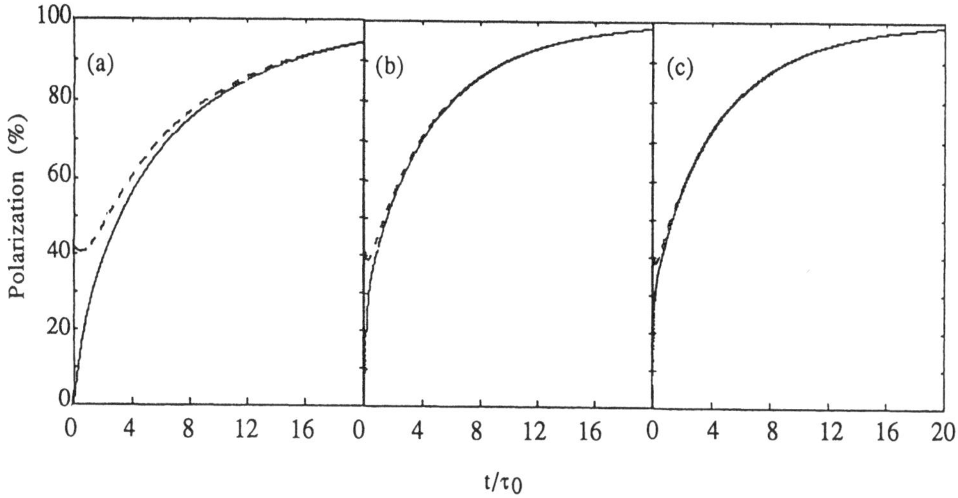

Fig. 2. Time dependence of the polarization of protons. $\tau_0(= A_{12})$ means the life time of the excited state. Figures (a),(b) and (c) are at the transition probabilities of A_{12}, $5A_{12}$ and $10A_{12}$, respectively. The solid lines refer to the ionization of beam atoms from both ground and excited states with a foil, the dashed lines from the excited state only with the laser ionization injection.

SCHEME

To obtain the high power laser flux, the interaction region is included in the optical cavity as shown in Fig. 3. In the optical cavity of the YAG laser with a $1kW$ output peak power, about $200kW/cm^2$ power flux will be expected even for beam size of $10mm^2$. This power flux is enough to ionize the hydrogen beam in the 2m interaction region.

We estimated the laser power of the FEL which is driven by a conventional linear accelerator and a helical wiggler as shown in Fig. 3 and Table 3, with one dimensional

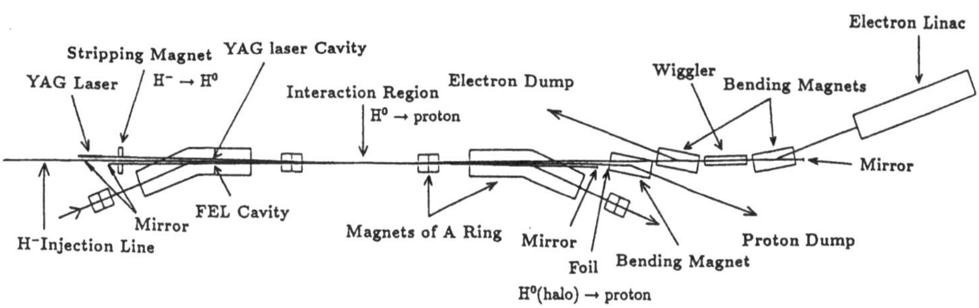

Fig. 3. Injection scheme by a FEL and YAG laser. The FEL system consists of an electron linac, two bending magnets, a wiggler and two mirrors. The electron beam, which is deflected by the magnet, passes through the wiggler which radiates photons. The photons are accumulated in the cavity between the two mirrors and excite the hydrogen atoms in the several meter interaction region. The excited hydrogen beam is ionized by the light in the YAG laser cavity.

Table 3. FEL parameters

Drive linac	
Energy	$52.7MeV$
Current	$10A$ (Bunch peak)
Normalized emittance	$\leq 3\pi cm \cdot mrad$
Energy spread	$\leq 0.2\%$
Phase angle of beam bunch	$2.5deg.$
Wiggler	
Wiggler length	$1.5m$
Wavelength of wiggler	$1cm$
Magnetic field	$4.22kG$
Wiggler type	helical
Beam size in the wiggler	$1mm$
Output radiation	
Wavelength	$470nm$
FEL gain	4.4%
Intercavity power flux	$2.4MW/cm^2$
Beam size in the interaction region	$3.6mm$

approximation model[3]. The maximum average flux in the cavity is expected to be $2.4MW/cm^2$, when the hydrogen beam size and the laser beam size are $10mm^2$. Even though, the estimated flux by the simplified one dimensional calculation may be too large by a factor of several times, it will be enough to get a high degree of polarization.

REFERENCES

[1] C.Ohmori, S.Hiramatsu and T.Nakamura, Proceedings of the 18th INS Symposium, March 1990, Tokyo, to be published.

[2] A.N.Zelenskiy et al.,Nucl. Instr. and Meth.,227(1984)429.

[3] P.Sprangle and R.A.Smith, Phys.Rev.A21(1980)293.

Higher Order Spin Resonances in High Energy Electron Storage Rings*

S.R. Mane

Brookhaven National Laboratory, Upton, NY 11973, USA

It was predicted by Sokolov and Ternov [1] that when electron and positrons circulate in a high energy storage ring, the particle spins become polarized by the emission of synchrotron radiation. In practice, the equilibrium degree of polarization is strongly affected by so-called depolarizing spin resonances. A formula for the equilibrium degree of polarization, including the resonances, was given by Derbenev and Kondratenko [2]. An algorithm to evaluate their formula was given by Mane, and coded into a computer program called SMILE [3]. The so-called SMILE algorithm treats arbitrary resonances, subject to the approximation of treating only linear orbital dynamics. Numerous formalisms for calculating synchrotron sideband resonances have also been published [4,5,6,7,8], and were all shown to be equivalent by Mane [8]. However, all of the above formalisms employ perturbation theory in one form or another, and the convergence of the various formalisms is still unresolved. The SMILE algorithm employs an expansion in powers of the orbital action variables, which becomes an expansion in powers of the beam emittances after averaging over the beam distribution. The other formalisms express the spin integrals of the resonances as a sum of Bessel functions, rather than a sum of powers of action variables.

This paper is intended to address the question of the convergence of the perturbation series for calculating the spin resonances. It does *not* address the question of developing techniques to model the imperfections in a storage ring, which is also an important question. Both analytical and numerical calculations were performed in the course of the investigation. The latter are mainly based on the SMILE formalism, as well as that in ref. [8]. The results can be summarized as follows. It turns out that the convergence of the SMILE perturbation series depends on the type of orbital modes being considered:

1. For resonances involving only horizontal and vertical betatron oscillations, the convergence of the SMILE perturbation series is rapid, even in storage ring models whose energy is as high as that of HERA or LEP.

2. For resonances involving only synchrotron oscillations, the convergence of the perturbation series depends on the spin tune modulation index α, defined by

* Work performed under the auspices of the U.S. Department of Energy

$$\alpha = \frac{\sigma_\epsilon^2(\gamma a)^2}{Q_s^2} \tag{1}$$

where σ_ϵ is the relative energy spread, Q_s is the synchrotron tune, γ is the average particle energy in units of rest energy, and $a = (g-2)/2$. If $\alpha \ll 1$, then the convergence is rapid, but if $\alpha \geq O(1)$, then the convergence is slow.

3. For synchrotron sidebands of parent betatron resonances, the convergence is again rapid or slow depending on whether $\alpha \ll 1$ or $\alpha \geq O(1)$.

Hence we see that the SMILE perturbation series is usually satisfactory, except in those cases involving synchrotron oscillations where the spin tune modulation index α is of order unity or more. In that case, we need to find an alternative method of summing the perturbation series. In fact, resonances involving synchrotron oscillations have been studied in detail, and formalisms for calculating such resonances, even when $\alpha \geq O(1)$, have been given in refs. [4] – [8], using a Bessel function expansion rather than a power series expansion in powers of the beam emittances. These formalisms suffer from the restriction that they treat only synchrotron sideband resonances, and do not treat other orbital modes. Hence, we can employ the following "hybrid" strategy: we use the SMILE algorithm for those resonances where it converges rapidly, and use the Bessel function expansions for those cases where the SMILE algorithm converges poorly. This strategy involves the neglect of certain spin integrals, as can be seen by the following example.

Note that the spin integral of a resonance of the form $\nu = Q_x + Q_s$+integer consists of two terms: (i) a horizontal betatron mode at first order and a synchrotron mode at second order, or (ii) a synchrotron mode at first order and a horizontal betatron mode at second order. The latter integral is included in the SMILE series, but not in the Bessel function expansions. Consequently, if the Bessel function expansion is used to calculate the resonance strength, the contribution of the second spin integral will be lost. It turns out, however, that this is usually a permissible approximation, because the contribution of that integral is much smaller than the first integral mentioned above.

The overall conclusion is that the use of a mixture of the SMILE and Bessel function formalisms can yield rapidly convergent expressions for the strengths of all spin resonances. This is achieved at the expense of occasionally neglecting certain spin integrals, which fortunately turn out, in practice, to be much smaller than those retained in the calculation.

The author thanks D.P. Barber, H. Grote, E. Keil and K. Yokoya for helpful comments and suggestions during the development of this work.

References

[1] A.A. Sokolov and I.M. Ternov, Dokl. Akad. Akad. Nauk SSSR **153** (1963) 1052 [Sov. Phys. Doklady **8** (1964) 1203].

[2] Ya.S. Derbenev and A.M. Kondratenko, Zh. Eksp. Teor. Fiz. **64** (1973) 1918 [Sov. Phys. JETP **37** (1973) 968].

[3] S.R. Mane, Phys. Rev. A **36** (1987) 120.

[4] Ya.S. Derbenev, A.M. Kondratenko and A.N. Skrinsky, Part. Accel. **9** (1979) 247.

[5] K. Yokoya, Part. Accel. **13** (1983) 85.

[6] C. Biscari, J. Buon and B.W. Montague, Il Nuovo Cimento **81** (1984) 128.

[7] J. Buon, Proc. 8th Int. Symp. on High Energy Spin Physics, Minneapolis (1988) (American Institute of Physics Conf. Proc. No. 187, vol. 2 (1989) p. 963).

[8] S.R. Mane, Nucl. Instrum. Meth., in press.

On Snake-Orbit Matching

Ya. S. Derbenev

Randall Laboratory of Physics, The University of Michigan,
Ann Arbor, MI 48109, USA

It is observed that Siberian Snakes express the substantial focusing effects on proton beams at low energies (about 1–10 GeV). The reduction of the transverse beam size in the Snake takes place during matching snake focusing with outside regular focusing of a beam and permits one to reduce the necessary aperture of snake magnets. Main limitations of the aperture and possibilities of snake-orbit matching are discussed.

INTRODUCTION

The Siberian Snake installed in an accelerator ring produces a few effects on particle motion in the ring. The most important effect is the distortion of the equilibrium orbit inside the snake section, which changes with energy, because the magnetic field of the snake is still constant. A second type of the snake's influence is a perturbation of focusing features of a machine lattice; its express in the betatron tune shift and also can increase effective beam emittance. Finally, the third kind of effects due to focusing effect of the snake are the excitation of vertical dispersion and perturbation of the horizontal one.

The outside effects go down with energy, as well as inside the orbit distortion and they are negligible or could be easily compensated in the rings with an injection energy above 10–20 GeV. A more serious question is how to suppress or compensate these effects at lower energies. We consider this phenomena and some possibilities to control them in the aspects which seem to be generally substantial and reliable.

1. Snake focusing effect

This effect is created by the correlation between the orbit distortion and the longitudinal snake field, B_z. Both are due to alternating the transverse field $\vec{B}_\perp$. We start with the perfect equation of particle motion:

$$\gamma Mc\dot{\vec{v}}_\perp = e\vec{v} \times \vec{B}_\perp + e\vec{v}_\perp \times \vec{B}_z,$$

here $\vec{v}_\perp$ is transverse (to z axis) particle velocity, v is longitudinal velocity. Transferring this equation to the derivation on z, will have:

$$\gamma Mc\vec{\nu}\,'_\perp = e\vec{i}_z \times \vec{B}_\perp + e\frac{\vec{\nu}_\perp}{v} \times \vec{i}_z B_z \tag{1}$$

here the unit vectors $\vec{i}_z$ notes z-direction. Due to the small value of z and the particle velocity oscillations in the snake's field, we can divide the transverse motion into two components: fast-oscillating (orbit distortions) and average smooth motion. Fast oscillations in the first approximation are given by the integral:

$$\tilde{\vec{\nu}}_\perp = \frac{e}{\gamma Mc}\vec{i}_z \times \int \vec{B}_\perp dz;$$

the equation for average motion was found by averaging equation (1):

$$\gamma Mc < \vec{\nu}\,'_\perp > = \frac{e}{v} < \vec{\nu}_\perp \times \vec{B}_z >$$

We can neglect the first term on the right side and also substitute $\nu_\perp \rightarrow \tilde{\nu}_\perp$, because the average value of B(z) is equal to zero. Also neglecting the distinction between v and the total particle velocity value, we get:

$$\vec{\rho}\,'' \approx (\frac{e}{pc})^2 < B_z \cdot \int \vec{B} dz >,$$

here $\vec{\rho} = (x, y)$ is the transverse two-dimensional coordinate. Using the equation

$$\mathrm{rot}\,\vec{B} = 0$$

and assuming $B_z\ (\vec{\rho} = 0) = 0$, will get a linear approximation of:

$$\vec{\rho}\,'' = (\frac{e}{pc})^2 < (\vec{\rho}\frac{\partial \vec{B}_\perp}{\partial z}) \cdot \int \vec{B}_\perp dz >= -(\frac{e}{pc})^2 < (\vec{\rho}\vec{B}_\perp)\vec{B}_\perp >$$

Further, we assume for simplicity that the average values of B_x^2, B_y^2 are equal and the interference from $< B_x B_y >$ is equal to zero. Finally, we get the equation for the average motion as follows:

$$\vec{\rho}\,'' = -\frac{1}{2}(\frac{e}{pc})^2 < B^2 > \vec{\rho} \equiv -\kappa^2 \vec{\rho}. \tag{2}$$

More rigorously, the parameter κ^2 can depend on z and also can be not quite the same in two planes; nevertheless, equation (2) could serve as a suitable model of snake focusing effect for qualitative consideration and estimations.

2. Matching of focusing

At first note that focusing the transformation inside the Snake is conservative, with the invariants:

$$\begin{aligned}(I_x)_{sn} &= \kappa x^2 + x'^2/\kappa, \\ (I_y)_{sn} &= \kappa y^2 + y'^2/\kappa;\end{aligned} \tag{3}$$

this transformation between points 1 and 2, which are coordinates of snake's ends, is as follows:

$$\begin{aligned}x_2 + x_2'/i\kappa &= (x_1 + x_1'/i\kappa)e^{i\psi} \\ y_2 + y_2'/i\kappa &= (y_1 + y_1'/i\kappa)e^{i\psi}\end{aligned} \tag{4}$$

$$\psi = \kappa\ell, \tag{5}$$

here ℓ is the length of the Snake.

We should compare forms (3) with the well-known general forms of invariants of focusing the lattice in a ring

$$\begin{aligned}I_x &= \frac{x^2}{\beta_x} + \beta_x(x' - \frac{\beta_x'}{2\beta_x}x)^2 \\ I_y &= \frac{y^2}{\beta_y} + \beta_y(y' - \frac{\beta_y'}{2\beta_y}y)^2,\end{aligned} \tag{6}$$

where the function $\beta_x(z)$, $\beta_y(z)$ describes the focusing parameters of a lattice and their modulation along the closed orbit. We assume for simplicity, that at the point $z_\pm$, which are an entrance and exit to the section with the Snake, respectively we have $\beta_x = \beta_y \equiv \beta_o$; $\beta_x' = \beta_y' = 0$; in such a case

$$I_x = \frac{x_+^2}{\beta_o} + \beta_o x_+'^2 = \frac{x_-^2}{\beta_o} + \beta_o x_-'^2, \tag{7}$$

and the same for the y-plane.

Moreover, suppose that we would be able to transform the particle motion from the ends of the Snake to the points $z_\pm$ using the canonical focusing transformation, similar to the crossover transformation (further we will call it extra crossover)

$$\begin{aligned}x_1 &= x_- F; \qquad & x_1' &= -x_-/F; \\ x_+ &= x_2' F; \qquad & x_+' &= -x_2/F\end{aligned} \tag{8}$$

After all of this substitution, we resume that the equality (7) will be fulfilled, i.e. the Snake will not increase amplitudes of particle oscillation, if

$$F = \sqrt{\beta_o/\kappa} \tag{9}$$

Under this condition, the variables (x_+, x'_+) and (x_-, x'_-) are connected as follows:

$$\begin{aligned} x_+ &= -x_-\cos\psi - \beta_o x'_-\sin\psi \\ x'_+ &= \frac{x_-}{\beta_o}\sin\psi - x'_-\cos\psi, \\ \psi &= \kappa\ell \end{aligned} \tag{10}$$

So the extra crossover (8) with the focusing parameter as (9) reduces the Snakes effect on focusing to the tune shift $\frac{\kappa\ell}{2\pi}$.

Note that the total tune shift in the section (z_-, z_+) is $\kappa\ell + \pi$; the contribution π is due to the transformation (8).

3. Extra crossover transformation

In view of the tune shifts ramping with energy, we cannot use a single quadrupole pair to provide the transformation (8) with the parameter (9). But it is possible to use two quadrupole doubblet on each side of the Snake in order to provide ramping parameter F at moveless magnets.

Let us consider such a lattice in the thin lens approximation. If we install one doubblet with the focusing parameter F at the point z_+ (right side) and another one with the focusing parameter $\hat{f}$ at some point $\hat{z}_+$ closer to the Snake, then we can calculate that f and $\hat{f}$ should be connected as follows (the point $z = 0$ is the middle of the Snake):

$$\begin{aligned} \frac{1}{f} + \frac{1}{\hat{f}} - \frac{z_+ - \hat{z}_+}{f\hat{f}} &= \frac{1}{F}, \\ 1 - \frac{z_+ - \hat{z}_+}{f} &= \frac{\hat{z}_+ - \ell/2}{F}, \end{aligned} \tag{11}$$

and we find how f and $\hat{f}$ should ramp with energy:

$$f = F\,\frac{z_+ - \hat{z}_+}{F - \hat{z}_+ + \ell/2}; \qquad \hat{f} = \frac{(z_+ - \ell/2)(z_+ - \hat{z}_+)}{z_+ - \ell/2 - F} \tag{12}$$

In practice it is not complicate to provide a positive value of f and $\hat{f}$, therefore, we will mind the limitations on change of F(p) as follows:

$$\hat{z}_+ - \ell/2 \leq F \leq z_+ - \ell/2 \tag{13}$$

4. Tune shift control

We also have to compensate the tune shift $\kappa\ell/2\pi$ ramping with energy during acceleration until it becomes negligibly small. In principle, the simplest way to provide this compensation is to make regular β-values of the lattice (i.e., without the Snake, but including the crossover considered above with $F = (z_+ - z_-)/2$ ramp with energy in order to fulfill the relation:

$$\int \frac{ds}{\beta(p,s)} + \frac{\kappa\ell}{z\pi} = \text{const}; \tag{14}$$

the integral here does not include the section (z_-, z_+). It seems easy to perform (14) if $\kappa_{inj}\ell << N$, here N is the number of lattice periods.

5. Snakes effects controlled at $\kappa\ell << 1$.

In the region $\kappa\ell << 1$ (i.e. the region where the Snake affects as thin lens with focusing parameter $1/\kappa^2\ell$) betatron tune shift can be easily compensated by small correcting quadrupoles. But generally, it does not mean that we can neglect by usage of extra crossover transformation which provides low β-function inside the Snake:

$$\beta_{in} \sim \frac{F^2}{\beta_o} << \beta_o\ , \text{ if } \beta_o\kappa >> 1.$$

Without the extra crossover, the betatron tune shift by the Snake would be

$$\Delta\nu \sim \beta_o \cdot \kappa^2\ell/4\pi;$$

(if we assume $z_+ \sim \beta_o >> \ell$) it will still not be small until

$$\beta_o\kappa^2\ell << 1. \tag{15}$$

On the other hand, during acceleration we achieve the region (15) and find we do not need extra crossover because the excitation of the particles free the oscillations by the Snake beams small in this region together with the tune shift. We can switch off the doublet $\hat{f}$ but should keep f equal to z_+ or switch on regular type of quadrupoles in the section in order to maintain constant ν-value with further acceleration. The betatron tune shift, which should be compensated by the usual correction technique is:

$$\Delta\nu_{sn} = \frac{1}{4\pi}\beta_{sn} \cdot \kappa^2\ell, \qquad (\beta_{sn} >> \ell)$$

6. On possible extensions of the method

We will now briefly discuss the possibility to extend the approach described above. Recall, that we assumed the entrance values of the β-function and matching focusing parameters F, f and $\hat{f}$ are equal for the two planes, (x, z) and (y, z). Obviously, such a method can be extended on the non-symmetric situation. We could then take into account that the points $\beta'_x = 0$ and $\beta'_y = 0$ could not coincide and we also could remove the condition of left-right symmetry. Generally, a perfect matching lattice can be more complicated in practice. At the time, it would be substantial to foresee in design the possibility to vary the configuration focusing magnetic field of the lattice in some area around the Snake, in order to use ramping focusing elements as matching elements during acceleration. (i.e. against the system of additional matching quadrupoles, as considered above, we could use system magnets, which provides snake-orbit matching at low energies and works as regular quadrupoles at high energies where the focusing effect of the Snake is small or even negligible.)

7. Examples of Snake focusing effects

Let us consider a few examples of the snake focusing effect at different injection energies. We normalize all the cases to helicoidal snake[1–3] and neglect the influence of the dipoles that match the spiral particle trajectory inside the Snake with outside sections. The radius of the spiral, r, and spin precession frequency in the Snake, ω, are connected with the field value B and its period λ by the formulas:

$$r = \frac{eB \cdot \bar{\lambda}^2}{pc}; \; (\bar{\lambda} = \lambda/2\pi).$$

$$\omega = \sqrt{\frac{\Omega_m^2}{\beta^2} + \left(\frac{1}{\bar{\lambda}}\right)^2} - \frac{1}{\bar{\lambda}};$$

$$\Omega_m = G\frac{eB}{Mc^2}; \qquad \beta = \frac{p}{(p^2 + M^2c^2)^{1/2}}$$

here e, M and p are the particle charge, mass and total momenta, c is light speed and G = 1.79. We assume $r << \bar{\lambda}$ and the perfect snake condition

$$\omega_m \cdot \ell = \pi, \tag{16}$$

here ω_m is ω at $p >> Mc$.

We also assume that B is constant during acceleration. Note, that at the condition of symmetricity of the snake field respectively the middle of the Snake (z = 0) as follows:

$$B_v(z) = -B_v(-z), \text{(vertical direction)}$$
$$B_h(z) = B_h(-z), \text{(horizontal direction)}$$

resulting axis of spin rotation lies in the horizontal plane (close to the particle velocity).

With the limitation on the injection value of r:

$$r_{in} \leq r_o$$

we have limitation on B:

$$B \leq \frac{p_o c}{e \lambda\!\!\!^{-2}} r_o,$$

here p_o is injection value of momenta. From (16) we have correspondent limitation on the snake length ℓ:

$$\ell \geq \frac{\pi \lambda\!\!\!^{-}}{[(\frac{Gp_o r_o}{Mc\lambda\!\!\!^{-}})^2 + 1]^{1/2} - 1}$$

and the resulting limitation of the snake focusing parameter $\kappa_o \ell$ at injection energy:

$$\kappa_o \ell \geq \frac{\pi}{\sqrt{2}} \cdot \frac{1}{[(\frac{Gp_o}{Mc})^2 + (\frac{\lambda\!\!\!^{-}}{r_o})^2]^{1/2} - \frac{\lambda\!\!\!^{-}}{r_o}}$$

Let us consider the numerical examples. Assume $\lambda\!\!\!^{-} = 20$ cm (i.e. $\lambda \simeq 1.25$ m) and $r_o = 3$ cm. Then for injection energy values (in GeV) 0.6, 1.5, 7.5, 12, and 70 we get the minimum values of $\kappa_o \ell$ as follows: $(\kappa_o \ell)$ min $\approx$ 3, 0.7, 0.1, 0.07, and 0.01. In fact, $(\kappa_o \ell)$ depends very weakly on r_o at injection energies above the value $(\lambda\!\!\!^{-}/Gr_o)Mc^2$. Note that the parameter, $\kappa_o \ell$, numerically can be larger at performance of the Snake as system of dipoles, but, of course, it will have the same order of value.

Finally, we recall that at $\kappa_o \ell << 1$ one could consider the Snake's focusing effect is small only if focusing the parameter $1/\kappa_o^2 \ell$ is larger than the machine β-function.

Conclusion

Let us resume our consideration of the Snake's effects on the focusing of the accelerator lattice.

1. Using Snakes in the energy region ~1–10 GeV, one should take care in matching the machines focusing lattice with the Snake as the focusing element; this matching should substantially ramp with the particle energy.

2. The general approach to this problem can be based on decreasing the β-function at the snake entrance to the optimum value excluding excitation of the free particle oscillations by the Snake. The beam compression also allows to substantially decrease the snake's magnetic aperture.
3. With this matching, one would also provide the constant value of the betatron tunes during acceleration, taking into account the tune shift by the Snake ramping down with the energy.

REFERENCES

/1/ Ya. S. Derbenev, A. M. Kondratenko, in Proceedings of the International Symposium on Polarization Phenomena in High Energy Physics, Dubna, USSR (November 1981), p. 281.

/2/ Yu. M. Ado et al., in Proceedings of the 8th International Symposium on High Energy Spin Physics, Minneapolis, Minnesota (September 1988), AIP V. 2, p. 1168–1173.

/3/ E. D. Courant, in Proceedings of the 8th International Symposium on High Energy Spin Physics, Minneapolis, Minnesota (September 1988), AIP V. 2, p. 1085–1092.

Reduction of the Spin Perturbations at Using the Intrinsic Framework

A.M. Kondratenko

Research Institute of Complete Electric Drive, SU-630090 Novosibirsk, USSR

An improved method to calculate the spin depolarizing perturabations in cyclical accelerators is presented including the rings with Siberian Snakes. The intrinsic orthonormal framework conjoined which precise particle velocity is used in order to reduce the expanded spin equations. The perturbation term are given in linear approximation on the particle declinations from equilibrium orbit.

1. Analysis of beam polarization precession in cylcical accelerators is being simplified in the intrinsic orthonormal framework conjoined with particle velocity. The idea of description bases on the well-known property which is that the spin projection on the velocity direction remains invariable in any magnetic fields for the particle without anomalous magnetic moment. Therefore, the spin rotation against velocity may be connected with the anomalous magnetic moment only. This property may be used in spite of the particle trajectories description is carried out in accelerating framework conjoined with equilibrium orbit.

The difference between the degree of polarization in the intrinsic framework and accelerating framework is connected only with ignorable spread of beam velocities in cyclical accelerators.

The simplification of description is useful in accelerators with different kinds of magnetic inserts (for example in accelerators with Siberian Snakes) in which the direction of equilibrium polarization is different from axial.

2. The most important thing for the description of the beam polarization precession is the linear approximation to the particle trajectories declination from the equilibrium orbit. Before the expression of spin perturbations in the linear approximation will be presented, we have to get an equation of the trajectory with introducing radial and longitudinal components of magnetic field on orbit (except, of course, axial field) which are needed for the beam polarization control.

Radius vector of particle $\vec{r}$ in the accelerating framework is written as:

$$\vec{r} = \vec{r}_e(\theta) + x\vec{e}_x(\theta) + z\vec{e}_z(\theta),$$

where x - is radial, z - is axial deflection from some ideal equilibrium orbit $\vec{r}_e(\theta)$. Particle position on orbit is longitudinal (along velocity) and it is determined by azimuth θ.

Periodic orth of accelerating framework are

$$\vec{e}_z, \vec{e}_x, \vec{e}_y = \frac{1}{R}\frac{d\vec{r}_e}{d\theta} \equiv \frac{\vec{r}\,'_e}{R} \tag{1}$$

and they satisfy the relations

$$\vec{e}\,'_z = K_z\vec{e}_y \,;\; \vec{e}\,'_y = K_x\vec{e}_z - K_z\vec{e}_x \,;\; \vec{e}\,'_z = -K_x\vec{e}_y \,,$$

where $2\pi R$ is the equilibrium orbit length. K_x, K_z - are dimensionless components of the equilibrium orbit curvature, which are expressed in terms of magnetic field components $\vec{H}$ on equilibrium orbit as following

$$K_x = H_x / < H_z > \quad ; \quad K_z = H_z / < H_z > ,$$

$$< H_z > = \frac{1}{2\pi}\int_0^{2\pi} H_z d\theta \quad ; \quad < K_x > = 0 \,; < K_z > = 1.$$

Writing the equation of particle movement in the magnetic field through the components of accelerating framework

$$\dot{\vec{v}} = \frac{q_0}{\gamma}\left[\vec{v}\vec{H}\right]$$

(γ - relativistic factor $q_0 = e/m(c = 1)$) we get the following equations in linear approximation, which are describing the transverse movement

$$\begin{aligned} x'' + g_x x &= g_{xz} z + K_y z' + \frac{\Delta\gamma}{\gamma} R K_z \,, \\ z'' + g_z z &= g_{zx} x - K_y x' - \frac{\Delta\gamma}{\gamma} R K_x \,. \end{aligned} \tag{2}$$

Here $\Delta\gamma/\gamma = (\gamma - \gamma_e)/\gamma$ are the relative deviation of energy from the equilibrium value, $K_y = H_y / < H_z >$,

$$\begin{aligned} g_x = K_z^2 + \frac{R}{< H_z >}\frac{\partial H_z}{\partial x} \quad &; \quad g_{xz} = K_x K_z - \frac{R}{< H_z >}\frac{\partial H_z}{\partial z}; \\ g_z = K_x^2 - \frac{R}{< H_z >}\frac{\partial H_x}{\partial z} \quad &; \quad g_{zx} = K_z K_x + \frac{R}{< H_z >}\frac{\partial H_x}{\partial x}. \end{aligned}$$

The relation between field derivatives follows from the equations $div\vec{H} = 0$; $rot\vec{H} = 0$:

$$\frac{\partial H_x}{\partial \theta} = R\frac{\partial H_y}{\partial x} + K_z H_y \,; \; \frac{\partial H_z}{\partial \theta} = R\frac{\partial H_y}{\partial z} - K_x H_y$$

$$\frac{\partial H_x}{\partial z} = \frac{\partial H_z}{\partial x} \,; \; \frac{\partial H_x}{\partial x} + \frac{\partial H_z}{\partial z} + \frac{1}{R}\frac{\partial H_y}{\partial \theta} = 0.$$

Slow synchronous oscillations of energy $m\Delta\gamma$ are being included in equations (2) adiabatically.

3. The equation of spin movement is written in intrinsic framework $(\vec{a}, \vec{\tau}, \vec{b})$ having a little difference from accelerating framework (1):

$$\vec{a} = \frac{[\vec{v}\vec{e}_z]}{|[\vec{v}\vec{e}_z]|} \simeq \vec{e}_x - \frac{x'}{R}\vec{e}_y \,; \; \vec{b} = [\vec{a}\vec{\tau}] \simeq \vec{e}_z - \frac{z'}{R}\vec{e}_y \,; \; \vec{\tau} = \frac{\vec{v}}{v} \simeq \vec{e}_y + \frac{x'}{R}\vec{e}_x \frac{z'}{R}\vec{e}_z \,. \tag{3}$$

Determining the angular velocity of orthonormal framework rotation

$$\vec{W}_b = \frac{1}{2}\left\{[\vec{a}\vec{a}'] + [\vec{\tau}\vec{\tau}'] + [\vec{b}\vec{b}']\right\} ,$$

we get the following equations for the spin precession velocity in linear approximation

$$\vec{S}' = \left[\vec{W}\vec{S}\right] \,, \; \vec{W} = \vec{W}_e + \vec{w}$$

on equilibrium orbit

$$(W_e)_1 = \vec{W}_e\vec{a} = \nu_0 K_x \,; \; (W_e)_3 = \vec{W}_e\vec{b} = \nu_0 K_z \,; \; (W_e)_2 = \vec{W}_e\vec{\tau} = \frac{q}{q_0}K_y \,; \tag{4}$$

for spin perturbation

$$\begin{aligned} w_1 \simeq \nu_0 \left[\frac{\Delta\gamma}{\gamma}K_x + \frac{z''}{R}\right] \; &; \; w_3 \simeq \nu_0 \left[\frac{\Delta\gamma}{\gamma}K_z - \frac{x''}{R}\right] , \\ w_2 \simeq \frac{q}{q_0}\left(-\frac{\Delta\gamma}{\gamma}K_y + K_x'\frac{x}{R} + K_z'\frac{z}{R}\right) &+ \frac{q_a}{q_0}\left(K_x\frac{x'}{R} + K_z\frac{z'}{R}\right) . \end{aligned} \tag{5}$$

In these equations q is the gryomagnetic relation, q_a - its anomalous part, $\nu_0 = \gamma_e q_a / q_0$.

The derivation of the equation (5) bas been done for the movement in magnetic fields and it will be generalized to the cases when both magnetic and electric fields take place.

If the longitudinal component of the magnetic field is of the same order of the transverse components, two spin perturbation transverse velocity components are discerned at high energies ($\nu_0 \gg 1$) before of all: along $\vec{a}$ and $\vec{b}$. On the contrary, only the longitudinal component of the precession frequency is not zero for the particle without anomalous magnetic moment ($q_a = 0$).

4. Let's write the expressions for the deviation of the precession frequency and equilibrium polarization direction for the particle moving near the equilibrium orbit. We have in linear approximation:

$$\begin{aligned} \Delta\nu &= < \vec{w}\vec{n} - \frac{1}{2} Im\, \vec{w}\vec{\eta} \int_{-\infty}^{\theta} \vec{w}\vec{\eta}^{*} d\theta >, \\ \Delta\vec{n} &= Im\, \vec{\eta} \int_{-\infty}^{\theta} \vec{w}\vec{\eta}^{*} d\theta \,. \end{aligned} \tag{6}$$

Here $\vec{n}(\theta) = \vec{n}(\theta + 2\pi)$ - is the equilibrium polarization direction. $\vec{\eta} = \vec{\eta}_1 + i\vec{\eta}_2$ two orthogonal solutions to $\vec{n}$, having the following property

$$\vec{\eta}(\theta + 2\pi) = \vec{\eta}(\theta) \cdot exp(-2\pi i \nu),$$

ν - is the generalized precession frequency of particle moving at equilibrium orbit.

5. Let's give a simple example for accelerators with one Siberian Snake inserted in straight section ($0 \div \theta_0$).

In this case the equilibrium polarization direction lies in the plane of orbit and it become longitudinal in opposite straight section, if the axis of spin rotation by Siberian Snake is directed along velocity. Let's assume the compensation conditions have been realized for this Snake. In this case in an ideal accelerator where particles orbits lie in the same plane (axial deviations are absent), the frequency of spin precession remains equal half ($\Delta\nu = 0$) and the precession axes spread are

$$\begin{aligned} \Delta\vec{n} &= \frac{\Delta\gamma}{\gamma}\phi \left[\vec{e}_x \cos\phi - \vec{e}_y \sin\phi\right] , \\ \phi &= \nu_0 \left[\int_{\theta_0}^{\theta} K_z d\theta - \pi\right] . \end{aligned}$$

Thus, in opposite straight section, where longitudinal polarization may be carried out, $\phi = 0$ and spread $\Delta\vec{n}$ is absent ($\Delta\vec{n} = 0$). It is interesting that beam polarization may be absent in other places of orbit at very high energy ($\Delta\gamma\phi/\gamma \gg 1$).

Measurement of Electron Polarization at TRISTAN

K. Nakajima, M. Arinaga, T. Kawamoto, Y. Mizumachi, Y. Mori, A. Ogata, and K. Yokoya

National Laboratory for High Energy Physics, 1-1 Oho, Tsukuba-shi, Ibaraki-ken, Japan

Measurement of electron beam polarization has been made in the TRISTAN e^+e^- storage ring using a laser polarimeter detecting an asymmetry in the Compton scattering of circularly polarized laser beam on polarized electrons. In the experiment to test operation of the polarimeter we observed a large asymmetry corresponding to a 40% polarization level at 29 GeV TRISTAN energy.

Introduction

In e^+e^- storage rings electrons (positrons) are transversely polarized anti-parallel (parallel) to the guiding magnetic field by the Sokolov - Ternov mechanism[1]. A natural polarization of electron beam has been observed in relatively low energy storage rings to determine a precise beam energy of the machine[2]. As energy increases it is expected that the degree of polarization becomes low due to strong depolarization effects. These are crucial problems to achieve high level polarized beams for e^+e^- and ep physics experiments in high energy storage rings exceeding the 30 GeV range.

A fast precision polarimeter is an essential tool to study polarization phenomena in e^+e^- storage rings. We have developed a monitor to measure polarization of electron beam, based on the Compton scattering of laser photons on electrons. Features of our polarimeter design are to minimize a systematic error as well as a statistical error and to allow for measurement of the transverse and longitudinal polarization simultaneously. We report the polarimeter system and results from measurement of polarization at TRISTAN made to test the system.

Electron Beam Polarization in TRISTAN

In the presence of depolarizing effects originating from magnetic field imperfections and misalignments in a real storage ring, the buildup of the beam polarization can be expressed in terms of the effective buildup time τ as

$$P_e(t) = \tau/\tau_p P_0(1 - \exp(-t/\tau)), \tag{1}$$

where P_0 is the ideal asymptotic polarization level, $P_0 = 8/(5\sqrt{3}) = 92.38\%$, and τ_p is the natural polarization buildup time. The buildup time of the polarization for the beam energy E in GeV is given by

$$\tau_p = 98.66(\rho^3/E^5)(R/\rho)\,\text{sec}, \tag{2}$$

where ρ is the bending radius in meters and R is the machine mean radius. In TRISTAN the natural polarization buildup time would be $\tau_p = 2.88 \times 10^9/E^5$ sec, typically 2 minutes for 30 GeV. An observable buildup time depends on the asymptotic degree of the polarization determined by depolarization effects. The absolute polarization level is expected to be very low around the depolarization resonance. It occurs when the spin tune $\nu = E/0.44065$ satisfies $\nu = n + n_x\nu_x + n_y\nu_y + n_s\nu_s$, where ν_x, ν_y, ν_s are the horizontal, vertical and synchrotron tunes, respectively and n, n_x, n_y, n_s are any integers. A polarization time will be 42 sec for a 30 % polarization level at 30 GeV in TRISTAN.

Compton Polarimetry

The principle of polarimeter is based on the asymmetry in the Compton scattering of circularly polarized photons from a laser on an electron beam. The differential Compton scattering cross section[3] of circularly polarized photons on polarized electrons is written in the electron rest frame as

$$d\sigma/d\Omega = (r_e^2/2)(q/q_0)^2\big[(1+\cos^2\theta) + (q_0-q)(1-\cos\theta) \\ \pm P_{eT}P_\gamma \cos\phi(1-\cos\theta)q\sin\theta \\ \pm P_{eL}P_\gamma(1-\cos\theta)(q_0+q)\cos\theta\big] \tag{3}$$

where q_0 and q are the incoming and outgoing photon energies and θ and ϕ are the scattering angles. In the laboratory system a transversely polarized beam gives rise to an up-down asymmetry in the backscattered γ rate, depending on the left-right photon polarization P_γ. Then the transverse asymmetry can be calculated by

$$A_T = \frac{(N_{Lu} - N_{Ld}) - (N_{Ru} - N_{Rd})}{N_{Lu} + N_{Ld} + N_{Ru} + N_{Rd}} \tag{4}$$

where N_{Lu},N_{Ld},N_{Ru} and N_{Rd} are the respective numbers of γ events above and below the midplane of the vertical distribution of γ rays for lefthand and righthand circularly polarized photon beams. A longitudinal polarization of electron beam can be detected as a difference in the energy distribution of γ events. Similarly the longitudinal asymmetry is given by

$$A_L = (N_L - N_R)/(N_L + N_R) \tag{5}$$

where N_L and N_R are the numbers of events in the distribution of scattered photon energy. The degree of polarization $P_{eT,L}$ can be determined by the asymmetry of the measured γ rate, the analyzing power Π and the photon polarization: $P_{eT,L} = A_{T,L}/\Pi P_\gamma$. The analyzing power is defined by the ratio of the spin-dependent term of cross section to the non spin-dependent term, calculated from the cross section integrated over the kinematic acceptance of the detector. The absolute degree of polarization should be determined by measuring the time dependence of the asymmetry and deriving the effective buildup time from eq. (1).

Laser Polarimeter

In order to measure an up-down asymmetry and a left-right asymmetry of γ rays backscattered on the polarized electron beam simultaneously, we use the single photon counting method[2], in which the electron beam is illuminated by a relatively low power, high repetition rate laser pulse. In this technique it is essential to measure both position and energy of each backscattered photon with high resolution.

We use a cavity-dumped Ar ion laser producing the 514.5 nm (2.41 eV) line. The cavity-dumped pulses are synchronized with the beam revolution frequency to illuminate the electron beam at each crossing. The laser pulse has a width of 15 ns (FWHM) and a peak power of up to 100W. The linearly laser light is converted to either right or left

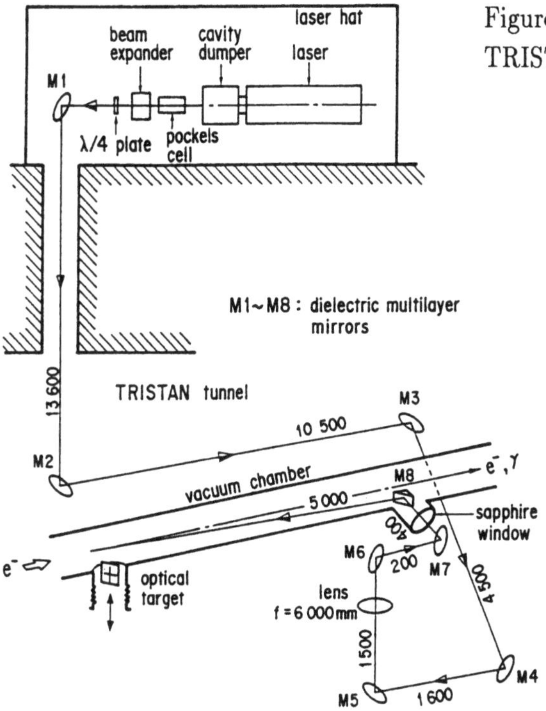

Figure 1: The laser and optical system of TRISTAN polarimeter

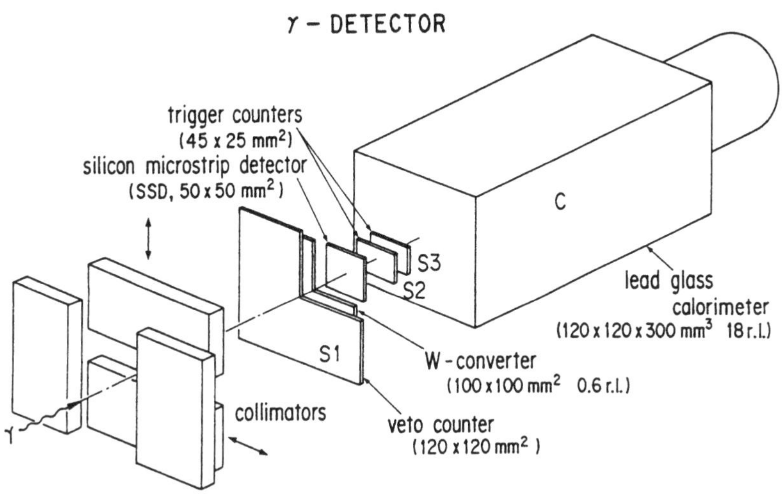

Figure 2: A schematic view of the γ detector arrangement.

circularly polarized light with a KD*P Pockels cell and a quarter wave plate. The laser beam is guided through the transport optics consisting of 8 mirrors and a final focusing lens towards the interaction point where the laser light crosses the electron beam at an angle of 8 mrad in the horizontal plane as shown in Figure 1. The laser beam is focused at this point with a rms spot size approximately less than 1 mm. The backscatterted γ rays travel along with the electron beam, then leave the beam pipe at the end of the first main dipole and finally reach the γ detector located about 38 m downstream from the interaction point.

The γ detector is schematically shown in Figure 2. A tungsten converter is followed by a position sensitive detector, trigger counters and a calorimeter. The photon trigger is provided by scintillation counters and a lead glass counter. The vertical position of a converted photon is determined from hit channels of the silicon microstrip detector with the position resolution 126 μm. The energy of a photon can be measured by the lead glass calorimeter with the energy resolution $\sim 6/\sqrt{E_\gamma(\mathrm{GeV})} + 1\%$.

Asymmetry Measurements

Asymmetry measurements have been made to test validity of the polarimeter operation. The polarimeter operation begins with alignment of the laser beam searching for optimum overlap of the laser spot and the electron beam in both space and time so as to maximize the backscattered Compton rate. Particularly the vertical alignment has to be achieved within 0.5 mm with the vertical rms beam size 0.2 $\sim$ 0.3 mm at 29 GeV. For accurate asymmetry measurement it is of importance to increase the ratio of the Compton rate to the background event rate originated from the beam-gas bremsstrahlung in the 200 m long straight section. A typical Compton rate was 2kHz over a 1kHz background rate for a 2 mA TRISTAN beam. A large amount of background was subtracted from collected events by taking only background events in laser-off runs besides left-right polarization runs. These three laser modes were alternately switched at an approximate rate of 2 Hz limited by a data acquisition speed of the computer.

In order to check an asymmetry artificially produced by a systematic error, asymmetry measurement was made at the beam energy 29.083 GeV with an integer spin tune 66 at which the depolarization resonance occurs. This measurement verified no transverse and longitudinal asymmetry within a statistical error 0.2%. The observation of natural polarization was carried out at 29.3 GeV with the spin tune 66.5. In this energy asymmetry measurements were made for the solenoid-off and -on condition to investigate effects of

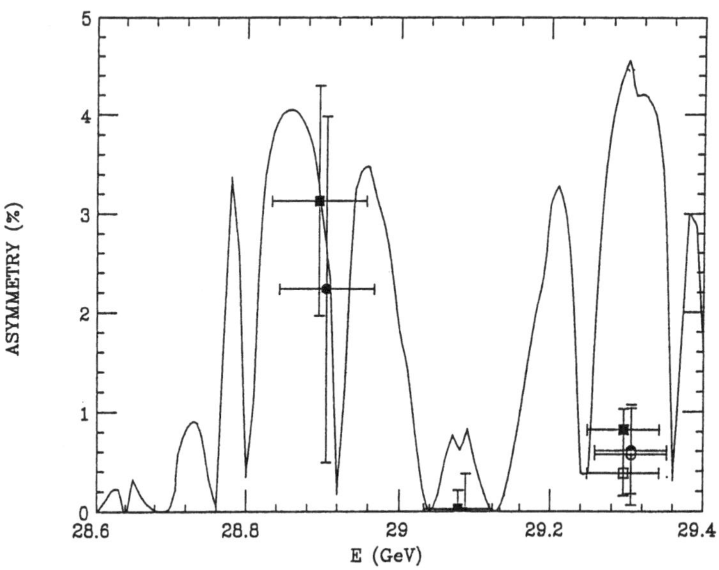

Figure 3: Asymmetry measurements as a function of beam energy. The plots show transverse asymmetry(•) and longitudinal asymmetry(■). The blank plots denote the solenoid-off measurement. The solid curve shows the expected polarization reduced by a factor 10.

three large solenoids installed in TRISTAN for the collider experiment. In the solenoid-on condition the longitudinal asymmetry showed twice higher value than in the solenoid-off condition, while the transverse asymmetry did not change. The result implys that the solenoid field causes the polarization vector to be tilted rather than depolarized. A large asymmetry has been measured in e^+e^- colliding beams at 28.9GeV. The measured asymmetry derives a polarization level of 40 % assuming the analyzing power of transversely 6% and longitudinally 13 %. Figure 3 shows results of transverse and longitudinal asymmetry measurements at three TRISTAN energies along with the expected polarization caliculated by the computer code SLIM[4].

Conclusion

We have constructed a polarimeter for measurement of electron beam polarization at TRISTAN and started measurements around 29 GeV. The measured asymmetry indicated a relatively high level polarization even in colliding beams with solenoid fields. The study on the electron beam polarization at TRISTAN will be continued.

References

[1] A.A. Sokolov and I.M. Ternov, Sov. Phys. Dokl. 8 (1964) 1203.
[2] D.B. Gustavson et al., Nucl. Instr.Meth. 165 (1975) 177.
[3] F.W. Lipps and H.A. Toelhoek, Physica XX (1954) 85, 395.
[4] A. Chao, PEP note 263, 1978. SLAC-PUB-2564, 1980.

List of Participants

Barber,D.P.	DESY /Germany
Bongardt,K.	Juelich /Germany
Courant,E.D.	Brookhaven /U S A
Derbenev,Ya.S.	Novosibirsk /USSR
Fieguth,T.	SLAC /U S A
Gianfelice-Wendt,E.	DESY /Germany
Golubeva,N.J.	Moscow /USSR
Grote,H.	CERN /Switzerland
Kondratenko,A.	Novosibirsk /USSR
Koutchouk,J.P.	CERN /Switzerland
Kreiser,H.	Hamburg /Germany
Krisch,A.D.	Michigan /U S A
Mane,S.R.	Brookhaven /U S A
Nakajima,K.	KEK /Japan
Ohmori,C.	Tokyo /Japan
Prasuhn,D.	Juelich /Germany
Ratner,L.G.	Brookhaven /U S A
Rossmanith,R.	CEBAF /U S A
Underwood,D.	Argonne /U S A
Wienands,U.	TRIUMF /Canada

Part III

Polarized Gas Targets

Recent Developments in Laser-Driven Polarized Sources

L. Young, K.P. Coulter, R.J. Holt, E.R. Kinney, R.S. Kowalczyk, D.H. Potterveld, and A. Zghiche

Physics Division, Argonne National Laboratory, Argonne, IL 60439, USA

ABSTRACT: Recent progress in the performance of laser-driven sources of polarized hydrogen and deuterium is described. The current status of the prototype source, I = $2.5 \times 10^{17} s^{-1}$, polarization = 0.29 (including atomic fraction), is comparable to classical Stern-Gerlach sources. A scheme to improve source performance by approximately an order of magnitude, using a combination of optical-pumping spin-exchange and RF transitions, is outlined.

INTRODUCTION

The development of high-density targets of polarized hydrogen and deuterium is central to the study of spin-dependent structure in nuclei. While such targets are of general interest, two experiments in particular motivate our development of polarized internal gas targets of hydrogen and deuterium: a) measurement of tensor analyzing power in electron-deuteron elastic scattering [1] and b) measurement of spin-dependent structure functions of the neutron and proton [2] . In both experiments, the desired figure of merit for the target is $np_{zz}^2 \approx np_z^2 \geq 10^{14}$nuclei/cm^2. In order to achieve this thickness, the proposed scheme is to feed a windowless storage cell [3] by a polarized source. At present, two options for the polarized source are being actively pursued, namely, the Stern-Gerlach source [4] at Heidelberg and the laser-driven source at Argonne [5].

The laser-driven polarized source is based on the principle of spin-exchange optical pumping. In this method, optical pumping creates a spin-polarized alkali atom, which, via spin-exchange collisions will polarize the electron of the hydrogen or deuterium atom. The nucleus of the H or D atom becomes polarized at low magnetic fields via the hyperfine interaction. This technique has great potential, which can be understood by considering that the photon flux available from a standard Ti-sapphire laser at the potassium D1 resonance line is 3-4 watts. A conversion efficiency of only 10% would give polarized H/D atoms at a flux $\geq 10^{18} s^{-1}$, far exceeding the flux available from conventional sources, $3 \times 10^{16} s^{-1}$. In this paper, recent developments (using both potassium and sodium as the spin-exchange intermediate) which show the laser-driven source to be comparable to the benchmark Stern-Gerlach source [6] will be discussed.

EXPERIMENTAL

The prototype spin-exchange optical pumping system is shown schematically in Figure 1. An Ar^+-pumped dye laser system provides the radiation required to optically pump the alkali, either Na or K. The spectral density of the laser is tailored to match

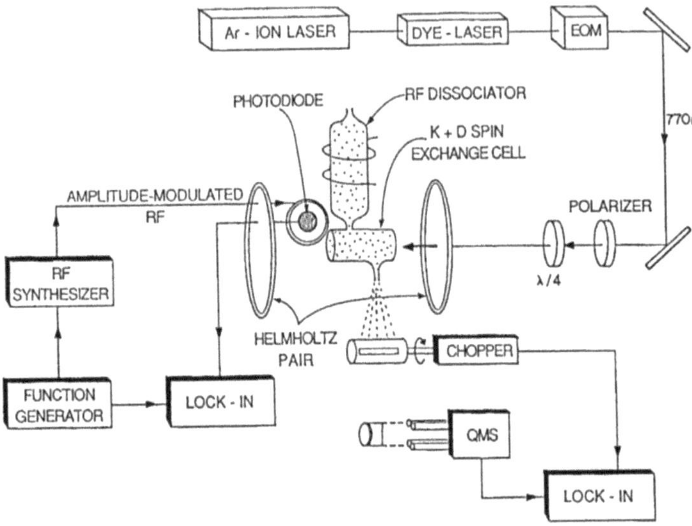

Figure 1. Schematic of the laser-driven source.

the Doppler-broadened absorption profile of the alkali vapor, circularly polarized, and sent to the spin-exchange cell located near the center of the diagram.

The spin-exchange cell is constructed of pyrex as an integral unit with the rf-dissociator and alkali reservoir. The interior is coated with drifilm [7]. The entire cell is heated to $\sim 250°C$ to prevent alkali condensation and placed in a static holding field of 10G. The H/D flux is controlled and measured with a servo-driven needle valve. The alkali density is independently controlled by the reservoir temperature and quantified by measuring the integrated absorption coefficient with a probe laser. Typically the ratio of Na to D atoms is 0.1 to 0.3%. The mixture of alkali and D atoms exits the spin-exchange cell through a spout, after which a chopper and quadrupole mass spectrometer are used to determine the fraction of D atoms in atomic form (AF).

The equilibrium polarization of the atoms is measured by optical detection of magnetic resonance transitions between adjacent Zeeman sublevels [8]. That is, increased fluorescence from the optically-pumped alkali atoms is observed as a result of altering the relative populations of the magnetic sublevels, either directly or via spin-exchange collisions with deuterium atoms. A typical signal using the Zeeman technique is shown in Fig. 2. The asymmetry between $\sigma+$ and $\sigma-$ pumping reverses as expected. The polarizations shown on the figure are deduced using a spin-temperature model, where the populations of the magnetic sublevels are given by $N_{m_f} \propto e^{\beta m_f}$, where β is the spin temperature. From the figure, there appears to be an incomplete equilibration between the polarization of the alkali and deuterium atoms.

The average deuterium polarization is expected to be measured well, since, the mean free path between D-alkali collisions is much larger than the cell dimensions ($\lambda = 1/\sigma_{se} n_A \approx 10^3$ cm). However, the alkali polarization may be overestimated, since regions of the cell with high laser intensity are preferentially sampled. This is confirmed by probe laser measurements of differential $\sigma+, \sigma-$ absorption as a function of position in the cell. The calculated polarization transfer efficiency is thus an underestimate, and

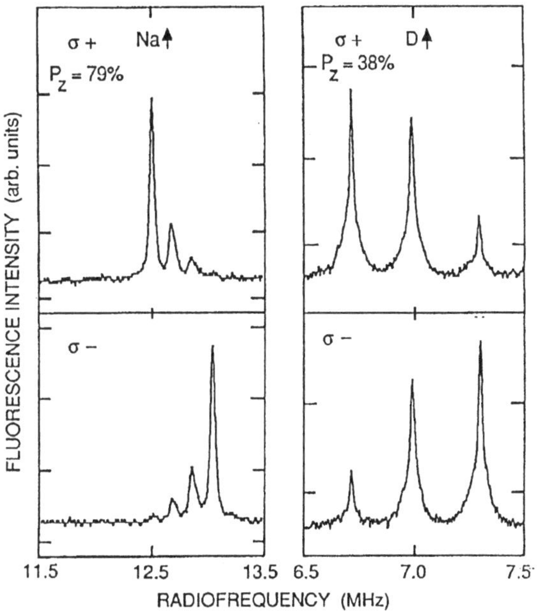

Figure 2. Typical fluorescence signal using Zeeman technique.

improvements in deuterium polarization can be expected with uniform high-power laser coverage of the cell.

RESULTS AND DISCUSSION

In this section we discuss the evolution of the laser-driven source. Only deuterium will be discussed in detail, although tests have shown that hydrogen polarization essentially equivalent to deuterium vector polarization, p_z, can be obtained. Initial studies (ANL'88) were conducted using potassium as the spin-exchange intermediate. This choice was motivated by the ease with which one could distribute the output of a cw laser over the absorption profile of the K vapor ($\approx$1.5 GHz Doppler + hyperfine) relative to other alkalis. The switch to Na (ANL'90) as the spin-exchange intermediate can be best understood by considering the processes which determine the deuterium polarization.

The processes in the spin-exchange cell can be modelled using a rate-equation approach, which leads to the following relationship for the efficiency of polarization transfer from the alkali to the D atom.

$$P_D = \frac{\gamma_{se} P_A}{\gamma_{se} + \Gamma}\left(1 - e^{-(\gamma_{se}+\Gamma)t}\right)$$

where γ_{se} = spin-exchange rate = $\sigma_{se}\bar{v}n_A$, Γ =loss rate (depolarization + flow rate) and P_D, P_A = electron polarization of the deuterium and alkali, respectively.

In order to enhance the efficiency of polarization transfer, one should maximize the spin-exchange rate, i.e. n_A, and minimize the loss rate. However, due to radiation trapping, n_A can not be increased arbitrarily without a corresponding decrease in P_A. Radiation trapping at low B-fields is a less severe problem for Na than for K. The critical density (where the alkali polarization is degraded by roughly 50%) is more than a factor of two higher for Na than for K.

Table 1 shows the performance figures for the laser-driven source configurations at low fields using K (ANL'88) and Na (ANL'90) as the spin-exchange intermediates. In addition, it shows the corresponding parameters for the benchmark classical atomic beam source (Bonn). The first three columns concentrate solely on source performance p_z, I and atomic fraction (AF). The final two columns relate to the expected figure of merit ($F = Ip_z^2\mathrm{AF}^2\epsilon$) for internal target applications involving a storage cell, as indicated in the introduction. The injection efficiency, ϵ, into the storage cell (fourth column) is unity for the laser-driven source since the source and cell can be directly coupled. The figure of 0.5 given for the Bonn source is estimated from the experience of the ANL-Novosibirsk collaboration [1]. It may be an underestimate, but is included to indicate the non-trivial nature of efficient injection into the storage cell.

Table I. Performance figures for polarized H/D sources.

Source	p_z	Intensity ($\times 10^{17} s^{-1}$)	Atomic Fraction	Injection Efficiency	F (s^{-1})
ANL'88	0.25	0.6	0.80	1.0	0.24×10^{16}
ANL'90	0.38	2.5	0.75	1.0	2.0×10^{16}
Bonn	0.95	0.3	1.00	0.5	1.4×10^{16}

As can be seen, the switch from potassium to sodium increased the deuterium polarization, p_z, from 0.25 to 0.38, for the reasons previously discussed. Another very marked improvement is the source intensity. This is due primarily to the increased laser power. The incident photon fluxes for the K and Na cases were $\approx 8 \times 10^{17} s^{-1}$ and $\approx 3 \times 10^{18} s^{-1}$, respectively.

FUTURE PLANS

Although substantial gains have been posted with the change in alkali and additional laser power, the laser-driven source has not yet met the design goals: I $= 4 \times 10^{17} s^{-1}, p = 0.5$. In particular, the p_{zz} obtained thus far, 0.15, yields a figure of merit similar to the existing source at Novosibirsk and must be improved an order of magnitude for the electron-deuteron elastic scattering experiments.

In order to increase the polarization (p_{zz}) and the intensity of the source, we plan a two-step process: optical-pumping spin-exchange with K at high field followed by RF transitions in a moderate field. The added complication of RF transitions subsequent to the spin-exchange is offset by the advantages of high-field optical pumping, namely, greatly reduced radiation trapping problems and better wall depolarization properties. The p_{zz} expected in this scheme is $p_{zz} = \epsilon_{24}\mathrm{AF}\ P_D$, where ϵ_{24} is the efficiency of the $4 \leftrightarrow 3$ and $3 \leftrightarrow 2$ RF transitions. Monte Carlo simulation of the process gives $\epsilon_{24} \approx 0.9$. Based on previous experience, we estimate a $P_D = 0.7$ and AF $= 0.8$ to be achievable, leading to a $\mid p_{zz} \mid= 0.50$. A recently-acquired Ti-sapphire laser provides an increase of a factor of 4 in available photon flux ($1.2 \times 10^{19} s^{-1}$). Thus, deuterium atom fluxes of $\sim 1 \times 10^{18} s^{-1}$ should be achievable.

SUMMARY

Over the past two years, an increase of approximately a factor of 8 in figure of merit has been realized for the laser-driven source of polarized deuterium. The gains have been due to a change in the spin-exchange intermediate from potassium to sodium and increased laser power. In its present state, $Ip_z^2\ \mathrm{AF}^2 = 2.0 \times 10^{16}\,\overrightarrow{\mathrm{D}}/s$, the laser-driven source is already comparable to classical atomic beam sources. Further improvements (a factor of ~ 12) are expected with increased laser power and a new scheme employing high-field optical-pumping spin-exchange followed by RF transitions.

ACKNOWLEDGEMENTS

We thank J. Gregar for his state-of-the-art glassblowing, as well as C. Kurtz and R. Amrein for technical assistance. Finally, we thank D. K. Toporkov and D. M. Nikolenko for assistance during some of the experiments, and R. A. Gilman for helpful discussions. This work was supported by the U. S. Department of Energy, Nuclear Physics Division, under Contract No. W-31-109-ENG-38.

REFERENCES

[1] R. A. Gilman et al., Phys. Rev. Lett. **65**, 1733 (1990).
[2] "A Proposal to Measure the Deep Inelastic Spin Dependent Structure Functions of the Nucleon at HERA", the HERMES collaboration, DESY-PRC-90-01 (1990).
[3] Barker, M. D. et al., in *Polarization Phenomena in Nuclear Physics - 1980*, edited by G. G. Ohlsen et al., AIP Conference Proceedings No. 69 (American Institute of Physics 1981), p. 931.
[4] G. Graw et al., in Proceedings of the IV LEAR Workshop, Villars, 1987, ed. C. Amsler et al., Harwood Acad. Publ., Chur (1988) p. 249.
[5] L. Young et al., in Proceedings of the Topical Conference on Electronuclear Physics with Internal Targets, Stanford, CA, 9-12 January 1989, ed. R. Arnold (World Scientific, Singapore) p. 125; Nucl. Phys. **A497**, 529c, (1989).
[6] H. G. Mathews et al., Nucl. Instru. and Meth. **213**, 155 (1983).
[7] L. Young et al., Nucl. Instru. and Meth. **B24/25**, 963 (1987); D. R. Swenson and L. W. Anderson, Nucl. Instru. and Meth. **B29**, 627 (1988).
[8] W. E. Bell and A. L. Bloom, Phys. Rev. **107**, 1559 (1957); S. G. Redsun et al.,Phys. Rev. **A42**, 1293 (1990).

Progress on the Michigan-MIT Ultra-Cold Polarized Hydrogen Jet*

R.S. Raymond and T. Roser

Department of Physics, The University of Michigan,
Ann Arbor, MI 48109, USA

This project exploits the fact that at reasonably low temperatures (0.3 K) atomic hydrogen in the two lower ground-state hyperfine states can be trapped by reasonable magnetic fields (5–7 T). If microwaves of the proper frequency (140 GHz for 5 T) are applied, forbidden transitions to the upper hyperfine states are driven. The flipped atoms, now on a potential hill, will accelerate out of the field, forming a beam/1/. The test apparatus, shown in Fig. 1, has a vertical dilution refrigerator with a horizontal mixing chamber, a superconducting solenoid, and feed systems for atomic hydrogen and microwaves.

At Minneapolis we showed signals of pulsed extraction, obtained by ramping the magnetic field through the values for resonance/2/. Since then we have refined our measurements and made many improvements in the jet itself and diagnostics. A schematic layout of the new set-up is shown in Fig. 2. An improved H° feed system was built, with a more stable dissociator and improved temperature control. Well-cooled, He-coated baffles were also installed to improve low-temperature accommodation.

We have made tests with a 7.5 T magnetic field and 213 GHz microwaves. The microwave source for these tests was a Ga-As tripler with a maximum output power of 7 mW. Tests were sufficiently encouraging that we have ordered a 213 GHz tube.

A 5 mm diameter aperture was installed in the output end of the mixing chamber. This type of aperture will be essential for us to have an atomic beam of sufficiently low emittance.

The film burner was moved from the still/heat exchanger area to the line between the vertical section of the refrigerator and the horizontal

---

* Collaborators: D. G. Crabb, W. A. Kaufman, A. D. Krisch, J. A. Stewart, B. S. Van Guilder, and B. Vuaridel from the University of Michigan; G. R. Court, T. J. Greytak, D. Kleppner, and A. Yu from MIT; Yu. M. Melnik, and A. F. Prudkoglyad from the Inst. for High Energy Physics, USSR; and V. G. Luppov and M. Mertig of JINR, USSR.

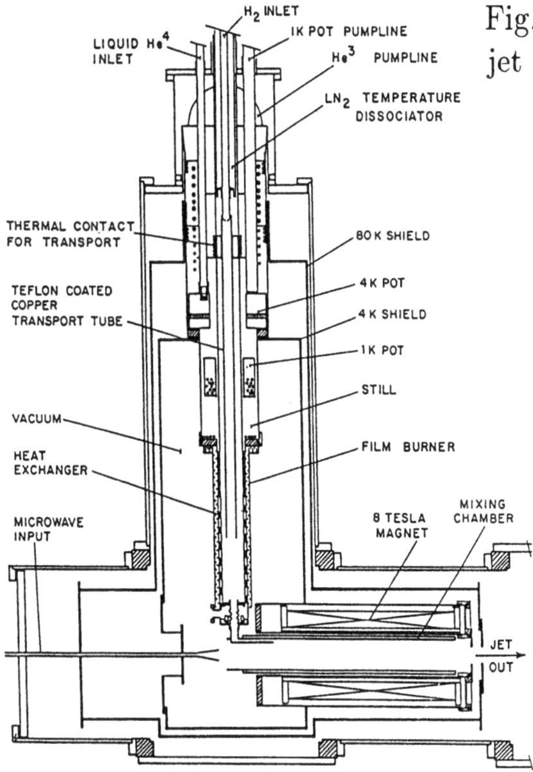

Fig. 1 Layout of the prototype jet apparatus.

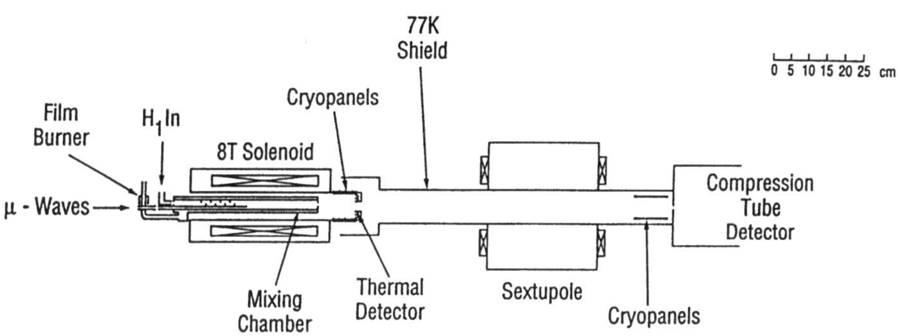

Fig. 2 Layout of the new storage cell and extracted beam detectors.

mixing chamber. With this change, the insulating vacuum pressure is the same with He film as with no film.

A sextupole magnet was built with a bore of 7.5 cm, a length of 30 cm and a maximum pole tip field of 0.3 T. A compression tube was assembled with a movable 1 cm diameter aperture and a quadrupole mass spectrometer.

With the improved jet we have observed a signal corresponding to 1.2 x 10^{16} atoms/s DC. We measured a temperature rise caused by re-

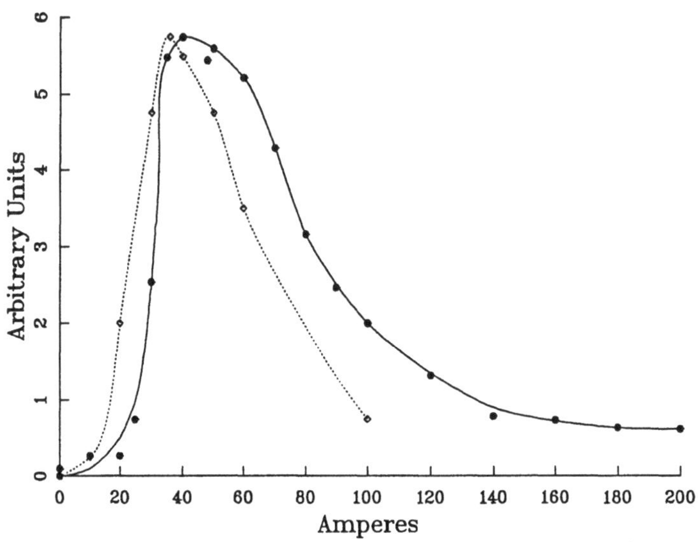

Fig. 3 Flux of atomic hydrogen as measured by the compression tube vs the sextupole current. The experimental results are shown by the solid dots and the open circles are a simulation assuming a 100% electron spin-polarized monoenergetic beam. The curves are to guide the eye.

combination energy (4.5 eV per recombined molecule) and assumed complete recombination and complete absorption of the energy. The response of the detector to applied power was measured with an electrical heater. The detector intercepted only about 15% of the expelled beam, so that the total expelled beam was about 8 x 10^{16} atoms/s. We believe this beam consisted of both *c* and *d* (1 and 2) state atoms and ray-tracing calculations indicate that more than 30% of extracted atoms may be focused to our desired spot size, so this measured flow corresponds to a focused flow of 1.5 x 10^{16} *d* (1) state atoms/s, or a target thickness of 1.5 x 10^{12} atoms/cm^2.

Measurements with the sextupole magnet and compression tube are shown in Fig. 3. The solid line is to guide the eye and the dashed line is a ray-tracing calculation, normalized so that the magnitude of the calculated maximum matches the measured maximum. There is qualitative agreement in the shapes, but, unfortunately, for these measurements the flow measured by the compression tube was lower than expected, from calculations and a thermal detector near the mixing chamber, by a factor of 6. This disagreement is not understood, but may be due to scattering by background gas. Cryopumping is now being improved to reduce background gas.

Design and construction of an improved jet for the experiment NEPTUN/NEPTUN-A at UNK[3] is now underway. Until the new jet is ready for testing, we will continue to work on improving the throughput and our understanding of the limitations of long-term operation.

This work was supported by a grant from the U.S. Department of Energy.

References

/1/ T. O. Niinikoski in Proc. of the International Symposium on High Energy Physics with Polarized Beams and Polarized Targets (Basel and Stuttgart 1981) p. 91; D. Kleppner and T. J. Greytak in Proc. of the 5th International Conference on High Energy Spin Physics, AIP Conf. Proc. **95**, 546 (1983).

/2/ R. S. Raymond et al. in Proc. of the 8th International Symposium on High Energy Spin Physics, (Minneapolis 1988), AIP Conf. Proc. **187**, 1311 (1989).

/3/ A. D. Krisch et al., Status of NEPTUN-A Experiment, in Proc. of the Workshop "Physics at UNK", (IHEP, Protvino, USSR 1989) p. 152.

Polarized Jet Target of Hydrogen and Deuterium Atoms

V.P. Ershov, V.V. Fimushkin, K.V. Kiselev, M.V. Kulikov, Yu.K. Pilipenko, V.B. Shutov, and I.V. Zhigulin

Joint Institute for Nuclear Research, Dubna,
Head P.O. Box 79, SU-101000 Moscow, USSR

The Accelerating and Storage Complex (UNK) of the Institute of High Energy Physics, Serpukhov has repeatedly been reported and discussed/1,2/. The first experiment that is planned to do at this accelerator is a polarized experiment with the internal beam (NEPTUN - NEPTUN-A). For this purpose the internal target area is being built at the accelerator ring. This report is about one of two polarized jet targets proposed for the experiment NEPTUN. It is being created at the Joint Institute for Nuclear Research, Dubna.

For successful operation of the accelerator, the target should be reliable and require the minimum of maintenance, especially just after the accelerator start.

From the physics point of view at an accelerator beam intensity of $6*10^{14}$ it is necessary to have a target thickness of $1*10^{14}$atoms$*$cm^{-2}. But for some experiments a target thickness of $1*10^{12}$ atoms$*$cm^{-2} would be sufficient. It could be obtained with a relatively simple design of the target. Taking this into account, we decided to develop the target based on the well-known atomic beam method.

A general view of the target is shown in fig.1. It is supposed to obtain a hydrogen polarized jet and a deuterium one. Molecular hydrogen (deuterium) is passed to the dissociator (1) working in the DC mode. It is planned to use two types of dissociator:

- a HF dissociator with water cooling, a generator frequency of 100 MHz and a power of up to 1 kw,
- a SHF dissociator with a generator frequency of 2.43 GHz and a power of up to 150 w.

When designing the dissociator, particular attention should be given to the safety operation of the target at the accelerator and to fast disassembling for cleaning.

The dissociator nozzle is cooled to 30 - 40 K. The jet is formed by a skimmer (2) and a collimator (3), which diameters are 4 mm and 12 mm, respectively. Spatial separation of the atomic beam in electron spin is realized in two sextupole magnets (6,8). The

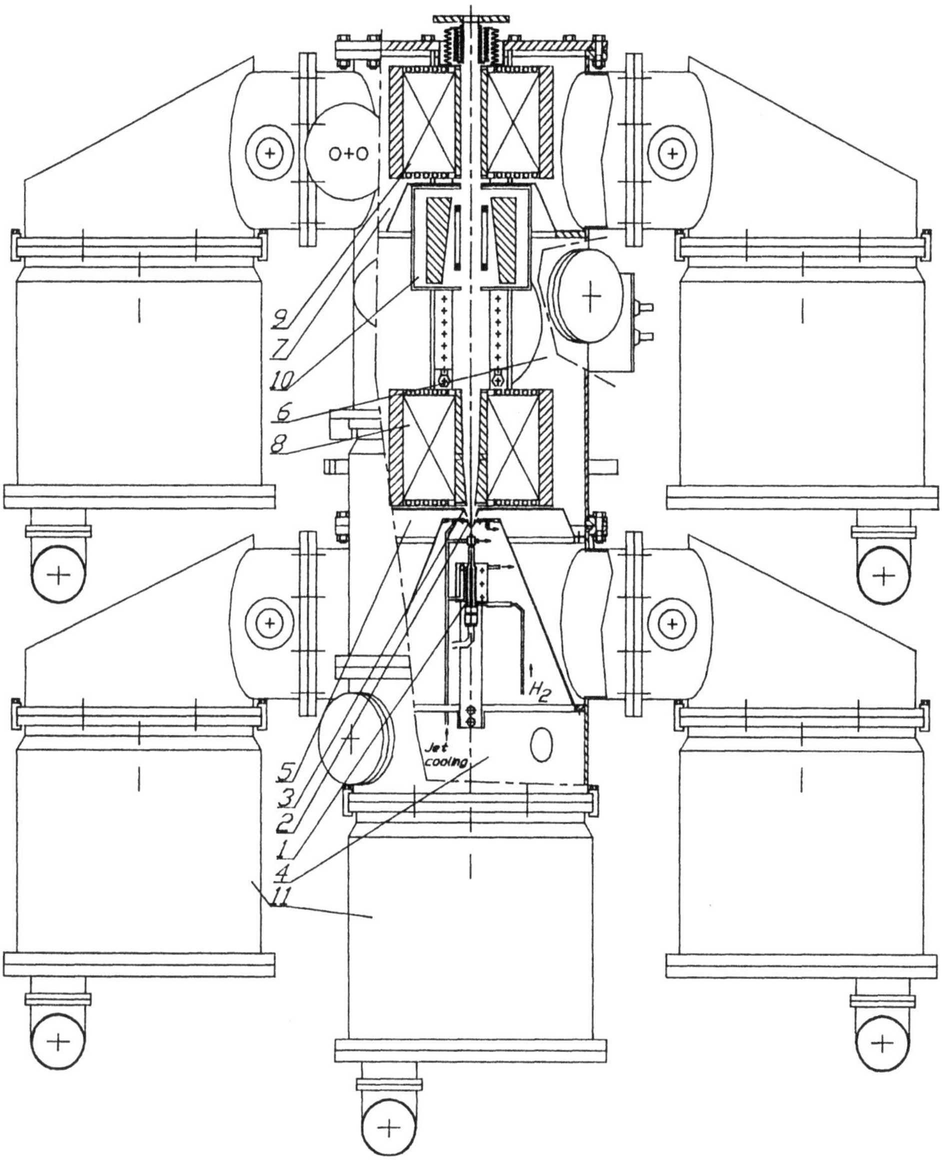

Fig.1. Polarized jet target of hydrogen and deuterium atoms.
1 - dissociator, 2 - skimmer, 3 - collimator, 4 - nozzle chamber, 5 - skimmer chamber, 6 - first sextupole chamber, 7 - second sextupole chamber, 8 - first sextupole magnet, 9 - second sextupole magnet, 10 - RF cell, 11 - turbomolecular pump.

geometry of the magnets (lengths and apertures) for the most probable velocity of atoms (1000 m/s) is optimized by the program /3,4/ for computing trajectories of hydrogen (deuterium) atoms in

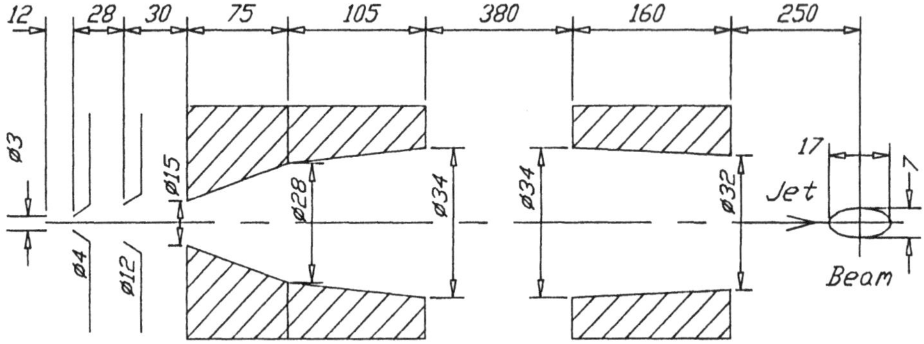

Fig.2. Magnetic structure of the polarized jet target.

gradient fields of multipole magnets. The accepted magnetic structure is shown in fig.2.

The nuclear polarization of the beam is made by the radiofrequency method. The RF transition cell is placed between the magnets. For hydrogen, transition 2 - 4 will be realized (Pz=+1). In the case of deuterium it is supposed to use a combination of two RF cells for obtaining the vector and tensor polarization. Changing the sign of the polarization will be realized by reversing a weak magnetic field in the interaction region.

The typical gas flow through the nozzle will be ~ 1 torr*l/s. To decrease the spectrometer background, vacuum in the interaction region of the jet with the beam should be better than $1*10^{-8}$ torr. For this, four-stage differential pumping of the target is suggested:

- nozzle chamber (4) - the vacuum is ~ $5*10^{-4}$ torr,

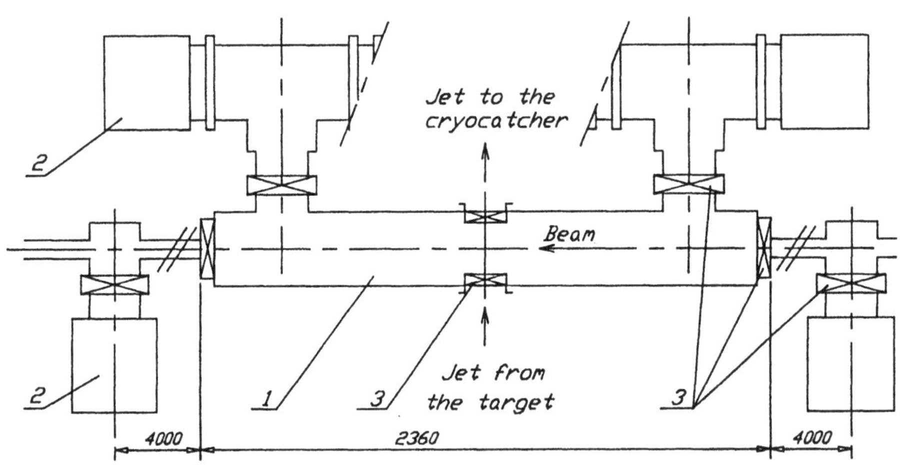

Fig.3. Vacuum system of the main accelerator ring ionguide at the target area.
1 - box, 2 - ion pump, 3 - vacuum valve.

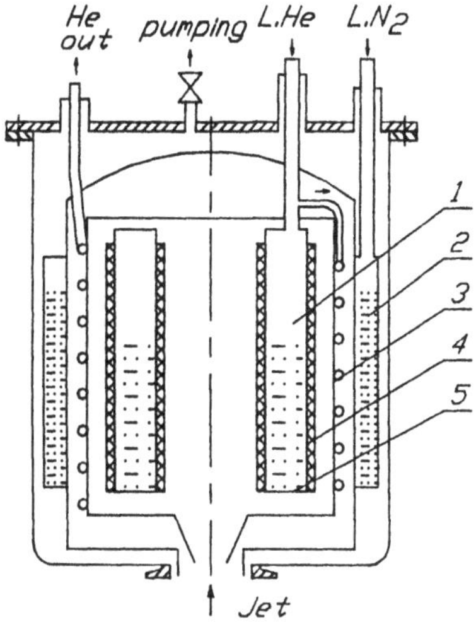

Fig.4. Cryocatcher.
1 - L. He vessel,
2 - L. nitrogen bath,
3 - He cooling shield,
4 - adsorbent,
5 - heater.

- skimmer chamber (5) $\sim 1*10^{-5}$ torr,
- first sextupole chamber (6) $\sim 5*10^{-7}$ torr,
- second sextupole chamber (7) $\sim 2*10^{-7}$ torr.

The chambers are pumped out by seven 2500 l/s turbomolecular pumps. The last stage can be pumped by cryopumps, which are being designed. The vacuum system of the accelerator ring at the target area is shown in fig.3. The required vacuum around the jet will be reached by 400 l/s ion pumps. The jet will be trapped by a specially developed cryocatcher (see fig.4).

References

1 V. I. Balbekov et al. Proc. of XII Intern. Conf. on High Energy Accel. - Batavia, 1983, p. 40.
2 A. I. Ageev et al. Proc. of XIII Intern. Conf. on High Energy Accel. - Novosibirsk, 1987, v. 2, p. 332.
3 Glavish H. F. Proc. Fourth Int. Symp. on Polarization Phenomena in Nuclear Reactions, Zurich, 1975, p. 844.
4 Kulikov M. V. et al. JINR Communication 13-88-470, Dubna, 1988.

Accumulation of Hydrogen Atoms in a Low Temperature Storage Cell of a Polarized Hydrogen Gas Jet Source

M. Mertig, A.V. Levkovitch, V.G. Luppov and Yu.K. Pilipenko

Joint Institute for Nuclear Research, Dubna,
Head P.O. Box 79, SU-101000 Moscow, USSR

Immediately after the first successful experimental low temperature stabilization of polarized atomic hydrogen it was proposed [1,2] to employ this technique for the creation of an intensive low velocity atomic hydrogen beam with a narrow velocity distribution. The beam will be formed by microwave-induced extraction of stored H↓-atoms from a low temperature ($T < 0.5$ K) stabilization cell (SC), which is placed in a high magnetic field ($B \sim 10$ T). This method was experimentally demonstrated up to a flux of extracted atoms of $\sim 10^{15}\ s^{-1}$ [3]. A further improvement of this value requires a detailed study if the processes of high effective transport of H-atoms to the SC and an effective microwave extraction. Although ESR-technique in principle allows to pump into a certain nuclear polarized state, due to spin exchange processes the extracted beam is expected to be a mixture of both upper hyperfine states [4].

At JINR, Dubna, an apparatus was designed [5], to develop this technique for the creation of a polarized hydrogen gas jet target. The schematic diagram of the apparatus is shown in Fig. 1 . The atomic hydrogen is produced in a pulsed microwave discharge dissociator. The atoms are transported from the dissociator to the SC with a volume of ~1 ccm. The SC is open to one side. The SC is placed at the center of the horizontal lying superconducting solenoid with a maximum magnetic field of 7 T. The SC is cooled by a continuously working He-3-evaporation refrigerator with a minimum temperature of 0.35 K and a cooling power of 25 mW at a temperature of 0.5 K. The presence of H -atoms in the He-covered cell is detected by standard bolometer technique.

In a first stage we focused our experiments on a high loading speed of the low temperature gas storage cell, what is an essential prerequisite for a high intensive gas jet source. The flux after transport if the atoms from the dissociator through a 4 K copper accommodator was determined by means if a calorimetric measurement using the uncoated SC as a recombination detector. In these

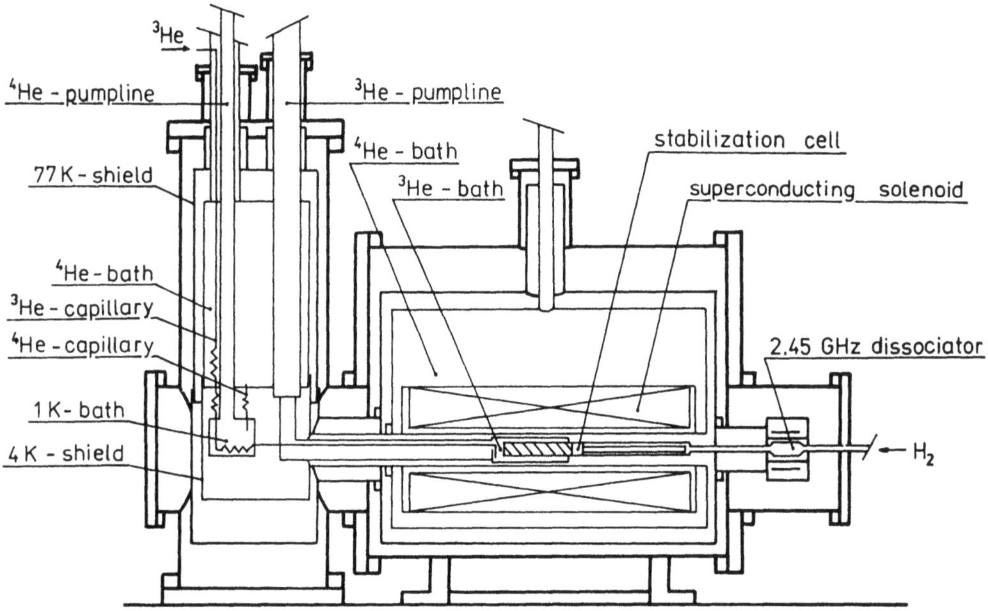

Figure 1. Schematic diagram of the apparatus

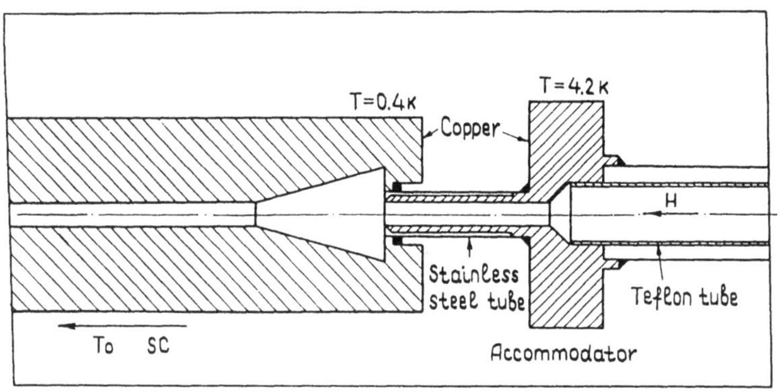

Figure 2. Design of the 4 K -accommodator

experiments by optimizing the dissociator action, a 4 K -cooled H-atom flux up to $8*10^{16}$ s^{-1} has been achieved. The efficiency E of the transport of the atoms through the critical temperature region from 4K to lower than 0.5 K was determined by measuring the flux of cold polarized H-atoms into the He-covered SC and compare it to the 4 K -flux at the same experimental conditions. For a design with an abrupt transition from 4 K to 0.5 K [5] a stable flux up to $1*10^{15}$ atoms$*s^{-1}$ has been observed with an efficiency of E ~ 10%. This result could be recently improved by a new design of the transition region, which is shown in Fig. 2. In this construction the He-film evaporates outside of the accommodator and the refluxing He-vapor carries the incoming H-atoms into the SC. The "pumping" action of the He could be experimentally confirmed by measuring an enhancement of the magnetic confinement in dependence of the He-film thickness up to a compression factor of 90. With this design a maximum stable flux of $5*10^{15}$ H↓ -atoms s^{-1} has been achieved. The maximum density of the accumulated gas was $7*10^{16}$ cm^{-3}.

At present, we are developing a new method to establish an electron-polarized H-beam without microwave extraction. This method is based on separation of H-atoms due to their electron spins in the gradient of a high magnetic field (B ~ 10 T), like proposed by D. Kleppner in 1981 [2], but without a build-up of a high density H -background gas [6]. The separation is achieved by cooling the atoms to temperatures T < 1 K at a field value of approximately half of the maximum field, where their magnetic energy is larger than thermal energy. Consequently, spin-down state atoms are attracted into the high field, where they are forced to recombine in order to prevent the density build-up. The spin-up atoms are accelerated to the low field region, where they leave the open cell and form a polarized beam. A new SC (see Fig. 3) to investigate this method is being build and under test.

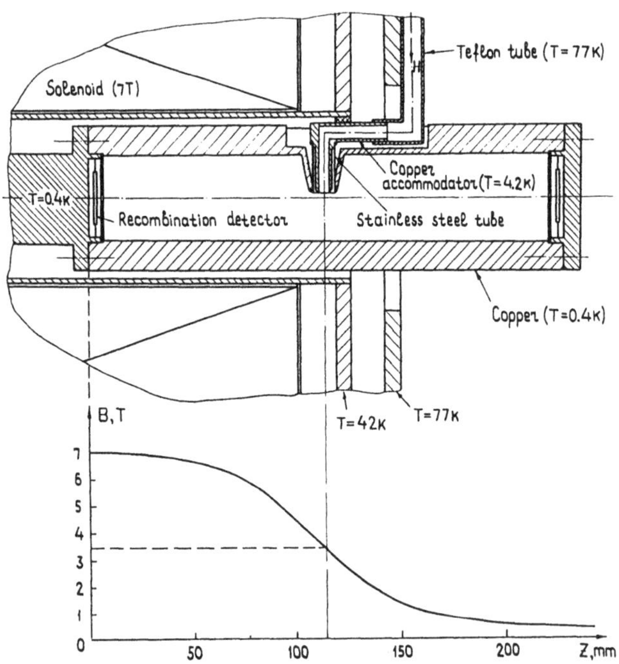

Figure 3. Design of the new storage cell and its position in the magnetic field

Acknowledgement

We would like to thank Prof.J.T.M. Walraven for many helpful discussions during developing the spin-separation technique.

References

1. T.O. Niinikoski, Proc. Intern. Symp. on High-Energy Physics with Polarized Beams and Polarized Targets, Lausanne (EXS 38, Birkhauser, Basel 1981) p. 191.
2. D. Kleppner and T.J. Greytak, Proc. of the 5-th Intern. Conf. on High Energy Spin Physics, Brokhaven, Ed. by G.M. Brunce, AIP Conf. Proc. No 95,546 (1983).
3. T. Roser et al. , to be published in Phys. Rev. Lett.
4. A.P.M. Mattheij, J. van Zwol, J.T.M. Walraven, I.F. Silvera, Phys. Rev. B 37, 4831 (1988).
5. M. Mertig, V.G. Luppov, Yu.K. Pilipenko, Proc. of the Conf. of Spin Polarized Quantum Systems, Torino, June 1988, Ed. by S. Stringari (World Scientific, Singapore, 1989) p. 353.
6. M. Mertig, Materials of the Neptun-Project-Workshop, Protvino, USSR, Sept. 1989

Intensity Measurements on the FILTEX Atomic Beam Source

W. Korsch

Universität Marburg and
Max-Planck-Institut für Kernphysik, Postfach 10 39 80,
W-6900 Heidelberg, Fed. Rep. of Germany

In order to increase the luminosity in storage ring experiments with polarized targets the development of internal targets has become of more and more interest for the last years. At the MPI für Kernphysik in Heidelberg our group is developing a dense polarized hydrogen target based on the storage cell technique[1]. This target is designed for the FILTEX experiment which intends to study the spin dependence in $p\bar{p}$- interaction with a polarized $\bar{p}$-beam and a polarized p-target at LEAR[2] and for the HERMES-experiment which is going to measure the spin-dependent structure functions of nucleons at HERA[3]. In the case of the FILTEX-experiment the internal target will act as a spinfilter, if the total cross section in $p\bar{p}$- interaction is spin dependent, as well as a polarized target for the determination of the spin correlation coefficients in $p\bar{p}$-scattering. To perform these experiments a mean areal density of about 10^{14} $\vec{H}/cm^2$ is needed. This can only be achieved with an intense H-atomic beam in combination with a storage cell since the target density is proportional to Q'_H/C, where Q'_H is the injected $\vec{H}$-flux and C is the conductance of the storage cell. Both quantities have to be optimized. This talk will be about the intensity optimization of the $\vec{H}$-beam on the new Heidelberg-Marburg-München polarized source.

The technique applied for the new atomic beam source is based on the working principles of a conventional atomic beam source which is described elsewhere. The idea to get a high intensity source was, to design a source which can be operated at high H_2-flows and to do systematic measurements with regard to the optimization to the intensity limiting parts and elements. The intensity which can be extracted from a source and injected into a target cell can be generally expressed by the following formula:

$$Q'_{\rm H} = \frac{1}{4} \cdot Q_{\rm H} \cdot \frac{\Delta\Omega}{\Omega} \cdot A \cdot T \ .$$

Here 1/4 stands for one hyperfine substate, $Q_{\rm H} = 2 \cdot \alpha \cdot Q_{\rm H_2}$ is the total atomic flux which is given by the product of the degree of dissociation α and the total H_2-flowrate $Q_{\rm H_2}$, $\Delta\Omega/\Omega$ is the relative, geometric, solid angle accepted by the first sextupole magnet, and is determined by the peaking of the beam and the magnet dimensions, A describes the attenuation of the beam due to residual gas scattering, and T is a transport-factor which expresses how many atoms in the focussing state can be transported through the magnet system and injected into a compression-tube or entrance-tube of a storage cell. From this formula follows that in such a factorisation four parameters contribute to the intensity. All these factors were investigated separately and have been tried to be maximized.

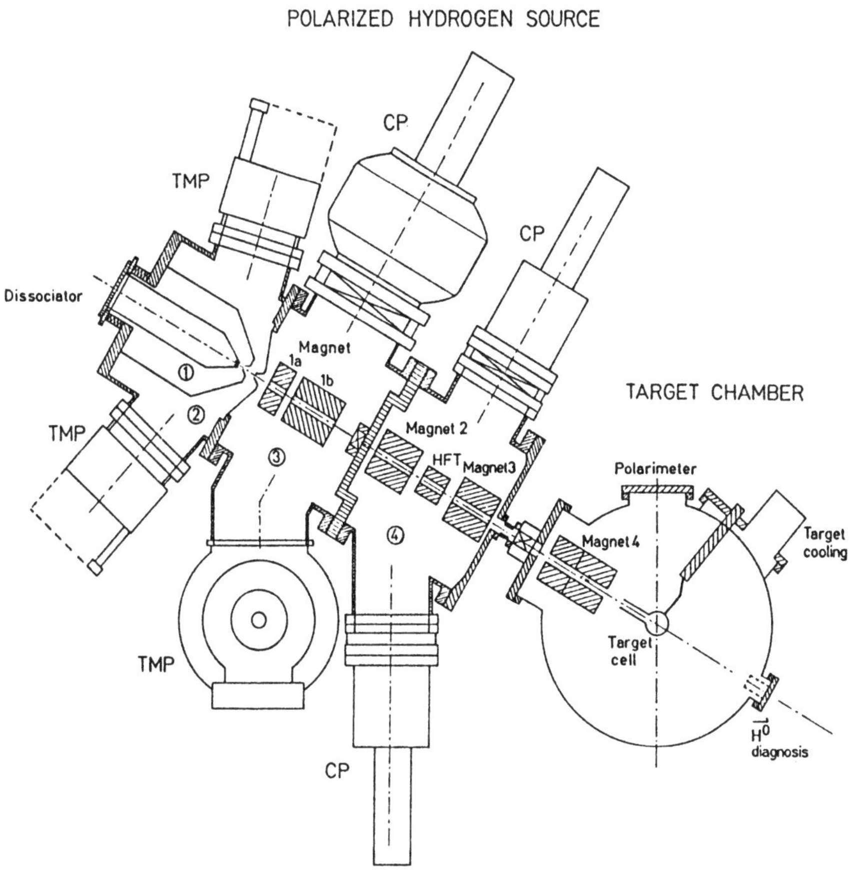

Figure 1 *The FILTEX atomic beam source.*

One of the basic requirements in order to be able to work at high H_2-flowrates is a powerful two stage differential pumping system. The system developed for the FILTEX-source is shown in figure 1. It has a total pumping speed of $\sim$ 19000 l/s and is equiped with a combination of turbomolecular- and cryopumps. This allows to work with H_2-flows of up to 5 mbar l/s without significantly attenuating the beam intensities at a nozzle temperature of 300 K (see figure 4a).

Furthermore, the degree of dissociation has to be maximized. The dissociator in the FILTEX-source consists of a linear discharge Pyrex glas tube. The needed HF-power is about 300 W at a frequency of 16 MHz. The HF is capacitively coupled to the plasma. The nozzle is sealed with indium on the glas tube in order to have a good thermal contact to the tube. Different nozzle materials were tested (Pyrex glas, gold, aluminum of different purities). The highest values for α could be obtained with an oxidized Al-nozzle (Al-purity $\geq$ 99.5 %). Figure 2 shows the data points for α as a function of Q_{H_2} at a nozzle temperature of 150 K. α was determined by chopping the beam, detecting the particles with a quadrupole-mass-spectrometer, and using 'lock-in' technique for signal processing. The values for α were obtained by adding small amounts of water vapour to the discharge ($Q_{H_2O} \leq 1\%$ of Q_{H_2}). The gain factor compared to a pure H_2-discharge was 1.8.

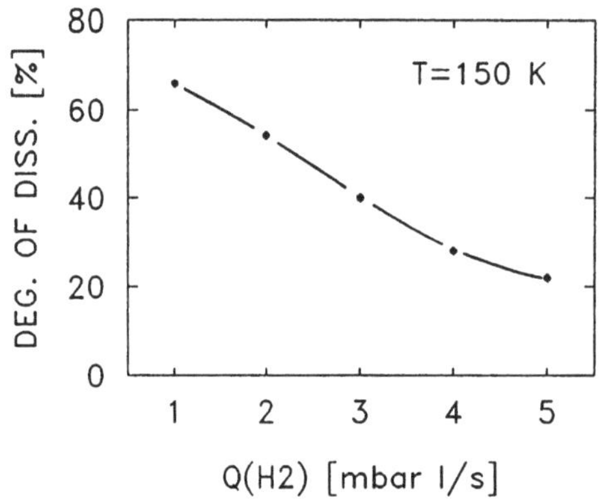

Figure 2 *The degree of dissociation as a function of Q_{H_2}.*

To study the attenuation of the beam due to residual gas scattering in the magnets several magnet dummies of realistic geometry were installed. These dummies were made of aluminum tubes with different inner diameters and conductance limiting endplates in order to reduce the effective pumping speed. Intensity measurements with a pure H_2-beam (without discharge) were performed with a compression tube. The experimental set-up is shown in figure 3.
Figure 4 shows the measured data points for two different nozzle temperatures.

The total magnet system, also shown in figure 1, was optimized with a Monte-Carlo code. A precise knowledge of the H-velocity distribution was needed for the optimization. Therefore the velocity distribution had been measured with a TOF-method. The extracted drift-velocity and the beam-temperature yielded large differences as compared to other sources (up to 30% higher resp.up to 70% lower). The sextupole-magnets are of segmented permanent type[4]. They have a cylindrical inner bore of 2.4 cm inner diameter and a pole tip field of 1.47 T. To minimize the scattering on the residual gas the first sextupole magnet is split into two in order to enhance the pumping speed in this region.

At the time of the presentation a first measurement of the $\vec{H}$ intensity with the complete magnet system a flux of $3 \cdot 10^{16}$ polarized atoms/s in two hyperfine substates could be reported on. The intensity was measured with a compression tube 4 cm behind the last magnet in the target chamber. The dimensions of the compression tube were 1 cm i.d. and 10 cm in length. In order to determine an absolute flux a calibration of the compression tube together with the ion gauge for the pressure reading was necessary. Therefore a small chamber of volume V, filled with H_2-gas of several millibar was connected at the end of the compression tube. By means of a needle valve H_2-gas could be injected in the tube. Then $V{\cdot}\Delta p/\Delta t$ was measured for small changes

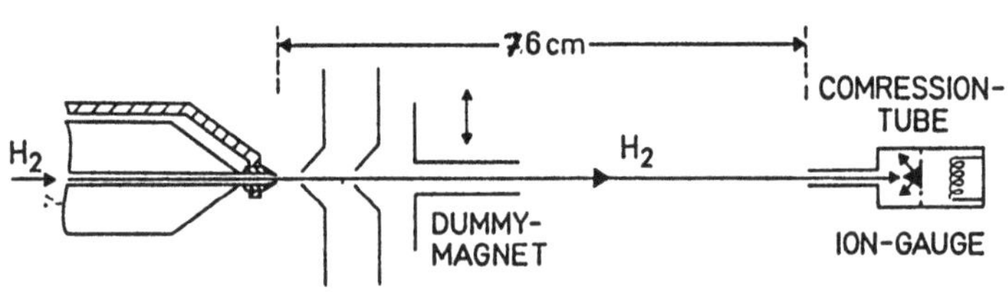

Figure 3 *Experimental set-up for the intensity measurements.*

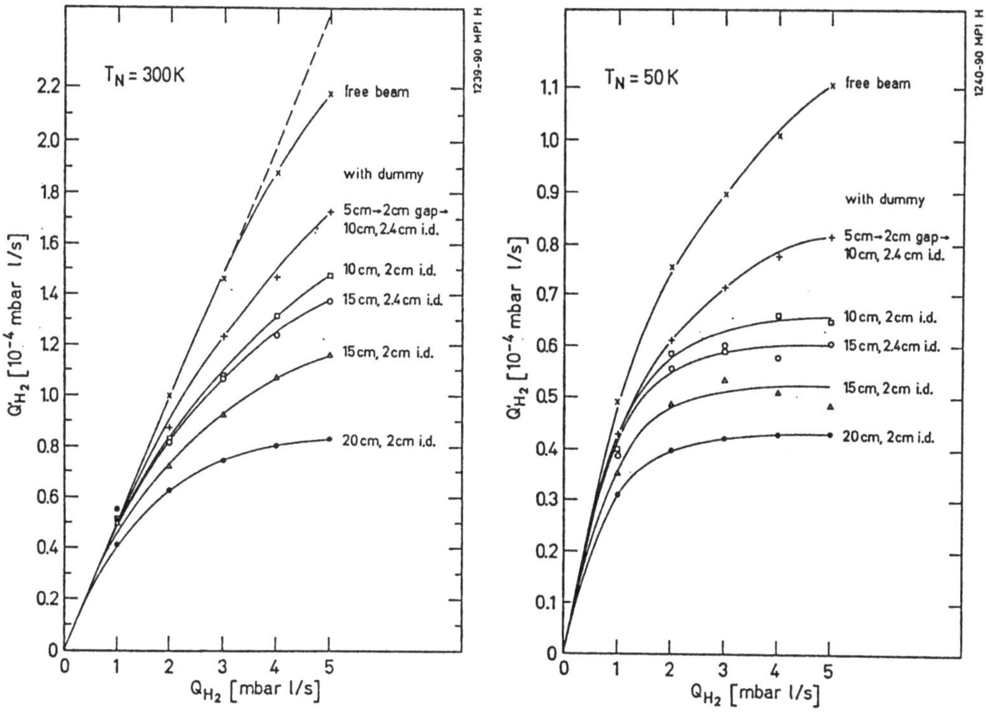

Figure 4a (left) and 4b (right) *Measured H_2 beam intensity for different dummy magnet geometries and two different nozzle temperatures.*

in Δp. Since the injected Q_{H_2} is equal to $(p_1-p_2)\cdot C \sim p_1\cdot C$, where p_1 is the pressure reading in the compression tube, p_2 the pressure in the target chamber, and C the conductance of the tube. Since Q_{H_2} was known absolutely, the product $\Delta p\cdot C$ could be determined.

Meanwhile, after improving the nozzle cooling (LN_2-cooling) the intensity could be increased to $3.6\cdot 10^{16}$ $\vec{H}$ in two substates. Figure 5 shows the temperature dependence of the intensity for an H_2-flow of 2 mbar l/s.

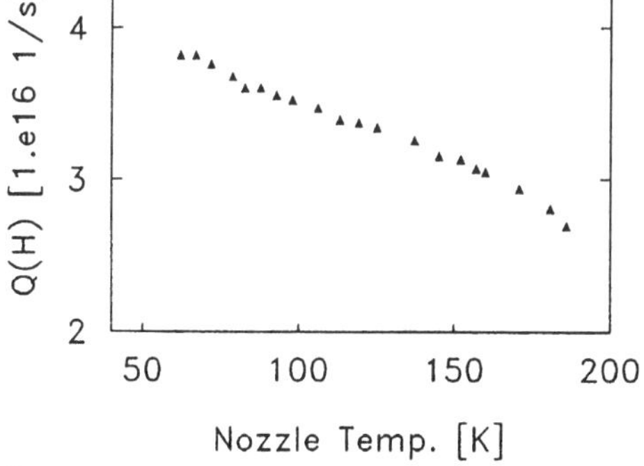

Figure 5 *Temperature dependence of the atomic beam intensity.*

[1]T. Wise and W. Haeberli, Workshop on Polarized Targets in Storage Rings 1984, Argonne Nat. Lab. ANL-84-50
[2]E. Steffens et al., Proc. to the 3rd LEAR Workshop, Tignes 1985
[3]The HERMES Collaboration, Proposal to Measure the Spin-Dependent Structure Functions of the Neutron and the Proton at HERA, DESY PRC 90/01, 1990
[4]G. Graw et al., Proc. of the International Workshop on Polarized Gas Jets, KEK Report 90-15, 1990

The Pumping Speed Limitations of the Atomic Beam Intensity

A.V. Sukhanov and D.K. Toporkov

Institute of Nuclear Physics, SU-630090 Novosibirsk, USSR

The limitations of the atomic beam intensity by the finite pumping speed are discussed. The beam intensity dependence on the pumping speed U_n was shown to be less then $U_n^{1/2}$. The experimental results confirming this fact are presented.

1. Introduction

The increased interest for the experiments with polarized gas jet targets has stimulated investigations of intense atomic beams [1,2]. In a largest part of previous works [see for example 3,4,5] the influence of the geometrical parameters (such as the distance between a nozzle and a skimmer, the nozzle geometry etc.) and the full gas flow on the atomic or molecular beam intensity has been carefully investigated. But the influence of pumping speed on the maximal achievable intensity hasn't been investigated. The main attention in presented work has been paid to this problem.

2. Layout of experiment and results.

The layout of experimental setup is shown in fig.1. Two cryopumps with capacities of 3500 and 1200 *l/sec* were used for pumping the nozzle and the skimmer chambers correspondingly. It was possible to decrease the pumping speed of the nozzle chamber

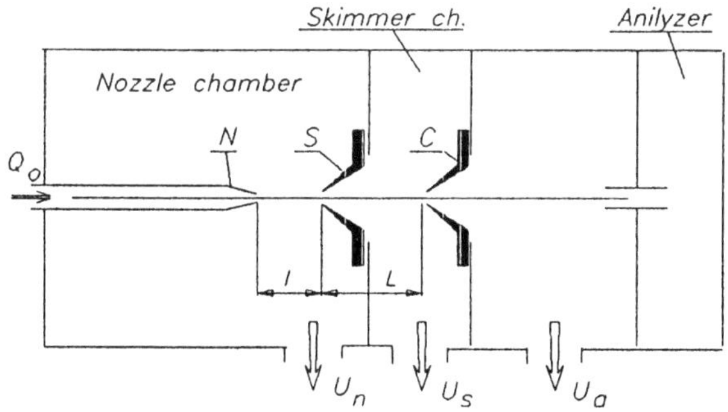

Fig.1. The layout of a common scheme for beam formation.

by using movable shields. The sublimation titanium pump with a capacity of ~ 5 10^3 *l/sec* has been installed to pump the collimator chamber and the ion pump NORD100 - to pump the analyzer chamber. The capacity of the latest one has been calibrated to measure an absolute intensity of the beam. The distance between the nozzle and the skimmer was variable and their sizes were chosen close to that used in common polarized sources: r_n=1 *mm*, r_{sk}=1 *mm*, r_{coll}=1 *mm*, L_{sc}=80 *mm*. Between the collimator and the analyzer chambers on the beam axis the tube with an internal diameter of 7 *mm* and the length of 100 *mm* has been installed. The distance between it and the collimator was equal to 70 *cm*. It was possible to cool the nozzle down to liquid nitrogen temperature. The beam intensity has been measured by the difference of vacuum in the analyzer chamber during the beam breaking by the scrapers in the skimmer chamber.

The measured beam intensities for maximal pumping speeds are shown in figs.2 and 3. For each full flow through the nozzle there is an optimal distance between the nozzle and the skimmer that provides the local maximum of the intensity. For given pumping speed there is also an optimal full flow provides the absolute maximal intensity of the beam.

Figure 4 shows the beam intensities as functions of full flow through the nozzle for different values of pumping speed of the nozzle chamber. The shown distances l_{ns} are corresponded to the absolute maximum of the intensity at each speed. Figure 5 shows the maximal achieved intensity as a function of pumping speed of the nozzle chamber. The maximum maximorum achieved intensity at room temperature was equal to 9 10^{18} */ster sec*.

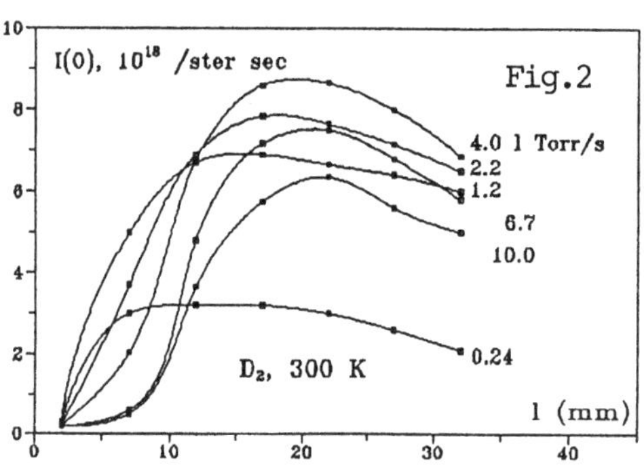

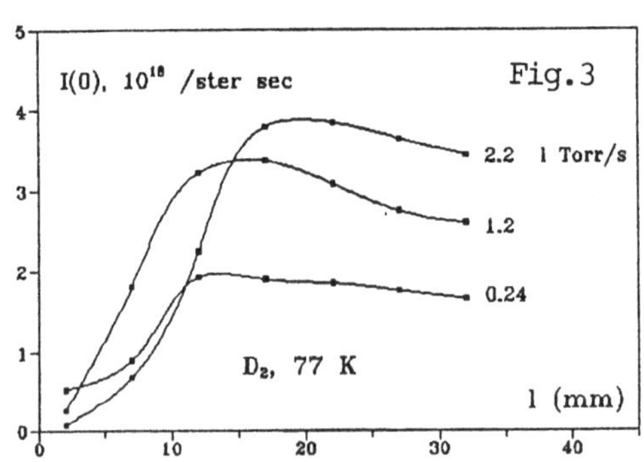

Fig.2,3. The beam intensity as a function of distance between the nozzle and the skimmer at different full gas flows.

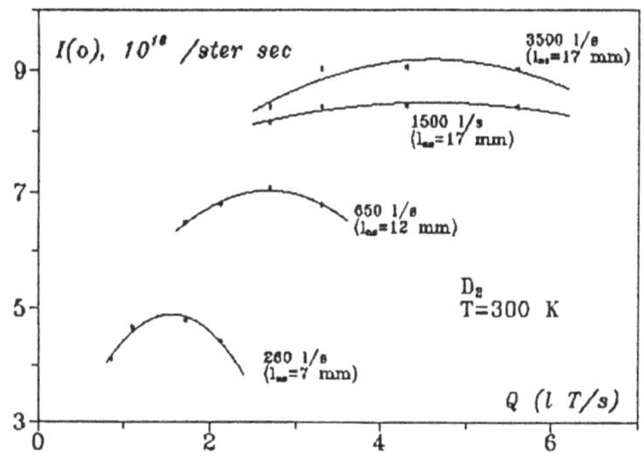

Fig.4. The beam intensity as a function of full flow through the nozzle for different pumping speeds.

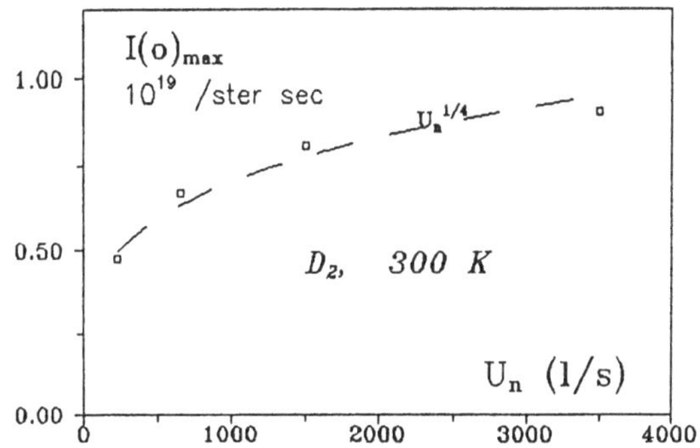

Fig.5. The maximum maximorum intensity as a function of pumping speed in the nozzle chamber.

3. Discussion of results

Let's assume that the nozzle forms the beam in a free molecular regime with an angular distribution [6]

$$\frac{dI}{d\Omega} = n\,\frac{Q_o}{\pi}\cos^{2n-1}\Theta \qquad (1)$$

here Q_o - is the flow through the nozzle, Θ - the angle between the beam axis and the direction to an observer. Let's also assume that the largest part of the beam is attenuated in the nozzle and the skimmer chambers by the scattering on a background gas, the density of which is fully determinated by corresponding pumping speed. In addition to the homogeneous background gas in front of the skimmer there is an inhomogeneous cloud of molecules scattered from the skimmer surface. At small sizes of the nozzle and the skimmer and the cosine angular law of molecules scattered from the surfase the intensity of the beam can be written in such form

$$\frac{dI(o)}{d\Omega} = n\,\frac{Q_o}{\pi}\exp\left\{-\sigma\int n\,dl\right\} = n\,\frac{Q_o}{\pi}\exp\left\{-\sigma\,Qo\left[\frac{\ell}{U_n} + \frac{L_{sc}}{U_s}\frac{r_s^2}{\ell^2} + \frac{A_n}{V\ell}\right]\right\} \qquad (2)$$

here σ (cм^2) is the scattering cross section, U_n и U_s (cм^3/sec) - pumping speeds of the nozzle and the skimmer chambers correspondingly, V(cm/sec) - average velocity of molecules scattered from the skimmer surface. The first and the second terms in the exponent factor describe the effects of beam interaction with homogeneous background in the nozzle and the skimmer chambers correspondingly, and the third term - the previously mentioned effect of scattering on the inhomogeneous preskimmer cloud. The numerical coefficient A_n can be easily calculated for chosen model of free molecular gas expanding from the nozzle. For $1 \leqslant n \leqslant 3$ it lays in an interval of $.19 \leqslant A_n \leqslant .82$.

The expression (2) qualitatively describes the experimentally observed effects, for instance:

- the existence of local beam intensity maximum at some distance between the nozzle and the skimmer for each given flow through the nozzle;
- the existence of the absolute maximum of the intensity at fixed pumping speed of the nozzle chamber;
- the increasing of beam intensity with the increasing of the nozzle temperature in accordance with $\sigma^{-1} = \sigma^{-1}(T)$ [7].

The suggested model predicts that the growth of the maximal achieving intensity has to be proportional to $U_n^{1/2}$. But the experimental data (see fig.5) are better described by some slower dependence ~ $U_n^{1/4}$. Such behavior of the beam intensity can be explained by an existence of some other beam-skimmer interaction havn't been taken into account in our simple model. For instance it can be the shock wave formation at the skimmer. If indeed the conditions of our experiment enabled this formation the beam at the nozzle was quite dense and preserved the central part of itself from the interaction with the background gas. To verify this statement we have studied the influence of the additional D_2 background was blown into the different chambers. The obtained beam attenuations in the nozzle and skimmer chambers are shown in fig.6. The extracted effective mean free path in the nozzle chamber was found approximately two times larger than in the

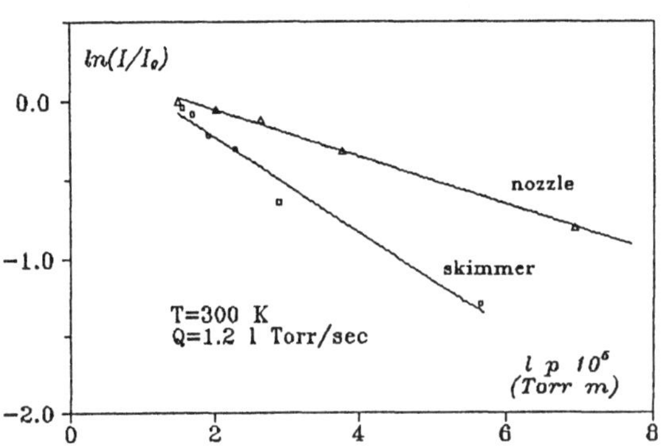

Fig.6.
The beam attenuation on the additional background in the nozzle and skimmer chambers.

skimmer chamber. So it really could means that the jet formed by the nozzle is quite dense in the nozzle chamber, and outer layers of the jet screening the background gas penetration to the jet axis.

4. Conclusions

Molecular beam with an intensity of 9 10^{18} */ster sec* was produced in the experimental setup using 3500 *l/sec* cryopump in the nozzle chamber. The further increasing of pumping speed is not the best way to produce more intense gas beam. The sources of beams that have the pumping speed of ~ 10^3 *l/sec* can produce the beams with the intensity close to practical limit caused by the beam-skimmer interaction.

Acknowledgments

The authors wish to express a cordial acknowledgments to Yu.S.Popov for the help in experimental equipment installation and to D.M.Nikolenko for stimulating discussions

References:

1. D.K.Toporkov et al., Particles and Fields Series 37, High Energy Physics, VIII Int. Simp., Minneapolis, Vol.II (1989)1286.
2. The proposal to measure the spin-dependent structure functions of neutron and the proton, The HERMES collaboration, January 1990.
3. W.Korsch, Dissertation, Univ. Marburg/Lahn, 1990.
4. A.V.Evstigneev et al., NIM A238(1985)12.
5. K.Teshima,Y.Yasunaga, Jap.J.of Appl.Phys. 22(1983)1.
6. R.D.Weaver and D.R.Francl, Rev.Sci.Instrum. 58(1987)2115.
7. A.Roth "Vacuum technology", North Holland publ.Co. (1976).

Compression of Polarized ^{3}He

W. Heil, G. Eckert, M. Leduc, M. Meyerhoff, P.J. Nacher, E.W. Otten, Th. Proschka, L.D. Schearer, and R. Surkau

Institut für Physik, Universität Mainz,
Staudingerweg 7, W-6500 Mainz, Fed.Rep.of Germany

This article gives a status report on the polarized ^{3}He target development work at Mainz. It is based on the optical pumping of the $^3S_1$ state of ^{3}He with subsequent metastability spin exchange to the ground state atoms. This pumping scheme is very efficient at ^{3}He pressures of about 1 Torr; for high density targets, however, a mechanical compression of the polarized gas is needed in a second step. First results with a Toepler compression pump look promising and will be discussed in the following.

1 Introduction and Physics Motivation

The technique of polarizing ^{3}He by spin exchange with optically pumped metastable ^{3}He-atoms is by now almost 30 years old and is well understood and well established [1]. The realization of highly polarized dense ^{3}He targets, however, had to await the advent of intense tunable pumping lasers in the particular infrared spectral region [2]. Since this basic problem has been solved, a fast increasing number of experiments is under preparation at several places. At the new 100 % duty factor electron accelerator MAMI at Mainz [3], scattering experiments with polarized beams and polarized targets are a major part of the future physical program. The measurement of the neutron electric form factor G_E^n in the reaction $^3\vec{H}e$ ($\vec{e}$, e'n) is one example of experiments for which polarized ^{3}He targets apply [4]. In the particular case of the quasifree scattering kinematic, the polarized ^{3}He nucleus can be regarded in a good approximation as a free polarized neutron, as the nucleons are predominantly in a s-state with the two proton spins paired off [5]. The measured quantity is then an asymmetry A_{exp} given by:

$$A_{exp} = P_e \cdot P_n \frac{-2\sqrt{\frac{\tau}{1+\tau}}\tan(\frac{\vartheta}{2})G_E G_M \sin(\beta) + 2\tau\sqrt{\frac{1}{1+\tau} + \tan^2(\frac{\vartheta}{2})}\tan(\frac{\vartheta}{2})G_M^2\cos(\beta)}{\frac{G_E^2+\tau G_M^2}{1+\tau} + 2\tau\tan^2(\frac{\vartheta}{2})G_M^2} \quad (1)$$

depending on the angle β between the target spin $\vec{S}$ and the three momentum transfer $\vec{q}$. With A_{exp} measured in parallel ($\vec{S} \parallel \vec{q}$) and perpendicular ($\vec{S} \perp \vec{q}$) kinematic one can determine the ratio G_E^n/G_M^n given by

$$\frac{G_E}{G_M} = -\sqrt{\tau + \tau(1+\tau)\tan^2(\frac{\vartheta}{2})} \cdot \frac{A_{exp}(\vec{S} \perp \vec{q})}{A_{exp}(\vec{S} \parallel \vec{q})} \quad (2)$$

We aim at a 10% measurement of G_E^n/G_M^n in the momentum transfer range of $5 \leq q^2 \leq 15$ fm^{-2} accessible with MAMI.

2 Nuclear Polarization of ^{3}He gas

The optical pumping is applied to $^3S_1$ metastable atoms created by a discharge in a ^{3}He gas. The density at which the pumping is efficient is set by the requirement of an homogeneous weak discharge in the gas and a long lifetime of the metastable atoms. This correponds to a ^{3}He pressure of about 1 Torr. The density of the metastable atoms [^{3}He*] is of order 10^{10} cm^{-3}. The advantage of the powerful solid state LNA lasers generating the desired wavelength $\lambda = 1.083$ μm is not only the larger power available but also their narrow bandwidth and tunability. This allows selecting pumping on only one component of the fine and hyperfine structure of the transition (usually the 2^3S_1 (F= 3/2) $\longrightarrow$ 2^3P_o component is chosen). Fig. (1) shows the optical pumping and detection schemes.

The ^{3}He gas is excited by a weak RF-discharge which produces the metastable 2^3S_1-atoms. The cell is immersed in a magnetic field B_o; the value of B_o is not critical and a field of a few Gauss is convenient. The only drastic requirement is the homogeneity of B_o: to avoid nuclear relaxation the field gradients must, in practice, be kept at about 1mG/cm over the entire volume where the ^{3}He atoms are confined. The laser beam is directed along the B_o field. To measure the nuclear polarization P an optical detection method is used, which gives a continuous reading of P and is not destructive [6]. The advantage of this ^{3}He pumping scheme lies in the very large reaction constant $< \sigma \cdot v >$ of about 10^{-11} cm^3/s for the polarization transfer from the metastable state to the nuclear spin in the ground state atom. With the resulting rate constant of $\gamma = < \sigma \cdot v > \cdot [^3He^*] \simeq 10^{-1}$/s, the sample can therefore be polarized within a typical time constant T_p of about $T_p \simeq 10$s. Fig. (2) shows the build-up of the polarization signal P in a 1 l cell filled with 2 Torr of pure ^{3}He

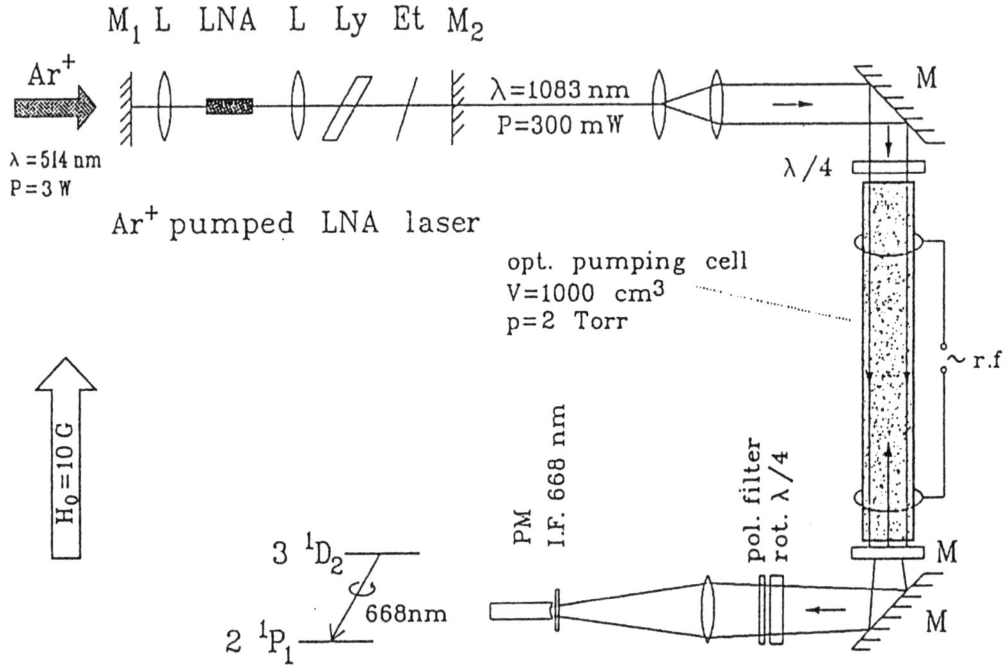

Figure 1: Optical pumping and detection schemes

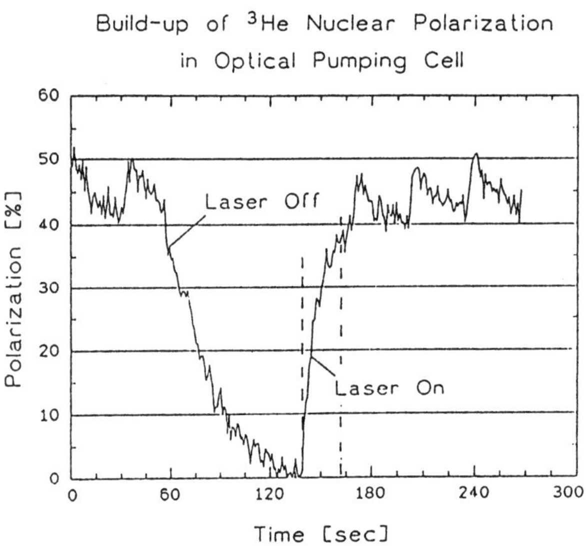

Figure 2: Build-up of ^{3}He nuclear polarization in the optical pumping cell

(N = 7.1· 10^{19}) when irradiated with the LNA-laser of fig.(1) . The production rate given by R = N·ΔP/T_p is about R = 2·10^{18} $^3\vec{H}e$/s.

Together with the absorbed laser power of 200 mW which corresponds to 1.1·10^{18} photons/s , this results in a transfer rate of 0.9 h into the nuclear system per absorbed photon, that makes this pumping scheme so efficient.

3 Increasing the density of optically pumped ^{3}He

For many applications especially in nuclear and high energy physics, dense polarized ^{3}He targets are needed. If this technique is to be used for high density targets, however, high density must be achieved either by mechanical compression, a way which has been followed first by J. Daniels and collaborators [7], or by cooling a side arm in a double cell configuration to low temperatures which is the domain of the Paris group [8]. If one aims at still higher densities than can be achieved by cooling [9], one has to apply some kind of mechanical compression which poses a number of technical problems: 1) avoiding any relaxing material, like stainless steel for instance, 2) achieving high compression factors and 3) maintaining high purity of gas. Our Mainz-Paris collaboration has undertaken the task and constructed a prototype of ^{3}He compressor which followed the ideas of the Toronto group and which is sketched in fig. (3).

Its an all glass apparatus in which a mercury piston is activated periodically to compress gas from the pumping cell (bottom part of fig. (3)) into the storage- or target cell. The compression cell is linked to the storage cell by a small tube containing a mercury valve on top of a section of sintered glass. The storage cell is leaking through a capillary tube calibrated so that a convenient pressure equilibrium is reached in each part of the apparatus. With an average transfer rate of 7·10^{17} $^3\vec{H}e$/s (P_o = 50 %), 120 polarization and compression cycles (T_c = 35 s) were needed to fill the target cell of V = 120 cm^3 at a ^{3}He pressure of 685 Torr. The build-up time of the polarization is monitored by NMR and shown in fig. (4), reaching a maximum polarization in the target cell of P_T = 30 %. To account for the polarization loss in the compression cycle, several causes of relaxation have been identified so far, which will hopefully be suppressed in the new version of this type of compression pump.

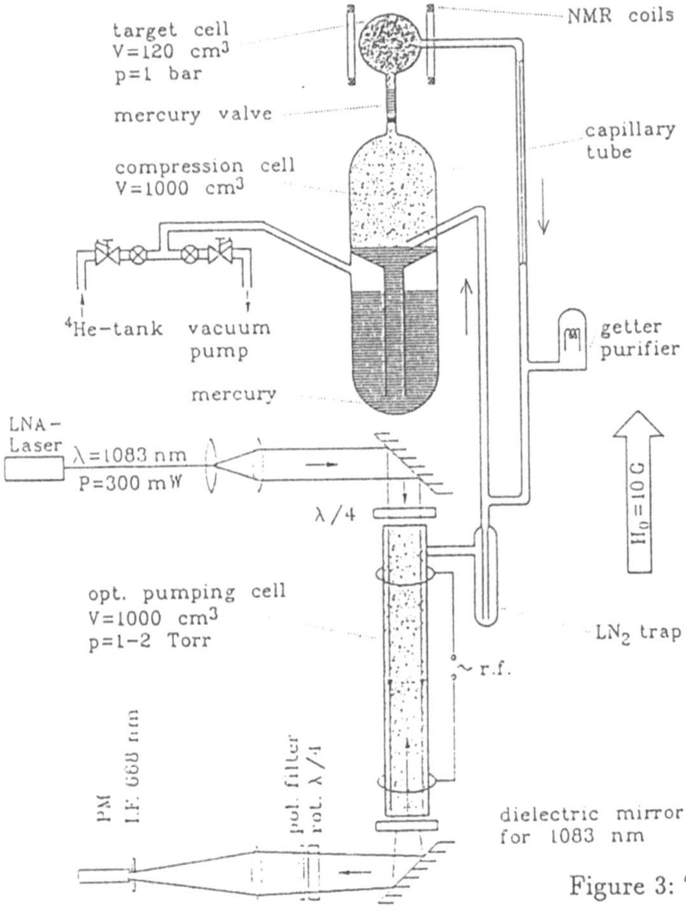

Figure 3: The Toepler-compressor

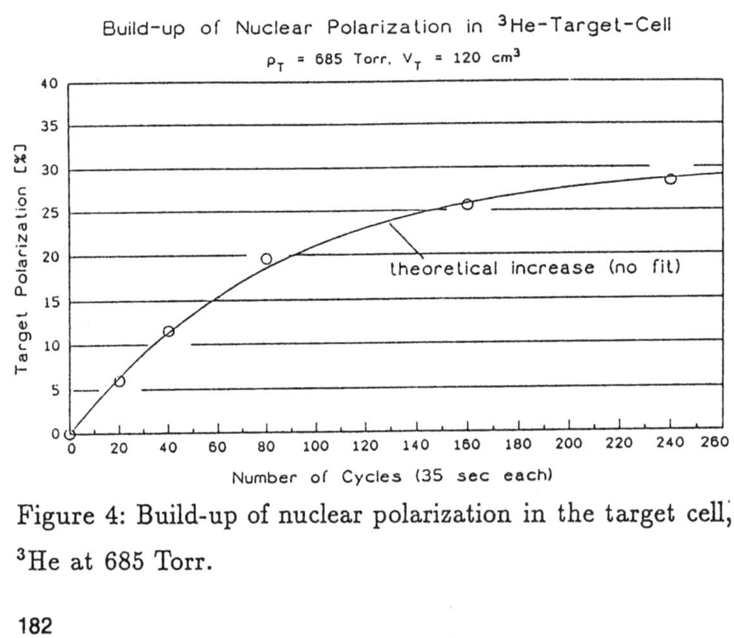

Figure 4: Build-up of nuclear polarization in the target cell, containing 120 [cm^3] ^3He at 685 Torr.

4 Future target development work

Since about two years, long a-axis grown LNA-crystalls are commercially available which can be pumped by Krypton arc-lamps [10]. So far these lasers can deliver 3 Watt of continuously tunable power at $\lambda = 1{,}083\ \mu m$ within a bandwidth of 2 GHz. The jump in the laser power by a factor of 10 as compared to previous state of affairs, can directly be transfered in increased production rates and higher steady state nuclear polarization, provided one is able to absorb the total amount of incident laser power in the pumping cell. Generally, this requires large pumping volumina, as the laser has to pump the fairly thin vapour of metastable atoms. What concerns the mechanical compression of the polarized gas, a new two stage design of the piston compressor is under construction, which provides a better adaptation of its pumping speed S_p to the increased production rates and which aims at higher endpressures. The apparatus consists of a commercially available oil free diaphragm pump with $S_p = 10$ l/min at p = 2 Torr compressing the gas to atmospheric pressure. In the second step a homemade piston compressor made of titanium and o-ring sealed copper valves should reach the endpressure of 10 bar. The latter has already been tested regarding relaxation and outgasing properties.

References

[1] F.D.Colegrove, L.D.Schearer, K.Walters Phys.Rev.132 (1963) 2561

[2] J.M.Daniels, L.D.Schearer, M.Leduc, P.J.Nacher J.Opt.Soc.Am.B4 (1987) 1133

[3] H.Herminghaus, Proc.Linear Accelerator Conf., Williamsburg VA, Oct.1988

[4] A3 Collaboration Mainz, Finanzierungsantrag an die DFG, Sonderforschungsbereich 201 (Mainz,1989)

[5] B.Blankleider, R.M.Woloshyn Phys.Rev.C29 (1984) 538

[6] M.Pinard, J.Van der Linde, Canad.Journ.of Physics 52 (1974) 1615

[7] J.M.Daniels, R.S.Timsit, Canad.Journ.of Physics 49 (1971) 525

[8] D.S.Betts, M.Leduc Ann.Phys.Fr.11 (1986) 267

[9] R.G.Milner, R.D.McKeown, C.W.Woodward, Nucl.Instr.Meth. A257 (1987) 286

[10] C.G.Aminoff, C.Larat, M.Leduc, F.Laloe Revue de Physique Appl.24 (1989) 824

Progress with High Density Polarized ^{3}He Targets

T.E. Chupp

Harvard University Physics Laboratories, Cambridge, MA 02138, USA

High density polarized ^{3}He targets have, in the last year, been employed in actual nuclear physics measurements at TRIUMF and at MIT-Bates. These measurements have shown the technical feasibility of two different techniques to produce targets and begin to provide experimental study of the spin structure of the ^{3}He nucleus important to efforts poised to study the structure of the neutron in quasi-elastic and deep inelastic scattering. One target, based on spin exchange with optically pumped Rb atoms has produced a target of length 4 to 8 cm and density of 1-3x10^{20} ^{3}He/cm^3 with polarization up to 60% and has been used in a 7 μa electron beam and a 20 na proton beam. The other technology, based on metastability exchange with optically pumped metastable 2^3S_1 ^{3}He atoms has produced a target with up to 30% polarization in a 11 μa electron beam with a length of 10 cm and density of 4x10^{18} ^{3}He/cm^3.

Introduction

Polarized ^{3}He targets have been developed for a variety of fundamental, high energy and nuclear physics measurements. These measurements include studies of parity and time reversal invariance violation in resonance neutron absorption[1], polarized neutron beta-decay asymmetry measurements[2], muon capture studies[3], measurement of the neutron's spin dependent deep inelastic structure function[4,5] and test of the Bjorken sum rule[6], the neutron's elastic form factors[7,8] and studies of elastic scattering [9] and the nuclear structure of ^{3}He[10]. Experiments with hot and epithermal neutrons have been underway for several years[11,12], however in the past year high density targets have been employed at TRIUMF and Bates reporting both technical successes and initial physics results.

Target Techniques and Performance

There are currently two techniques, both based on optical pumping of an intermediate species which transfers polarization to ground state ^{3}He atoms – either Rb atoms or metastable 2^3S_1 He atoms. In the target based on spin exchange with Rb,[13] the Rb is optically pumped by light at the D1 resonance (795 nm) produced by a Titanium-doped sapphire (Ti^{3+}:Al_2O_3) laser. At least 4 watts of laser power at 795 nm which is provided by the (Ti^{3+}:Al_2O_3) laser. This laser technology has only become available in the past year and rapid development and optimization of the lasers for these targets have led to the significant performance advances. The Rb atomic polarization, carried by the valence electrons, is transferred to the ^{3}He nucleus during collisions. The time constant for spin transfer is several hours because of the weakness of the hyperfine interaction that mediates the spin exchange and because the collision rate is limited by the Rb density of about 6×10^{14} atoms/cm^3. However, it is possible to achieve high polarization in densities of 1-3x10^{20} ^{3}He/cm^3 as long as a small fraction of N_2 (5x10^{18} N_2/cm^2) is included to prevent radiation trapping and to aid in recombination of ionic species produced by the incident beam.[14]

For an experiment at the MIT/Bates electron linac, a two cell target was employed. This geometry was chosen because the ionization produced by an $\approx 10\mu A$ incident electron beam destroys the Rb polarization. We have therefore separated the volume in which the Rb is optically pumped from the target volume through which the electron beam passes. The two volumes are connected by a transfer tube designed to have a 10 minute time constant, which is much less than the time for polarization build-up in the pumping cell (several hours) and much less than the relaxation times due to wall collisions and the ionizing effects of the electron beam in the target cell (several tens of hours). Thus the two cells are very strongly coupled and the ^{3}He polarization is essentially equal in the two volumes.

The entire target is made of alumino-silicate glass (Corning 1720 or Schott 8409) which has wall relaxation times of 20 hours or greater (greater than 100 hours has been observed). The target cell is 7.5 cm long and 2.2 cm inside diameter with glass windows thinned to 120 μm through which the electron beam passes. Magnetic spectrometers were employed and the acceptance of the spectrometers was limited to 4 cm ($\pm$ 2 cm) along the electron beam with 2 cm thick heavy metal target slits. These slits eliminated background produced by electrons scattering from the target windows. The effective areal partial densities of the target were 3.6×10^{20} ^{3}He/cm^3 and 3×10^{19} N_2/cm^3.

Polarization as high as 40% was achieved during bombardment but significant variation of the polarization can occur. The variation is the result of instability of the laser power mostly due to changes in the cooling water temperature. This can be easily remedied with a stabilized cooling system. Furthermore, an uncompromised experiment would allow easy access to the laser system for tuning. The maximum observed polarization of 40% was the consequence, in part, of the angle of incidence of the laser light on the glass pumping cell. In order to orient the ^{3}He polarization at the desired angles with respect to the vector momentum transfer ($\vec{q}$) of the scattered electrons, the light was at non-normal incidence which results in elliptically polarized light. With normal incidence, 60% polarization was achieved. The time constants for ^{3}He polarization relaxation were observed to be unaffected by electron beam currents up to 22 μA. This confirms the expectation that the N_2 in the target reduces the relaxation rate to a fraction of the rate at which ^{3}He atoms are ionized.[14,15] These results combined with previous measurements with a charged particle beam [15] demonstrate a detailed understanding of the ionization induced relaxation process in the Rb spin exchange targets. The ^{3}He target thickness was also monitored by measuring cross sections throughout the running period. One important result is that the gas target thickness was not appreciably affected by changing beam current.

The target built at TRIUMF [16,17] employs a single cell. This is feasible because the low beam currents used – 20 nA of 250 MeV protons – do not appreciably affect the Rb optical pumping. The target cells used in several recent measurements at TRIUMF were 8 cm long with 2.2 cm inside diameter and were made of Corning 1720 glass. The end windows were 100 μm thick. A vertex tracking chamber is used to separate protons scattered from the windows and those scattered from the gas in the target. The maximum polarization achieved in a target with 2400 torr (at 273 K) of ^{3}He was 63%. In a new development, the TRIUMF group has used cryogenic techniques to produce cells with 9000 torr (at 273 K) of ^{3}He in which 40% polarization has been achieved.

Polarization measurements for the spin-exchange targets have been based on Adiabatic Fast Passage NMR [18]. With this technique, the magnetization of the ^{3}He is measured in a set-up that is calibrated by measuring NMR signals from protons in a room temperature H_2O sample in an identical target cell. The precision of NMR polarimetry is limited to about 5% due to limitations of the signal-to-noise ratio for the proton signals (the polarization is less than 10^{-8}) and an accurate determination of the ^{3}He density in the cryogenically filled cells. Corroborative measurements have been provided by several nuclear physics experiments including:

1. measurement of polarization of neutrons at LAMPF[1] and Grenoble[12] with precision of 10%
2. measurement of the spin correlation parameter, A_{nn} for the $^3\vec{He}(\vec{p},\pi^+)^4$He reaction at TRIUMF with 10% precision, and
3. measurement of the helicity dependence in elastic scattering of polarized electrons from polarized ^{3}He at Bates with 40% precision achieved with analysis of part of the available data.

The meatastability exchange target is also a two cell arrangement.[19] The metastable ^{3}He atoms are produced in a weak room temperature discharge at 2 torr in a boro-silicate (Pyrex) pumping cell. The metastable 2^3S_1 ^{3}He atoms are polarized by optical pumping of the $2^3S_1 - 2^3P_{0,1,2}$ transition with 1.083 μm laser light. The laser used, a Nd:YAP crystal pumped by Kr arc lamps, is one of two lasers recently developed by Leduc *et al.*[20] which have inspired significant advances of performance of the metastability exchange technique.

For the target used at Bates, the target density is increased by cooling of the target cell to 17K at which the number density is 4×10^{18} ^{3}He/cm^3. The target cell is copper which is cooled by a temperature controlled cold finger. The windows through which the electron beam passes are $5 \mu m$ copper foils which are fastened with epoxy. The cold copper is a surface with very poor ^{3}He polarization relaxation properties. Therefore a layer of N_2 is condensed onto the cold surfaces which reduces the surface relaxation rate.

The polarization is measured in the glass pumping cell, and the polarization in the target cell can be inferred from polarization transients after the pumping cell polarization is destroyed. Once the ratio of target cell to pumping cell polarization is determined as a function of density, temperature, wall relaxation rate and beam current, a continuous monitor of the polarization is possible by measuring the circular polarization of a 668 nm fluorescence line from the discharge. The run at Bates provided about 11 μA of electrons and it was determined that the target cell polarization was 85% of that measured in the pumping cell. The significant difference of polarization is due to the effect of the ionization on the ^{3}He polarization relaxation rate in the target cell where several ionic species are present.[14] The polarization in the target cell typically ranged between 20% and 30%, depending primarily on the laser output power. A measurement of the helicity dependence in polarized-electron-polarized ^{3}He scattering, described below, suggests that the ^{3}He polarization in the target cell is consistent with expectations.[8]

Recent Experimental Results

The TRIUMF group has been most prolific, completing measurements of elastic scattering and quasi-elastic scattering (nucleon knock-out) of polarized and unpolarized protons from the polarized ^{3}He target. The elastic scattering results are described in the proceedings of this symposium.[16] This experiment probes the spin structure of the bound ^{3}He nucleus by comparison of the knock-out reactions with previously measured nucleon-nucleon spin correlations.[21] The spin correlations A_{no} with A_{on}, for polarized beam or polarized target respectively, were compared and found to be very different which shows that a the T-matrix has a sixth, non vanishing (f) amplitude. This amplitude vanishes in nucleon-nucleon scattering due to isospin symmetry, but gives a significant contribution to proton scattering on ^{3}He.

The spin correlations in (p,2p) nucleon knock-out measurements are sensitive to small components of the ^{3}He wave function which contribute to the protons polarization in polarized ^{3}He. For fully polarized ^{3}He, the net proton polarization is calculated to be $P_p = -1.4\%$, that is the net proton spin is opposite that of the nucleus. The neutron polarization is calculated to be $P_n = 87\%$.[22] The fact that $P_p \neq 0$ and $P_n < 1.0$ is the consequence nuclear tensor force induced $D-$states and the isospin mixing induced S'-states. The TRIUMF experiment[17] at 250 MeV was most sensitive to the S'-states and shows that the S'-states contribute about 1.5% to the ^{3}He wave functions in agreement with the Faddeev calculations of Afnan and Birrell.[23] Data were also collected on the ^{3}He(p,pn) knock-out reaction which reveal spin correlations significantly less than expected. However the neutron data have limited statistics and are subject to errors due to proton contamination in the neutron channel. A 2% proton contamination correction would account for the discrepancy of experiment and theory.

Results from recent running at Bates have been published for the measurement at 574 MeV of the electron helicity dependence of polarized-electron-polarized ^{3}He scattering near the quasi-elastic peak. The ^{3}He target polarization was oriented nearly parallel or anti-parallel to the momentum transfer $\vec{q}$ so that the experiment was most sensitive to the contributions of the neutron. A helicity dependence of $-3.46\% \pm 1.23\% \pm 0.54\%$ was reported where the first uncertainty is due to statistical fluctuations and the second due to systematic corrections dominated by the beam and target polarization measurements. Current calculations of the expected electron helicity dependence are based on the model of quasi-elastic scattering as elastic scattering from the polarized nucleons in the polarized ^{3}He nucleus into quasi-elastic final states[7,22] but do not include final state interactions and meson exchange corrections. The expected electron helicity dependence is in the range -4.3% to 4.5%, consistent with the measured result. Analysis of the helicity dependence measured with the Rb spin exchange target is in progress and these data combined with the result quoted above should provide a precision of 0.5% on the helicity dependence.

With these first results, it seems clear that precise measurements with electrons and protons will soon reveal the detailed spin-structure of ^{3}He. More detailed calculations of these processes which include the final state and meson exchange effects are now needed.

Future High Density Polarized ^{3}He Targets

Significant advances in the volume and ^{3}He density in the Rb spin exchange target are planned for the near future. Reliable cryogenic techniques have been developed for cell filling and laser technology is rapidly improving. A target planned for deep inelastic scattering measurements at SLAC (experiment E-142) will be 30 cm long, have a target cell volume of 100 cm^3, ^{3}He density of 3×10^{20} atoms/cm^3 and be optically pumped with at least 10 watts of laser power.[5] The metastability exchange target should also realize significant increase in density when cooled to 4K and in polarization with improved cryogenic wall coatings and laser technology provided by lamp pumped long LNA crystals, both of which have been perfected by Leduc *et al.*.[24] Continued running at TRIUMF and Bates are planned along with several other experiments including E-142 at SLAC. The technology for a highly developed target will be in place when CEBAF comes on line.

References

1. K.P. Coulter, PhD. Thesis, Princeton University (1989).
2. E.G. Wasserman, T.E. Chupp, J.M. Richardson,J.F. WIlkerson, T. Bowles, R.G.H. Robertson, S. Elliot, G. Greene, M.S. Dewey, *private communication.*
3. G. D. Cates *et al.* Proposal to LAMPF PAC(1990).
4. R. Woloshyn, Nucl. Phys. **A 496**, 749 (1989).
5. R. Arnold *et al.* Proposal to SLAC EPAC **E-142**, (1989).
6. J.D. Bjorken, Phys. Rev. D **9**,1476 (1966); Phys Rev. **D1** 1376 (1966).
7. B. Blankleider, R. Woloshyn, Phs. Rev. **C 29**, 538 (1984).
8. C.E. Woddward *et al.*, Phys. Rev. Lett. **65**, 698 (1990).
9. R.H. Landau, M. Sagen, G. He, Phys. Rev. **C41**, 50 (1990).
10. O. Häusser *et al.* Proposals to TRIUMF Program Advisory Committee **E-541** (1988); and **E-616** (1990).
11. K.P. Coulter*et al.* NIM **A 288**, 463 (1990).
12. F. Tasset, Report to 3^{me}Souffle, Institute Laue-Langevin planning workshop, Grenoble, France (1990).
13. T.E. Chupp *et al.* Phys. Rev.**C 36**, 2224 (1987).
14. K. Bonin, T. Walker, W. Happer, Phys. Rev. **A 37**, 3270 (1988).
15. K.P. Coulter *et al.* NIM **A 276**, 29 (1989).
16. P.P.J. Delheij, *et al.*, in these proceeedings.
17. A. Rahav, O. Häusser, C.A. Miller, P.P.J. Delheij, W.P. Alford, T.E. Chupp, K. Fergerson, R. Henderson, C.D.P. Levy, K.P. Jackson, J. Mildenberger, B. Morrissette, M.C. Vetterli, R.M. Woloshyn, submitted to Phys. Rev. Lett. (1990)
18. T.E. Chupp, K.P. Coulter, Phys. Rev. Lett. **55,** 1074 (1985).
19. R. Milner, R. McKeown, C. Woodward, NIM. **A274**, 56 (1989).
20. C.L. Bohler *et al.* J. App. Phys **63**, 2497 (1988).
21. D.K. Haskel *et al.* Phys. Rev. **C34**, 236 (1986).
22. J.L. Friar, B.F. Gibson, G.L. Payne, A.M. Bernstein, T.E. Chupp, Phys. Rev. **C42**, 2310 (1990).
23. I.R. Afnan, N.D. Birrell, Phys. Rev. **C29**, 823 (1977).
24. C.G. Aminoff, C. Larat, M. Leduc, F. Laloë, Revue Phys. Appl., **24** 827 (1989).

Spin Relaxation on Surfaces

W. Widdra

Fachbereich Physik und Wissenschaftliches Zentrum für Materialwissenschaften, Philipps-Universität Marburg, Renthof 5, W-3550 Marburg/Lahn, Fed. Rep. of Germany

Spin relaxation via fluctuating fields and their magnetic field dependence are discussed. In particular the differences for relaxation mechanisms on surfaces based on spin diffussion compared to the mechanisms in the bulk are pointed out. Three magnetic field dependencies of alkali spin relaxation on surfaces - one for chemisorption and two for physisorption - are described.

Introduction

Spin relaxation on surfaces is an important problem of preparing polarized internal targets by using storage cells. To prevent strong depolarisation one needs to know about the relaxation mechanism on the surface. The interaction of spin polarized atoms with their environment on the surface can be described by a hamiltonian which has to include the interaction of the electronic and nuclear moments of the atoms with the local electric and magnetic fields. The fluctuating part of the hamiltonian can cause transitions between different states leading to equal occupied states, namely to spin relaxation. A magnetic dipole moment is sensitive to fluctuating magnetic fields. Some examples are the Fermi-contact-interaction with conduction band s-electrons of metals[1] or the dipole-dipole-interaction of diffusing spins[2] or a spin-orbit-interaction. On the other hand an electric quadrupole moment can cause relaxation due to fluctuating electric field gradients produced e.g. through adsorbat diffusion[3,4,5] or lattice vibrations (spin-phonon-interaction)[6]. This paper intends to concentrate on relaxation basing on fluctuating fields related to diffusive processes.

Spin relaxation via diffusion

The change of magnetic substate occupation probabilities of an ensemble of atoms - including spin relaxation - can be described by a master-equation [2]. The transition-probability (per unit time) between the substates is given by the product of the strength and the spectral density of the fluctuations at the transition frequency:

$$W_{mn} = \frac{1}{\hbar^2} \cdot G(0) \cdot j(\omega_{mn}).$$

From the autocorrelation function of the fluctuating part of the hamiltonian (ensemble averaged)

$$G(\tau) := \overline{\langle m|H(t)|n\rangle\langle m|H(t+\tau)|n\rangle^\star}$$

the spectral density can be calculated by fouriertransforming:

$$j(\omega) = \frac{1}{G(0)} \int_{-\infty}^{+\infty} G(\tau) \cdot exp(-i\omega\tau) d\tau.$$

An often used phenomenological Ansatz (known as BPP-Ansatz[7]) for the correlation function is given by

$$G(\tau) = G(0) \cdot exp\left(-\frac{|\tau|}{\tau_c}\right)$$

where τ_c is the correlation time of the fluctuations. In case of diffusion it is the inverse jumping rate. Fouriertransforming this Ansatz gives

$$j(\omega) = \frac{2\tau_c}{1 + (\omega\tau_c)^2} \longrightarrow \begin{cases} 2\tau_c & \text{if } \omega\tau_c \ll 1 \\ \frac{2}{\tau_c \omega^2} & \text{if } \omega\tau_c \gg 1 \end{cases}$$

For short correlation times compared to the inverse frequency (low frequency limit) the spectral density is independent of the frequency and therefor also independent of the magnetic field which determines the transition frequencies.

If the fluctuating fields are generated by spin diffusion it is possible to calculated the autocorrelation function[2,8]. In the low frequency limit $\omega\tau_c \ll 1$ the probability of finding a spin at time t on the place $\vec{r}$ (which has started at $\vec{r}(t = 0) = \vec{r}_0$) can be calculated from the diffusion equation. The result depends on the dimensionality d of the diffusion process:

$$P(\vec{r}, \vec{r}_0, t) = (4\pi Dt)^{d/2} exp\left(-\frac{(\vec{r} - \vec{r}_0)^2}{4Dt}\right)$$

where D is the diffusion coefficient which is inverse to τ_c Calculating the spectral density yields the following dimension dependent behaviour[5]

$$j(\omega) \stackrel{\omega\tau_c \ll 1}{\longrightarrow} \begin{cases} \tau_c & \text{for d=3} \\ \tau_c ln \frac{1}{\omega\tau_c} & \text{for d=2} \\ \frac{\tau_c}{\sqrt{\omega\tau_c}} & \text{for d=1} \end{cases} .$$

The spectral density depends on the frequency for two and one dimensional systems. As consequence for this systems the relaxation rate diverges by decrease of the magnetic field to zero.

Calculations for the high frequency limit $\omega\tau_c \gg 1$ show dimension independent the same result as the BPP-Ansatz mentioned above[8].

Experimental results

The theory in the previous section is applied first for chemisorbed atoms on metal surfaces. Therefore the nuclear spin relaxation of nuclear spin polarized ^{7}Li-atoms was studied on a oxygen covered tungsten-(110)-surface at a temperature about 1100K[5,11]. The mechanism of relaxation is believed to base on fluctuating electric field gradients due to adsorbat diffusion on the surface.

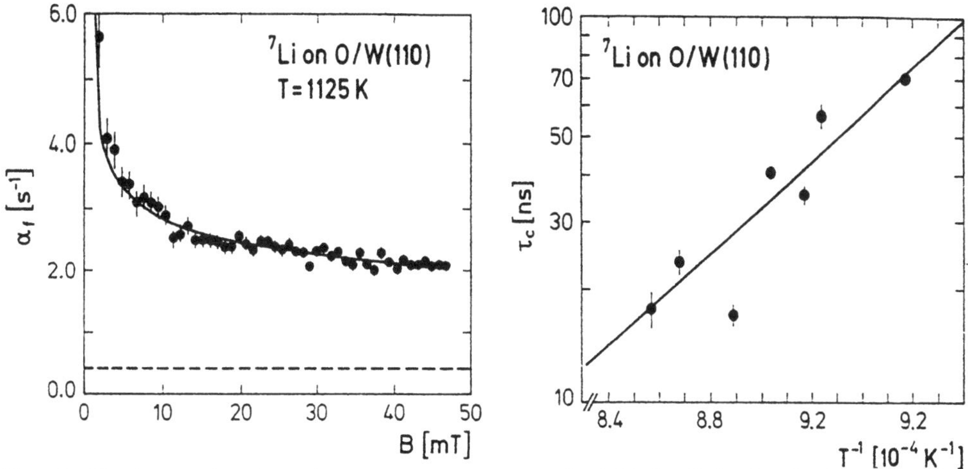

Fig. 1. Nuclear spin relaxation rate versus magnetic field for ^{7}Li adsorbed on a oxygen covered W(110) surface (left side). Extracted correlation times versus inverse temperature is shown on right side.

The measured relaxation rate is shown in figure 1. The dependence on the magnetic field is clearly visible. Using the model for 2-dimensional diffusion one can describe the data as marked by the solid line and extract a correlation time. The extracted correlation time is displayed on the right side of figure 1 for various surface temperatures. These values can be explained quit well with a thermal activated behaviour

$$\tau_c = \tau_c^0 \cdot exp\left(\frac{E_{Diff}}{kT}\right)$$

with $\tau_c^0 = 1.2 \cdot 10^{-13}s$ and a diffusion energy $E_{Diff} = 1.2eV$. For this system the relaxation rate and its temperature and magnetic field dependence can be understood with this model in the low frequency regime.

Two experiments which are dealing with physisorption of alkali metal atoms on surfaces will be now discussed. The short interaction time because of the weak Van-der-Waals-bonding to the surface is overcome in both cases by the use of storage cells. The spin relaxation is studied after several surface interactions. The relaxation rate is often described by

$$\alpha = const \cdot \frac{\tau_s}{\tau_s + \tau_v} \cdot G(0) \cdot \frac{2\tau_c}{1 + (\omega\tau_c)^2}$$

with a typical time of flight in the cell τ_v and the dwell time τ_s on the surface. The additional factor arises from the fact that an interaction can cause relaxation only during the fraction of time when the atom is adsorbed on the surface.

Figure 2 shows results for optical pumped Na-atoms in a pyrex cell[9]. The steady state electronic spin polarisation drops with decreasing magnetic field. This is explained by an increasing relaxation rate. The relaxation rate is calculated from the assumption that the correlation time is equal to the dwell time $\tau_c \approx \tau_s \approx 0.8 \cdot 10^{-10}s$. Therefore and because this experiment is sensitive to the electronic spin polarisation only one can use the high frequency limit of the spectral density. For the spectral density the BPP-

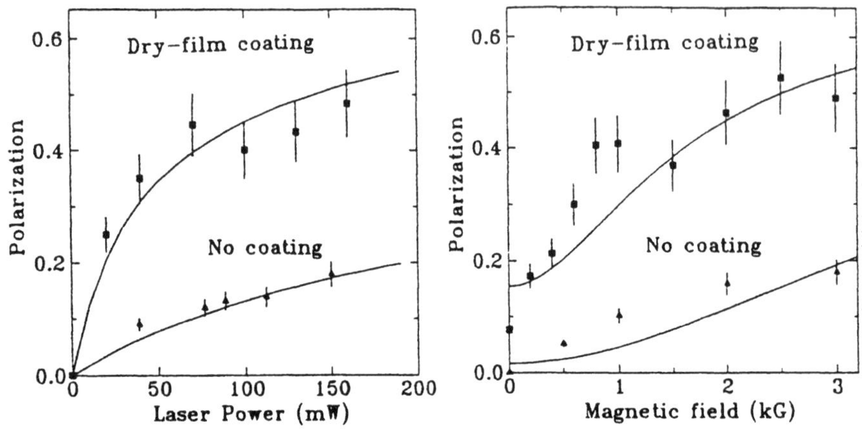

Fig. 2. Steady state electronic spin polarisation of Na in a pyrex cell versus magnetic field (right side) and versus laser power used for optical pumping.

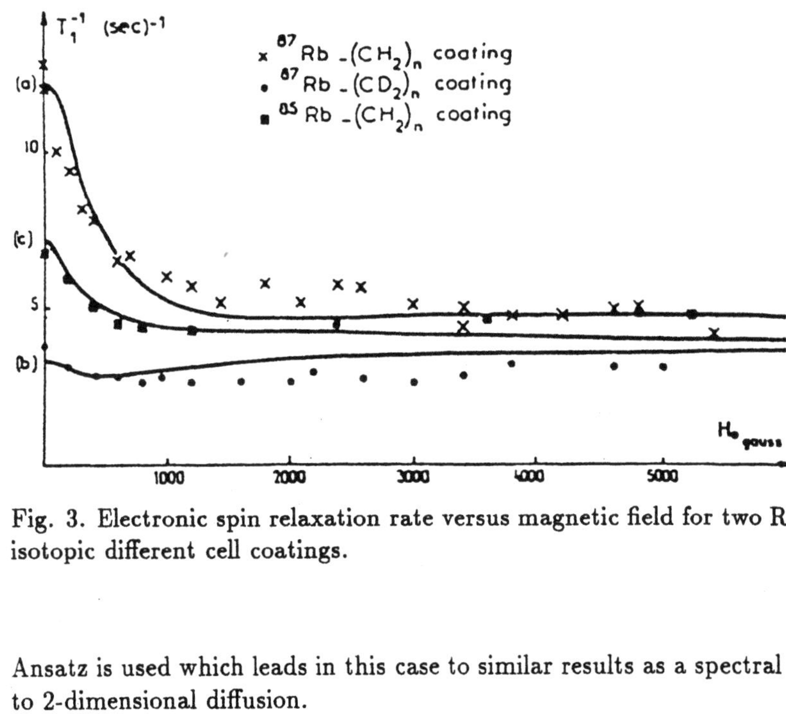

Fig. 3. Electronic spin relaxation rate versus magnetic field for two Rb-isotops and two isotopic different cell coatings.

Ansatz is used which leads in this case to similar results as a spectral density according to 2-dimensional diffusion.

The last example is a very early work and shows measurements of the electronic and nuclear spin relaxation of different Rb-isotopes in a paraffin-coated cell[10]. The measured electronic relaxation rates for the isotops ^{85}Rb and ^{87}Rb are displayed in figure 3 for the different cell coatings -$(CH_2)_n$- and -$(CD_2)_n$- .

From the different results the authors conclude to have relaxation based on magnetic dipole-dipole-interaction with the protons or deuterons of the coating. The solid lines represent calculations with two different correlation times using again the BPP-Ansatz. In this case the use of a spectral density which can describe 2-dimensional diffusion on surfaces could lead to slightly different results.

Conclusion

Often the measured increasing spin relaxation rate with decreasing magnetic field is explained by a frequency dependent spectral density of the fluctuations of the hamiltonian. The well known BPP-Ansatz from solid state physics which gives a similar spectral density as calculations for diffusion based fluctuations can describe the behaviour of the relaxation rate only in the high frequency limit of the spectral density. However calculation for diffusion on surfaces yields a different spectral density because of the lower dimensionality. This spectral density has a frequency (magnetic field) dependence even in the low frequency limit.

References:

[1] K.Waßmuth, D.Fick, Phys.Rev.Lett. 59 (1987) 3007

[2] A.Abragam, Principles of Nuclear Magnetism, Oxford 1961

[3] B.Horn, W.Dreves, D.Fick Z.Physik B48 (1982) 335

[4] U.Memmert, M.Bickert, M.Riehl, K.Waßmuth, D.Fick, R.F.Haglund Hyp.Int. 39(1988) 295

[5] M.Riehl-Chudoba, U.Memmert, D.Fick Surf.Sci. to be published

[6] I.von Kranendonk Physica 20 (1954) 781

[7] N.Bloembergen, E.M.Purcell, P.V.Pound Phys.Rev. 73 (1948) 679

[8] C.A.Sholl J.Phys. C14 (1981) 447

[9] M.Tanaka, T.Ohshima, K.Katori, M.Fujiwara, T.Itahashi, H.Ogata, M.Kondo Phys.Rev.A 41 (1990) 1496

[10] M.A.Bouchiat, J.Brossel Phys.Rev. 147 (1966) 41

[11] M.Riehl-Chudoba PhD-thesis Marburg 1989

Storage Cell Target for Polarized Hydrogen and Deuterium

W. Haeberli

Department of Physics, University of Wisconsin, Madison, WI 53706, USA

This paper describes recent wall depolarization studies, in which the degree of polarization of hydrogen atoms in a storage cell was measured for different wall coatings as a function of wall temperature. For the cells tested, the average number of wall collisions of atoms in the cell varied between 40 and 380. It is found that cells coated with a thin film of fluorinated polymers tolerate 380 wall collisions for wall temperatures between 120K and 300K without significant loss in polarization.

Introduction

The use of polarized atomic beams as internal targets in storage rings (so called polarized jet targets) has been considered for a long time but the density achieved in polarized jets is far below the 10^{14} atoms/cm^2 needed for most applications. Here we report recent tests of storage cells which show that the target thickness can be enhanced by at least two orders of magnitude.

Enhanced Target Thickness by use of Storage Cell

We have proposed[1] to increase the target density compared to that obtained with a polarized jet target by injecting the atomic beam into a cell. The cell has an opening for injection of the atomic beam, and apertures for entry and exit of the circulating beam. Rather than passing through the circulating beam only once, as they do for a jet target, the atoms undergo several hundred collisions with the wall of the cell before leaking out, and thus offer a much improved economy of polarized atoms, i.e. a much increased target thickness. We reported a first test of this idea[2] already in 1980.

The density of atoms in the storage cell is proportional to the number N of atoms injected per second, and inversely proportional to the combined gas conductance C of the openings of the cell: $\rho = N/C$. For example, three tubes of 10 cm length and diameter 7 mm have a combined conductance of 6.2 l/s for room temperature hydrogen. Thus $N = 6 \times 10^{16}$ atom/s injected into the cell yields a target density of 10^{13} atoms/cm^3. If the detection system can accept events from the central 10 cm of the target, the corresponding target thickness is 7.5×10^{13} atoms/cm^2, a gain of about two orders of magnitude compared to the corresponding jet target. The gain is higher or lower, depending on the size of the cell apertures required to admit the circulating beam and the atomic beam. The only other parameter available to affect the gas conductance is the wall temperature of the storage cell. Cooling the cell increases the target thickness since the gas conductance decreases in proportion to the velocity of the atoms in the cell.

Study of Wall Depolarization in Storage Cells

The test method in use at our laboratory has previously been described in some detail.[3] Briefly, it consists of passing a 50 keV D^+ beam through the target cell, where some of the D^+ ions pick up a polarized electron from the polarized target atoms. The resulting fast D^0 atoms acquire nuclear polarization by hyperfine interaction with the picked-up polarized electron. The resulting deuteron tensor polarization is detected by the anisotropy of neutron emission from the $^3H(d,n)^3He$ reaction initiated by the 50 keV D^0 beam. Careful measurements on a teflon coated cell with 120 wall collisions showed a constant target polarization of $(88 \pm 2)\%$ for wall temperatures between 150K and 300K. An additional error of about 4% must be added for possible systematic errors. It should be noted that all polarization values quoted here are relative to an ideal polarization value of $P = 0.5$ for the setup used, in which the atoms from an atomic beam source without RF transitions are inspected in a weak magnetic field.

Since the report[4] at the previous High-Energy Spin Conference in Minneapolis, measurements have been taken with an improved test system. The old system with diffusion pumps and elastomer gaskets was replaced with a new stainless steel system with metal gaskets, pumped by turbo- and cryo-pumps. In addition, the probe beam is now momentum analyzed so that the uncertainty arising from the D_2^+-component of the beam is eliminated. Finally, the new system allows cells to be cooled down to 25K to study the temperature dependence of the target polarization.

Measurements were made on cells with different diameters (6.3 mm, 10.0 mm, 15.9 mm) and different lengths (193 mm, 153 mm, 140 mm respectively). In all three cases the length of the entry tube for the atomic beam was about 150 mm. The diameter of the entry tube was chosen as 10 mm in order to accommodate the diameter of the atomic beam. In the average, the atoms in these three types of cells have suffered 380, 150 and 40 wall collisions previous to being hit by the beam. The number of wall collisions was calculated from a computer simulation of the wall collisions in the cell.

Figure 1 shows a sample of results. The number N_0 indicates how many wall collisions the atoms in the cell have suffered in the average. The results show that the

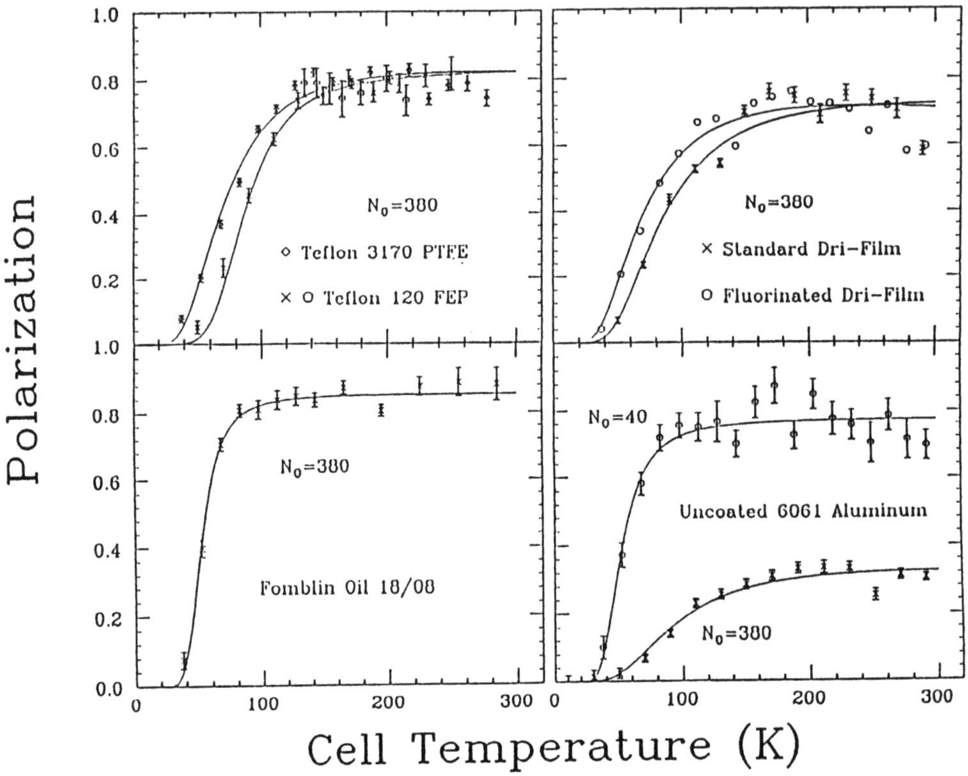

Fig. 1. Average polarization of atoms in storage cells as a function of wall temperature for different wall coatings. N_0 is the average number of wall collisions the atoms in the cell have undergone prior to being sampled by the beam.

performance of fluorocarbon polymer wall coatings can be improved by using a high temperature version of Teflon (Teflon 3170 compared to Teflon 120). A fluorinated version of Dryfilm is found to be superior at low temperatures to the regular form of Dryfilm. It is interesting to note that aluminum cells with no coating other than the natural oxide film show good polarization after 40 wall collisions and significant polarization even after 380 wall collisions. Cells coated with Fomblin oil permit large polarization down to about 80K, but the vapor pressure at room temperature ($\sim 10^{-8}$ mbar) is larger than for the other surfaces tested.

The curves in Fig. 1 are based on a simple model of wall depolarization.[4] The assumption is that in each collision the atoms stick to the surface for some length of time, with a mean sticking time $\tau = \tau_0 e^{E_B/kT}$, where E_B is the binding energy of the atom on the surface. The time dependence of the polarization of the atom is described by a relaxation time τ_R. With these assumptions one can derive the average polarization of a population of atoms which has made in the average N_0 wall collisions prior to being sampled, with the result

$$< P >= P_0\left[1 + N_0\left(\frac{\tau_0}{\tau_R}\right)e^{E_B/kT}\right] \tag{1}$$

The curve for Fomblin for instance is based on the parameters: $E_b = 17$ meV, $\tau_0/\tau_R = 1.3 \times 10^{-4}$.

Target Thickness

The target thickness traversed by the D^+ beam can be determined from the neutron count rate per microampere deuteron beam. Calibration of the count rate in terms of target thickness is accomplished by admitting a known pressure of gas to the storage cell vessel, and correcting the count rate for the different electron pick-up cross section of the H gas compared to the gas used for calibration.

The measured target thickness for a room temperature cell with 380 wall collisions (cell length 193 mm, cell diameter 6.3 mm) is $(1.0 \pm 0.1) \times 10^{13}$ atoms/cm^2 with 1×10^{16} atoms/s injected into the cell. For this cell, the gain in target thickness compared to using the same atomic beam as a jet target is about 200. The average dwell time of the atoms in the cell is calculated to be about 1.6 ms. The target thickness can be increased to 1.5×10^{13} atoms/cm^2 by cooling the cell walls to about 120 K, but for temperature below 100K all coatings tested show significant loss in polarization.

Conclusion

Polarized hydrogen atoms stored in cells coated with thin layers of fluorinated polymers retain their polarization after 380 wall collisions for wall temperatures down to about 100K. In comparing the present measurements to those reported by W. Luck et al.[6], it should be noted that their measurements are sensitive only to the degree of polarization of hydrogen atoms in the cell, but do not include the effect of depolarization by recombination.

The measurements reported here were made with only a weak magnetic field (0.5 mT) over the storage cell. Application of a magnetic field strong enough to decouple electron and nuclear spin is expected to reduce wall depolarization. Such tests are now in preparation.

References

[1] W. Haeberli, in Nuclear Physics with Stored Cooled Beams, AIP Conf. Proc. No. 128 (American Institute of Physics, NY 1984) p.251.

[2] M.D. Barker, G. Caskey, C.A. Gossett, W. Haeberli, D.G. Mavis, P.A. Quin, S. Riedhauser, J. Sowinski and J. Ulbricht, in Polarization Phenomena in Nuclear Physics 1980, AIP Conf. Proc. No. 69 (American Institute of Physics, NY), p.931.

[3] W. Haeberli and T. Wise, Polarization Phenomena in Nuclear Physics 1985, Suppl. J. Phys. Soc. Japan, Vol. 55 (1986) p.483.

[4] T. Wise, A. Converse and J.S. Price, in High Energy Spin Physics 1988, AIP Conf. Proc. No. 187 (American Institute of Physics, NY), p.1527.

[5] W. Haeberli, in Proceedings of Workshop on Polarized Beams and Polarized Gas Targets, (KEK, Japan) February 1990 (to be published).

[6] W. Luck, these Proceedings

The Measurements of the Atoms' Polarization in the Drifilm Coating Storage Cell

K. Coulter[1], *R. Gilman*[1], *R.J. Holt*[1], *E.R. Kinney*[1], *R.S. Kowalczyk*[1], *S.I. Mishnev*[2], *J. Napolitano*[1], *D.M. Nikolenko*[2], *S.G. Popov*[2], *D.H. Potterveld*[1], *I.A. Rachek*[2], *A.B. Temnykh*[2], *D.K. Toporkov*[2], *E.P. Tsentalovich*[2], *B.B. Wojtsekhowski*[2], and *L. Young*[1]

[1]Argonne National Laboratory, Argonne, IL 60439, USA
[2]Institute of Nuclear Physics, SU-630090 Novosibirsk, USSR

The polarization of the atoms in the storage cells covered by drifilm was measured in the presence of the intense circulating electron beam in the storage ring. The tensor polarization $P_{zz} = 0.572 \pm 0.053$ was obtained for the cell of the phase 1 of the experiment. For the active cell of the phase 2 of the experiment polarization was measured close to zero. The results are discussed.

1. Introduction

The experiments on elastic and inelastic electron scattering by polarized deuterons are in progress in Novosibirsk using a 2 GeV electron storage ring VEPP-3.[1,2]

The storage cells for polarized atoms were designed and applicated to increase the luminosity of the experiment as a result of the collaboration between the Institute of Nuclear Physics at Novosibirsk and Argonne National Laboratory.[3] The idea of the storage cell technique was discussed previously[4] and the reader is referred to them for details.

The most important reasons decreasing the polarization of the atoms in the cell are:

collisions of the polarized atoms with the cell walls;

the presence of the strong high-frequency magnetic field produced by circulating beam;

the dilution of the polarized atoms by unpolarized background or ions, confined by the electric field of the beam.

2. Experience with the phase 1 cell

Figure 1 displays the diagram of the experimental straight section of the storage ring. The passive storage cell installed in the vacuum chamber was an aluminum tube 940 mm in length and

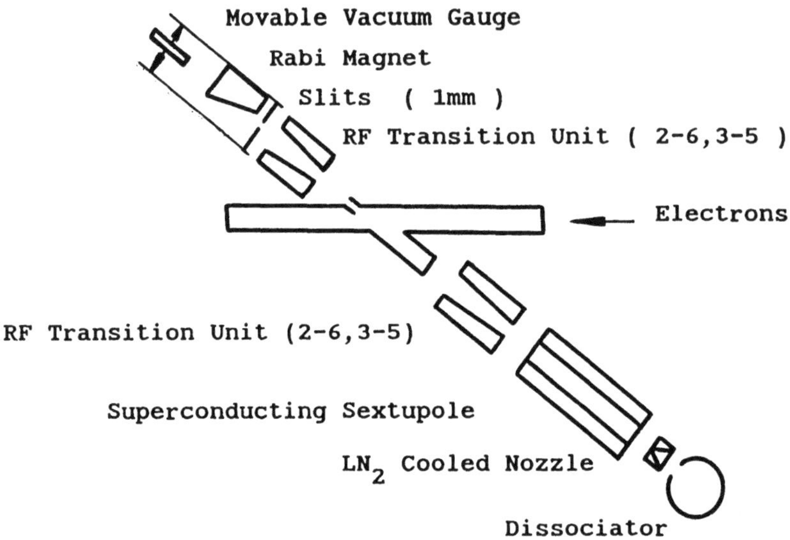

Fig. 1. Layout of experimental straight section of VEPP-3.

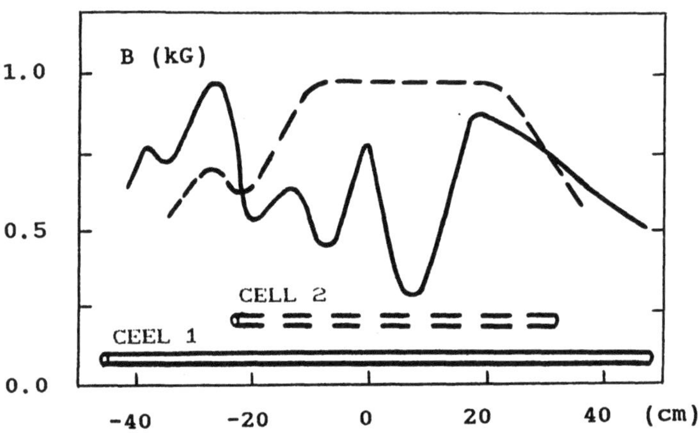

Fig. 2. Magnetic field distribution along the cell.

coated by drifilm[5] to give a surface of less than 0.001 depolarization probability. The relatively large cell cross section (46 x 24 mm elliptical shape) was determined by the aperture necessary for electron beam during injection. Monte-Carlo simulation of the atomic motion in this cell yield a dwell time of 7.5 ms and an average number of 470 wall collisions per atom. The polarized beam was injected through a 10 mm diameter, 60 mm in length tube, opposite to which is a small sampling port directed at the polarization monitor. A holding magnetic field was applied to the cell in order to keep polarization (Fig.2). The atomic beam source produced 10^{16} atoms per second

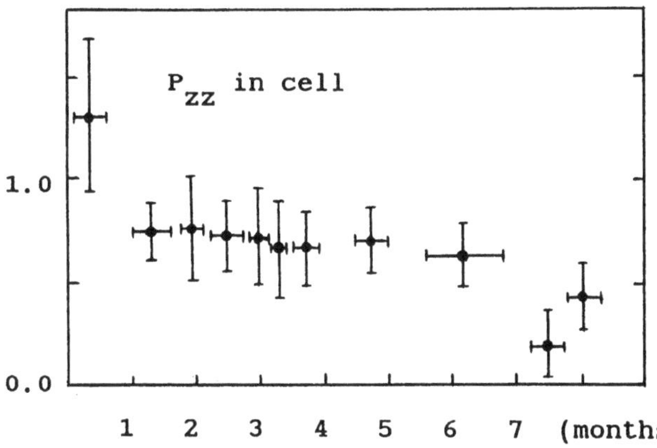

Fig. 3. P_{zz} in the phase 1 cell as a function of time

with tensor polarization P_{zz} close to unity.[6] The efficiency of the RF transitions was controlled continuously during the experiment.[3] The effective value of P_{zz} (see Fig. 3) for the target was extracted from the measurements of the asymmetry in elastic ed scattering at low momentum transfer where the disagreement between theories is less than ~ 5%. The extracted polarization $P_{zz} = 0.572 \pm 0.053$ is in agreement with the expected one.

3. Phase 2 storage cell

Figure 4 shows a diagram of active cell which consisted of U-shaped tube which could be opened and closed against a flat wall. The cell is 520 mm length with an opening of roughly 19 mm by 9 mm. The polarized atoms from the same source were injected

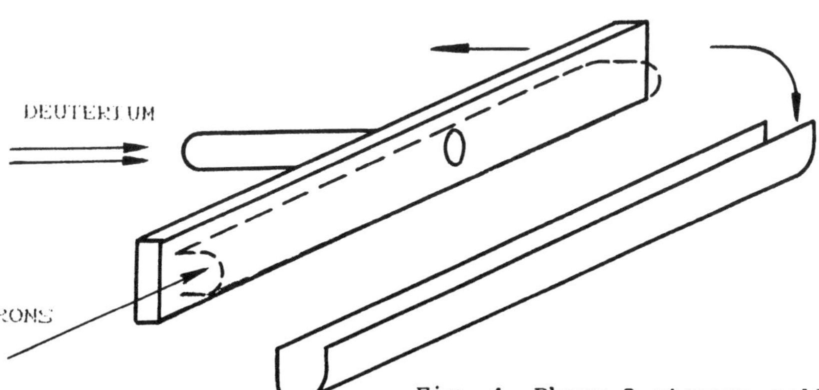

Fig. 4. Phase 2 storage cell.

through 12 mm diameter and 150 mm in length tube, opposite to which a is sampling port directed to the polarization monitor. These cell dimensions provide a useful thickness $5x10^{12}$ atoms/cm^2 (for 15 cm of the central part). The average number of wall collisions calculated to be 620 and dwell time 3.5 ms. The U-tube walls were slightly taped in order to decrease heating due to the synchrotron radiation and the movable stainless steel and copper blocks placed just upstream of the cell were provided additional shielding. More uniform and stronger magnetic field was applied along the length of the cell (Fig.2). Two electron-deuteron detection systems placed symmetrically in the vertical plane were used to extract the events of elastic electron scattering. For the first time the polarization measurement was produced at the 1 GeV electron energy. The time length of the electron bunch ~ 0.25 ns in this case. The asymmetry of the reaction A = 0.004 ± 0.062 was measured while the expected one is 0.147 for P_{zz} equal unity. In order to eliminate the resonant depolarization of atoms by electron bunch the polarization measurement was repeated at 0.9 GeV energy when only one RF generator of the storage ring was on. The time length of the bunch is 3.5 ns in this case and resonant depolarization is sufficiently suppressed[7]. The asymmetry A = -0.025 ± 0.043 was obtained that corresponded polarization close to zero.

4. Conclusions

These cells were surrounded by the accelerator environment approximately the time of one year. For the phase 1 cell the polarization obtained was close to expected one. The polarization close to zero was measured for phase 2 cell. The depolarization of the atoms by the electron bunch is improbable. The most probably reason is the destroying of the drifilm wall surface under the influence of the synchrotron radiation. When the cell was removed from the vacuum chamber it was found that synchrotron light attained the drifilm cover inner surface of the cell in the position when cell open. As the analysis have shown it might be synchrotron light reflected from the vacuum chamber of the accelerator at large angle to the beam axis and existing shielding was not able to eliminate radiation damage. The small mechanical damages of the drifilm cover were found as well. We plan to install more effective shielding from synchrotron light in the near future.

REFERENCES

1. V.F. Dmitriev *et al.*,*Phys. Lett.* **157B**,143 (1985); M.V.Mostovoy *et al.*, *Phys. Lett. B* **188**, 181 (1987).
2. B.B.Wojtsekhowski *et al.*,*JETP Lett.* **43, 733** (1986).
3. S.I. Mishnev *et al.*, *High-Energy Spin Physics,* AIP Conf. Proc. **187**, 1286 (1989).
4. See for example, R.Holt, *Proc. of Spectrometer Workshop,* Williamsburg,(1983); T.Wise and W.Haeberli, *Proc. of Workshop on Polarized Targets in Storage Rings,* Argonne National Laboratory, ANL-84-50, p.249 (1984).
5. L. Young *et al.*, *NIM* **B24/25**, 963 (1987); D. R. Swenson and L.W. Anderson, *NIM* **B29**, 627 (1988).
6. A.V.Evstigneev, S.G.Popov and D.K.Toporkov, *NIM* **A238**,12(1985).
7. E.R.Kinney *et al.*,Preprint ANL PHY-6202-ME-89.

Description of an Atomic-Beam Sampling Polarimeter

H.-G. Gaul

Max-Planck-Institut für Kernphysik, Postfach 103980,
W-6900 Heidelberg, Fed. Rep. of Germany

A polarimeter for thermal atomic hydrogen and deuterium beams is presented. The polarimeter is designed to have an accuracy of 3 % for the determination of both electron and nuclear polarisation. The measurements are done by exchanging certain hyperfine states by means of high frequency transitions. The beam is then guided through a sextupole system which serves as a filter for atoms with a positive electron spin projection. The beam intensity is measured with lock-in-technique by using a quadrupol mass spectrometer for the detection of the chopped atomic beam.

1. Introduction

The precise measurement of the spin-dependent structure functions of the proton and neutron in the HERMES experiment [HER] at DESY requires a polarized lepton beam - provided by the HERA electron ring - as well as a polarized hydrogen and deuterium gas target of high density. The newly developed technique of internal storage cell gas targets [HAE] which are fed with polarized atoms will allow target densities which are larger than the target densities of polarized gas jets [EKS]. Storage cells are successfully tested at VEPP-2 in Novosibirsk [TOP]. The price one has to pay for this enhancement is a possible depolarization of the atoms during their stay in the storage cell. Wall collisions with the cell material as well as the interaction with the electromagnetic fields of the electron bunches cause depolarization of the gas. Therefore it is absolutely necessary to monitor the target polarization to a good precision. The following sections will deal with the idea of the polarimeter and its physical and technical realisation.

2. Design of the Polarimeter

The polarization of the target will be measured in the following way:

A sample of the target gas will be extracted out of the cell. This is done by means of a tube which guides the atoms in the direction of the polarimeter. A nozzle at the end forms a **sample beam** which is analyzed by the polarimeter. One has to ensure that this sample beam represents the true polarization value of the target itself. Sample beam and target polarization are not neccessarely equal because additional wall bounces inside the extraction tube may lead to further depolarization. Monte-Carlo studies [BBR] on the polarization of the atoms in the storage cell have shown that one has to take care on a correction factor which has at least to be determined experimentally.

Once the sample beam is formed it can be analyzed with the polarimeter by means of **radio-frequency spectroscopy**. The goal is to measure the relative occupation num-

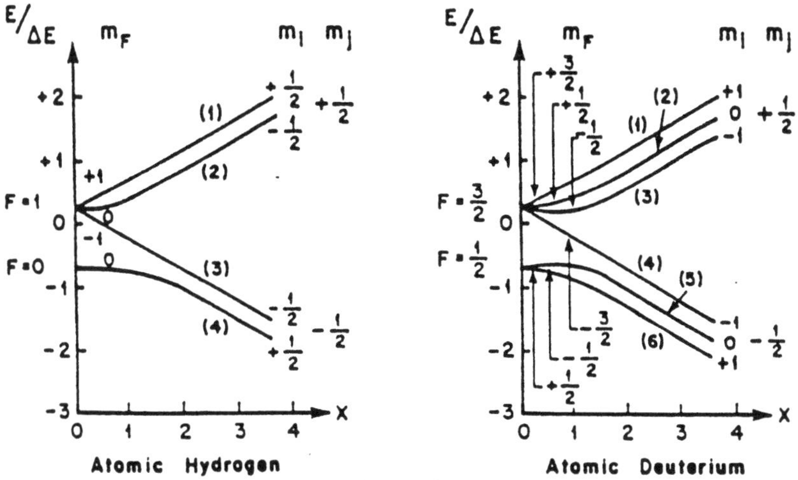

Fig. 1: Hyperfine structure splitting of ^{1}H and ^{2}H in an external magnetic field

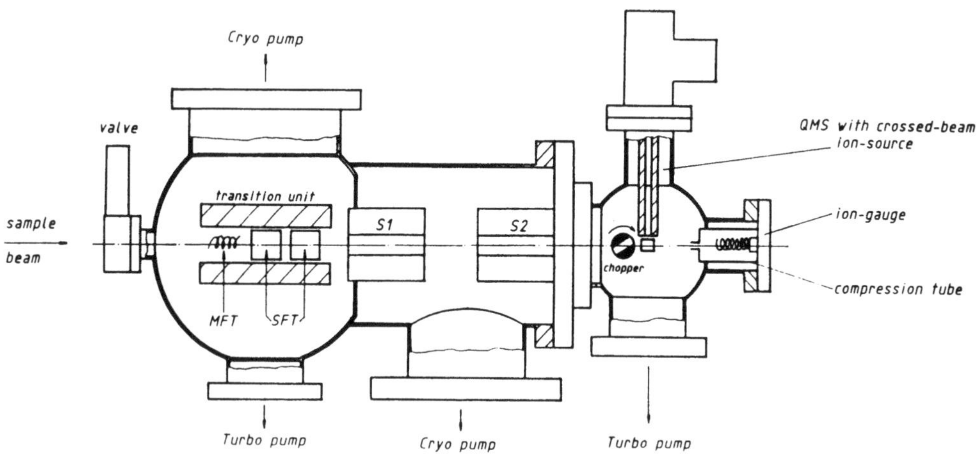

Fig. 2: Cross section of the HERMES polarimeter

bers n_i of the gas under study. From this numbers one can calculate the polarization. To make the description more clear the hyperfine structure splittings of ^{1}H and ^{2}H are shown in figure 1.

First the sample beam has to pass through a region where radio-frequency transitions (RFT) are located. They exchange certain hyperfine sublevels by inducing magnetic dipole transitions between hyperfine states of the ground state atoms. Figure 2 illustrates the set-up of the polarimeter.

Next the beam passes a magnetic sextupole system S1 and S2. This only focuses $|m_J = +\frac{1}{2}>$-states into the detector system while $|m_J = +\frac{1}{2}>$-states are rejected. $|m_J = \pm\frac{1}{2}>$ represents hyperfine states with either positive or negative eletron spin projection on a certain field direction.

The atoms as well as the molecules which are recombined in the storage cell are detected by a sensitive quadrupole mass spectrometer (QMS) equipped with a cross-beam ion source. The fraction of molecules has to be known because the measured target polarization must be corrected by this value. The sample beam is chopped and by lock-in technique the background will be suppressed.

3. Working Principle of the Polarimeter

The relative occupation numbers of the hyperfine states can be composed to an **occupation vector** $\vec{n}$ where the ordering of this vector is according to the order of states given in figure 1. The following example is given for hydrogen.

Let us start with the action of the sextupole magnet on $\vec{n}$. The states with $|m_J = +\frac{1}{2}>$ correspond to the upper two components of $\vec{n}$. Therefore they will be kept while the lower two components become zero when the ensemble of atoms is passing through the sextupole. The signal at the QMS looks like $I_{RFToff} = I_{beam} \cdot (n_1 + n_2)$. The population of the states involved will be exchanged if a RFT unit is swichted on. That means that the corresponding components are exchanged. The (2-4) - transition in hydrogen exchanges states 2 and 4 so that component 2 and 4 of the vector $\vec{n}$ have to be exchanged. If this newly formed ensemble is passing the sextupole magnet the signal looks like $I_{RFTon} = I_{beam} \cdot (n_1 + n_4)$. By taking the ratio of RFT on to RFT off one gets

$$S_{24} = \frac{I_{RFTon}}{I_{RFToff}} = \frac{n_1 + n_4}{n_1 + n_2}$$

Immediatley it becomes clear that this signal ratio differs from unity if n_2 and n_4 are not equally populated. The same arguments hold for the other transitions which involve different states. The following sequence illustrates the proccess described above.

$$\begin{pmatrix} n_1 \\ n_2 \\ n_3 \\ n_4 \end{pmatrix} \xrightarrow{\textit{no transition}} \begin{pmatrix} n_1 \\ n_2 \\ n_3 \\ n_4 \end{pmatrix} \xrightarrow{\textit{sextupole}} \begin{pmatrix} n_1 \\ n_2 \\ 0 \\ 0 \end{pmatrix}$$

$$\begin{pmatrix} n_1 \\ n_2 \\ n_3 \\ n_4 \end{pmatrix} \xrightarrow{2-4} \begin{pmatrix} n_1 \\ n_4 \\ n_3 \\ n_2 \end{pmatrix} \xrightarrow{\textit{sextupole}} \begin{pmatrix} n_1 \\ n_4 \\ 0 \\ 0 \end{pmatrix}$$

If one measures three signal ratios for ^{1}H and additionally uses the normalization condition $\Sigma n_i = 1$ a system of linearly independent equations in the unknown relative occupation numbers can be set up.

$$\begin{pmatrix} 1 - S_{13} & 1 & S_{13} & -1 \\ 1 & 1 - S_{23} & S_{23} & -1 \\ 1 & 1 - S_{24} & -1 & S_{24} \\ 1 & 1 & 1 & 1 \end{pmatrix} \begin{pmatrix} n_1 \\ n_2 \\ n_3 \\ n_4 \end{pmatrix} = \begin{pmatrix} 0 \\ 0 \\ 0 \\ 1 \end{pmatrix}$$

The real system behaves more complicated. If the efficiency ϵ of a transition is not equal to unity the states are not completely exchanged. In addition the sextupole magnet does not separate the states totally. Finally the factor I_{beam} which cancels out by taking the ratio of measurements must not be uniform in time. These effects cause errors which have to be included into the calculation. Therefore the matrix of coefficients must be written as

$$\begin{pmatrix} A(S_{13}-\delta)-B\epsilon & A(S_{13}-1) & B(S_{13}-\delta)-A\epsilon & B(S_{13}-1) \\ A(S_{23}-1) & A(S_{23}-\delta)-B\epsilon & B(S_{23}-\delta)-A\epsilon & B(S_{23}-1) \\ A(S_{24}-1) & A(S_{24}-\delta)-B\epsilon & B(S_{24}-1) & B(S_{24}-\delta)-A\epsilon \\ 1 & 1 & 1 & 1 \end{pmatrix}$$

where $\delta = 1 - \epsilon$.

The solution of this system gives the desired relative occupation numbers. Monte-Carlo studies have shown that the determination of the polarization of either ^{1}H or ^{2}H is precise to about 3 % if the errors on transition efficiency, incomplete separation and intensity fluctuations are each of the order of 1 %.

4. Physical and technical details

All transitions employed are so-called **Abragam-Winter** or **adiabatic passage** transitions [ABR].

These are the **strong-field transitions** (SFT) are very common. The induce transitions between states of different Zeeman multiplets. Two different types are used: σ-transition ($\Delta m_F = 0$) and π-transitions ($\Delta m_F = \pm 1$). The σ-type needs a rf field which is parallel to the static magnetic field while the π-type radio-frequency field is perpendicular to the static one. The SFT is made of a two-rod resonant line at $\frac{\lambda}{4}$ which is embedded in a cavity. One end of each rod is grounded while the other end forms a capacitor. This allows the tuning of the resonator. It is experimentally shown that up to 99 % efficiency are reachable with an input power of about 2 Watts. These efficiencies are measured using the hydrogen test source at Heidelberg.

The other type of transitions are so-called **medium-field transitions** (MFT) [JAE]. They induce $\Delta m_F = \pm 1, \Delta F = 0$ transitions within a multiplet. The MFT uses the fact that the transition frequencies of the hyperfine substates are no longer equidistant in a medium magnetic field of about 20 % of the critical field. There the resonance condition for a fixed frequency is met at different magnetic field values. By superimposing a gradient magnetic field parallel to the static one the transitions will take place at different positions. Using again the occupation vector the MFT for hydrogen has the following effect:

(i) positive gradient in direction of flight

$$\begin{pmatrix} n_1 \\ n_2 \\ n_3 \\ n_4 \end{pmatrix} \xrightarrow{2-3} \begin{pmatrix} n_1 \\ n_3 \\ n_2 \\ n_4 \end{pmatrix} \xrightarrow{1-2} \begin{pmatrix} n_3 \\ n_1 \\ n_2 \\ n_4 \end{pmatrix}$$

(ii) negative gradient in direction of flight

$$\begin{pmatrix} n_1 \\ n_2 \\ n_3 \\ n_4 \end{pmatrix} \xrightarrow{1-2} \begin{pmatrix} n_2 \\ n_1 \\ n_3 \\ n_4 \end{pmatrix} \xrightarrow{2-3} \begin{pmatrix} n_2 \\ n_3 \\ n_1 \\ n_4 \end{pmatrix}$$

Again the sextupole only transmits the upper two components in the sample beam which are in the first case n_1 and n_3 and in the second case n_2 and n_3.

Table 1 gives an overview on the transitions which are used in the polarimeter.

Hydrogen	MFT	MFT	SFT
transitions	2-3	1-3	2-4
frequency (MHz)	70	70	1430
B_z (Gauss)	50	50	50
gradient field	+	-	±
type	coil	coil	cavity

Deuterium	MFT	MFT	SFT	SFT	SFT
transitions	3-4	1-4	1-6	2-6	3-5
frequency (MHz)	22	22	375	355	355
B_z (Gauss)	24	24	24	24	105
gradient field	+	-		±	
type	coil	coil		cavity	

The vacuum system consists of two cryo pumps of 1500 l/s pumping speed each and two turbo-molecular pumps. The turbo pump of the detector stage is backed by an additional turbo pump to increase the compression ratio for hydrogen and deuterium.

Parts of the polarimeter are fabricated and already under test. The ^{1}H partial pressure in the last chamber is about $6 \cdot 10^{-10}$ mbar without gas inlet and $3 \cdot 10^{-9}$ mbar for a gas inlet which corresponds to the one expected for the HERMES experiment. The system has not been baked out up to now. The transition units are tuned and ready for test measurements.

5. Summary

Radio-frequency spectroscopy provides a useful tool for the determination of the relative occupation numbers of the hyperfine substates of both hydrogen and deuterium atoms in their ground states. As a first application of the polarimeter it is planed to measure the wall depolarization of hydrogen at different magnetic field values.

[ABR] A. Abragam, J.M. Winter, Phys.Rev.Lett 15 (1958) 374

[BBR] B. Braun, HERMES report 04-90, under preparation

[EKS] C. Ekström, CERN accelerator school 1989, CERN 90-04, 24 April 1990

[HAE] W. Haeberli, AIP Conference Proceedings No. 187 (1988) 1487

[HER] HERMES proposal

[JAE] H. Jänsch, D. Krämer, E. Steffens, Hyperfine interactions 22 (1985) 253

[TOP] D.K. Toporkov, this workshop

A Polarimeter for Polarized Proton Gas Targets[+]

W.S. Luck[1], *H.J. Jänsch*[1], *D. Fick*[1], *H.T. Duong*[2], *and E. Steffens*[3]

[1]Philipps-Universität Marburg, Fachbereich Physik,
W-3500 Marburg, Fed. Rep. of Germany

[2]Laboratoire Aime Cotton, CNRS F-914050 Orsay, France

[3]Max-Planck-Institut für Kernphysik, Postfach 103980,
W-6900 Heidelberg, Fed. Rep. of Germany

Experiments that use the storage cell technique[1] to increase the density of a polarized gas target can make use of a polarimeter to prepare the experiment and to monitor the target polarization during the runs. We developed such a device that is mechanically much more compact than already existing methods and can thus be installed into a storage cell target set up. So far our technique has been used to study the problem of depolarization of atomic hydrogen in storage cells[2]. The technique uses the property of hydrogen atoms in a weak magnetic field to have always identical nuclear and electron polarization since each of the four hyperfinestates contributes equally to the electron and the nuclear polarization (fig.1). Therefore a measurement of the electron polarization P_e is sufficient to determine the nuclear polarization P_n. To do so, the hydrogen atoms are excited by low energy electron impact ($E_e \approx 80$eV, fig. 2). This excitation is so fast that the depolarization of the atomic electron by the excitation can be neclected. The life time of the 3P excited state is long enough for hyperfine coupling to transfer spin polarization to angular momentum polarization. This polarization of the angular momentum leads to a circular polarization of the subsequently emitted fluorescence light of the atoms. So by measuring the circular polarization of the electron impact induced H_α-fluorescence light the nuclear spin polarization of the atoms before the excitation can be determined.

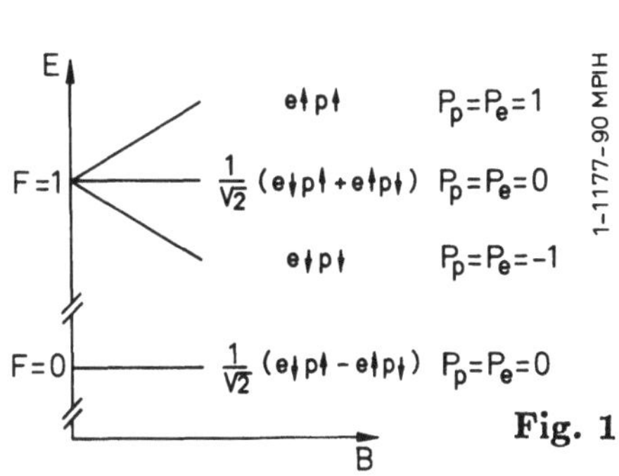

Fig. 1.

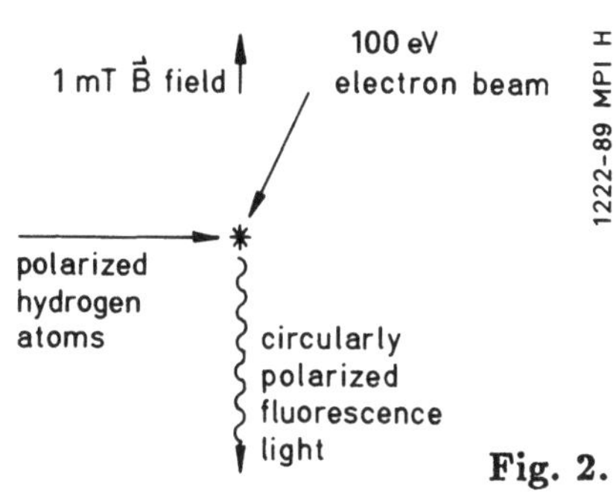

Fig. 2.

Fig. 1. The hyperfinestates of hydrogen atoms in a weak B-field.

Fig. 2. The principle of the electron impact polarimeter.

+) Supported by the Bundesministerium für Forschung und Technologie, Bonn under contract 06MR133I

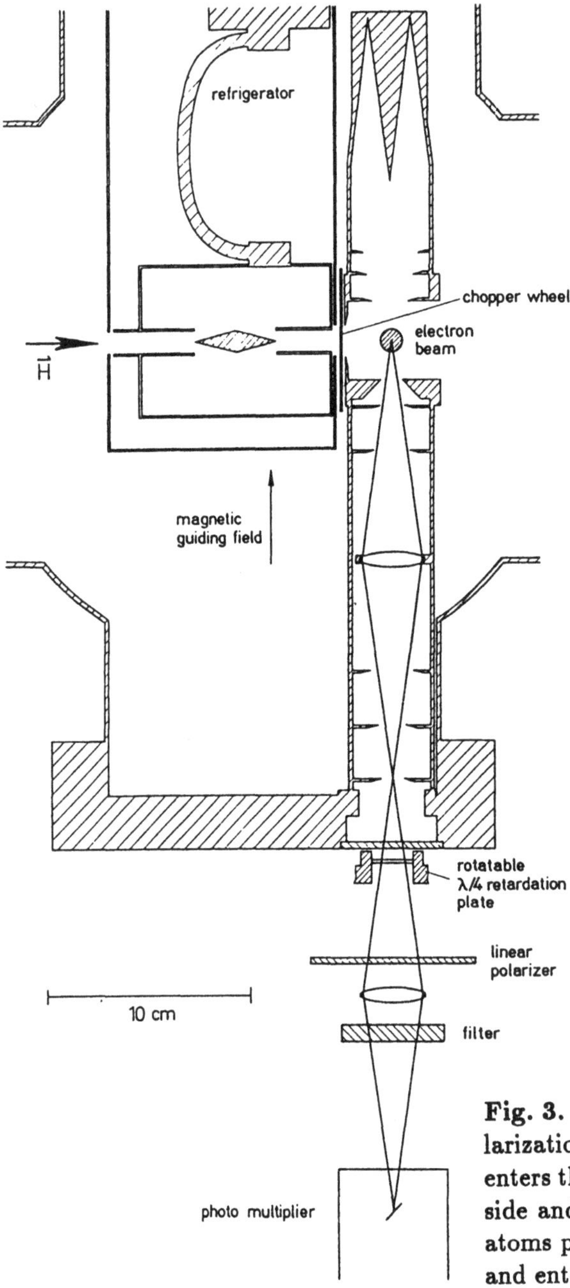

Fig. 3. Apparatus used to study wall depolarization in storage cells. The atomic beam enters the box-shaped storage cell at the left side and is stopped on a cone. The stored atoms partly leave the cell at the right side and enter the polarimeter.

Test Set Up

Polarized hydrogen atoms are produced by a conventional source, using a SIN-type rf dissociator and one sextupole magnet[3]. The flux of hydrogen atoms into the interaction zone of the polarimeter is around $2{\cdot}10^{15}$ atoms/s. In order to measure the background the hydrogen beam can be interrupted by a chopper wheel mounted close to the interaction zone of the polarimeter. The main source of background is the electron impact excitation of residual gas hydrogen molecules, leading to dissociation of the molecules with one of the two hydrogen atoms ending up in the n=3 state. The H_2 residual gas is generated almost entirely by recombination of the atomic hydrogen beam on the surfaces of the vacuum chamber, of the storage cell and the chopper wheel (when the chopper is closed). The gas conductances inside the vacuum chamber are high enough that the background counting rate varies only slightly with the position of the chopper wheel. Thus a measurement of the background can be performed with closed chopper. The total error introduced to polarization measurements of storage cell targets by this method is estimated to be less than 16%. This error can be reduced by increasing the pumping speed of the vacuum system.

A pair of Helmholtz type coils generates a magnetic guiding field in the interaction zone of 1mT to provide a quantization axis for the polarization. The electron beam is produced by a simple and compact arrangement consisting of an indirectly heated cathode (VALVO 30AX) and two electrodes. Electron currents in the mA range are obtained at energies around 80eV. The induced fluorescence light is imaged by two lenses to the cathode of a cooled single photon counting multiplier (RCA 31034). The H_α-line is selected by an interference filter (Schott MA3-0.3). The circular polarization is measured by means of a $\lambda/4$ plate and a linear polarizer.

Our test set up to study wall depolarization with the new polarimeter is shown in fig. 3. For the polarization measurements we used a storage cell which could be removed during measurements on the free atomic beam for calibration purposes. To determine the polarization of the stored hydrogen atoms the polarization of the atoms effusing out of an exit hole of the storage cell is measured. This way a meaningful determination of the target polarization is obtained since the atoms move randomly inside the storage cell. The storage cell measurements reported here were performed with a box shaped cell providing a mean number of wall collisions around 1000. If the storage cell is cooled the target atoms move slower between the cell walls and the target density is increased. In order to investigate to which temperatures storage cells can be cooled without a loss of target polarization our test storage cell is connected to a closed cycle helium refrigerator. The accessible temperature range is 25K-300K.

Measurements with the Test Set Up

The H_α photon counting rate induced by electron impact excitation of the free atomic beam with a polarization of $P_e = P_n = \frac{1}{2}$ is of the order of 1kHz at an atomic beam intensity around $2 \cdot 10^{15}$ atoms/s. This gives a sensitivity of the device of about 2 photons/10^{12} hydrogen atoms. If the background is subtracted a circular light polarization of P=0.172±0.002 (statistical error) is measured resulting in an experimentally determined analyzing power of the method around 0.34 . This analyzing power turns out to be independent on the energy of the electron beam in the energy range from 20eV to 80eV. Its dependence on the strengh of the magnetic guiding field is also weak.

We applied our method to measure the polarization of hydrogen atoms moving in storage cells. For a measurement on hydrogen atoms effusing from the storage cell the photon counting rate is of the order of 100Hz. Fig. 4 shows the polarization of the stored atoms normalized to the polarization of the free atomic beam that was always measured seperately. The flux of the hydrogen atoms out of the cell that reaches the polarimeter normalized to the atomic beam flux is also displayed. The measurements

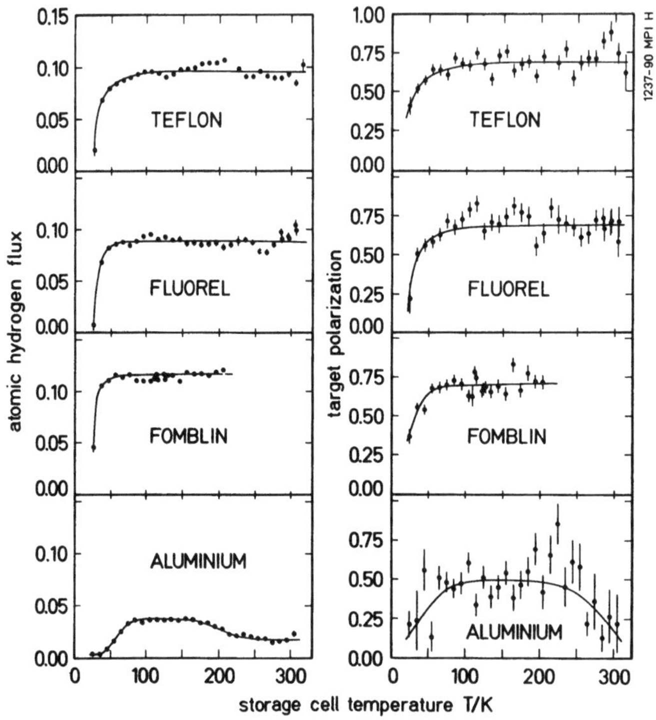

Fig. 4. Measured flux (normalized to the flux of atoms into the cell) and polarization of hydrogen atoms effusing out of the storage cell.

show that fluorocarbon coatings like TEFLON, FLUOREL and FOMBLIN can be used for storage cell coatings at temperatures down to 80K. At lower temperatures the target polarization and density decreases. Aluminum oxide coatings produce in all respects inferior results. The storage cells with fluorocarbon coatings showed target polarizations around 70% of the polarization of the free atomic beam. Remarkably enough no change of this value was observed when the storage cell was modified to give 500 instead of 1000 wall collisions. The 70% polarization of the stored atoms compared to the free beam is presently not understood. Further investigations to clarify this point are presently under preparation.

Discussion

The counting rate around 100Hz, the large analyzing power of 0.34, the weak dependency of this analyzing power on the electron beam energy and on the magnetic guiding field strength enables the application of the method as a useful polarimeter for storage cells.

The ratio of the flux of hydrogen atoms effusing from the storage cell into the polarimeter to the atomic beam flux is about 1/10. This can be understood considering the angular distribution of atoms effusing out of a tube[4]. The drop of flux out of the storage cell at low temperatures can be described by a simple model assuming physisorption of the hydrogen atoms at the storage cell surfaces with an activation enthalpy of 15meV and a probability to recombine to molecules proportional to the mean residence time of an atom at the storage cell surface. The decrease of the polarization at low temperatures can be described similarly as the shape of the intensity curve by assuming a depolarization probability at the storage cell surfaces proportional to the stay time at the surface, although the absolute value of the polarization is not understood at present.

The constant polarization of the stored hydrogen atoms in cells coated with fluorocarbons in the temperature range 80K-300K allows to cool the storage cell down to 80K, resulting in a 90% increase of the target density as compared to room temperature operation.

Outlook

A device, similar to the test set up described here, that fits into the target chamber of the FILTEX test to be performed at the storage ring TSR at Heidelberg is being built at present[5]. This polarimeter will measure the polarization of the hydrogen atoms effusing out of an exit hole of the target cell. Now first sensitivity measurements with the new device have been performed successfully.

References

1 T. Wise and W. Haeberli, Workshop on Polarized Targets in Storage Rings 1984, Argonne Nat.Lab.ANL-84-50, p.249

2 W.S. Luck et al., proc. 7th int. conf. pol. phen. nucl. phys., Paris 1990, J. Phys. (Paris) Colloq. in print

3 W. Korsch, diploma thesis, Universität Marburg 1987

4 J.A. Giordmaine and T.C. Wang, Journal of Applied Physics 31,3(1960)463-471

5 M. Metzger, diploma thesis University of Heidelberg, in preparation

Gas Dynamics Aspects of Internal Targets

F. Sperisen

Indiana University Cyclotron Facility, Bloomington, IN 47408, USA

Based on our experience with the Indiana Cooler storage ring some aspects relevant to the design of internal polarized gas target experiments are discussed. In particular, it is pointed out that, perhaps contrary to general perception, increasing the target thickness does not necessarily lead to a proportional gain in luminosity: In our studies with cooled beams of 100 to 200 MeV, interacting with internal H_2 targets, the luminosity as a function of target thickness shows a broad maximum in the range of 10^{14} to 10^{15} atoms/cm^2.

This report deals with some gas dynamics considerations of the interaction of internal gas targets with stored beams. It is largely based on our recent experience with the Indiana Cooler, a storage ring with electron cooling for medium-energy light ions.[1] Studies have been done with jets as well as extended, diffuse (unpolarized) targets of various gases,[2] exposed to cooled proton beams of up to 300 MeV.

Maximizing the luminosity. One of the most important performance parameters of internal targets is the average luminosity $\langle L\rangle = \langle I\rangle t$, the product of beam intensity $\langle I\rangle$ and target thickness t, that can be achieved in a particular experiment. Much of the current target development effort goes toward maximizing t, apparently with the general assumption that this will lead to a proportional gain in the luminosity. However, an increase in t results in a shorter lifetime of the stored beam, reducing $\langle I\rangle$ and thus offsetting to some extent the direct benefit through t. To gain more insight into the complex interplay betweeen beam, target and cooling, the concept of the "loss cross section" has been introduced.[3,4] This quantity is defined as $\sigma_L = (f_R t\tau)^{-1}$, where f_R is the revolution frequency of the stored beam ions and τ is the beam lifetime. It can be shown that $\langle L\rangle$ is essentially proportional to the filling rate and inversely proportional to σ_L. So, the luminosity is maximum when the target thickness is chosen to *minimize* σ_L. Fig. 1 shows a measurement of $\sigma_L(t)$ taken recently to optimize t for a $\vec{p}p$ elastic scattering experiment with a cooled beam of 185 MeV protons in the G straight section of the Indiana Cooler.[5] The loss cross section has a broad minimum, and, under this circumstance, a target thickness of $1 \cdot 10^{14}$ atoms/cm^2 (to take the typical design goal of present polarized target projects) would be considered optimal; increasing t beyond that value would not bring any gain in luminosity.

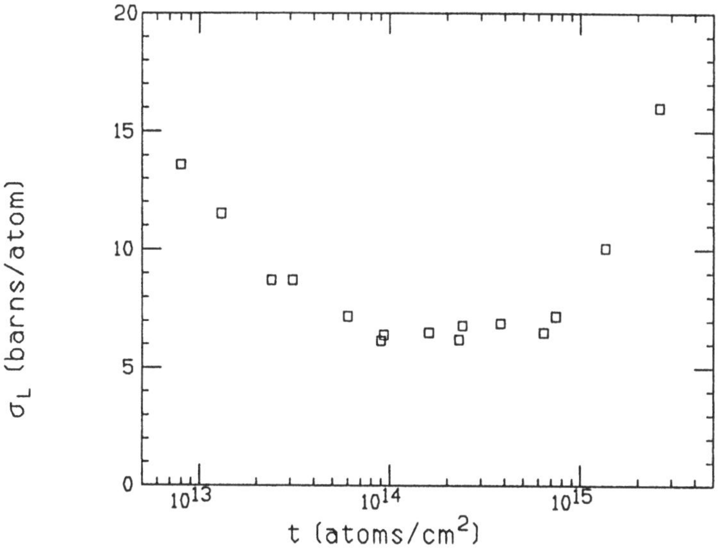

Fig. 1 - Loss cross section σ_L measured as a function of H_2 target thickness t with a 185 MeV cooled proton beam in the Indiana Cooler. (Preliminary results of the CE-08/17 collaboration.[5])

The loss cross section is expected to depend on several machine properties. Some of them may offer a way to decrease the loss cross section. Of particular interest here is the fact that the rise in σ_L towards small t is due to the beam interaction with the residual gas in the ring: For $t \to 0$, τ is constant (given by the residual gas pressure in the ring), and σ_L is thus proportional to t^{-1}. Therefore, in the thickness range where the residual gas interaction contributes significantly to σ_L (up to close to its minimum), improving the ring vacuum would help the luminosity.

One of the ring parameters to be optimized for maximum $\langle L \rangle$ is the transverse ring acceptance A. On the one hand, it determines the clearance the storage cell has to provide for the beam and is thus limiting the achievable target thickness (at a non-dispersed target location t is roughly proportional to $A^{-3/2}$). On the other hand, the beam lifetime is expected to increase linearly with A. Furthermore, the acceptance affects the luminosity via the beam injection efficiency. Experimental verification of the expected A dependence for various target gases and beam energies are planned in the near future.

Differential pumping. Another aspect I want to discuss briefly is that of differentially pumping the gas load of a polarized storage cell target. To be specific, consider a polarized hydrogen source feeding the cell with $1 \cdot 10^{17}$ atoms/s. If the first pumping stage, containing the cell, is being pumped with a speed of 5000 l/s, a pressure of about $4 \cdot 10^{-7}$ mbar could be maintained (assuming full recombination of the atomic hydrogen before pumping). This would be accomplished most efficiently (in terms of required flange size as well as capital cost) with a cryogenic pump. At a typical H_2 capacity of 15 std. liters, regeneration would only be needed every 2200 hours. The gas flow into the second stage would be best restricted by a tube with a cross section similar to that of the target cell. However, if reaction products have to be detected at small forward or large backward angles, the pumping stages may have to be separated by a thin foil with an aperture clearing the beam. Assuming this worst case, with an aperture area of 3 cm^2 (typical for our experiments in the Indiana Cooler G section) and again a pumping speed of 5000 l/s, the pressure would drop by a factor of 38 across such an aperture to nearly $1 \cdot 10^{-8}$ mbar. This should be sufficient to keep the average ring pressure below about $5 \cdot 10^{-9}$ mbar which, for a ring of 10^2 m circumference, would represent a thickness of $3.3 \cdot 10^{12}$ atoms/cm^2, small compared with envisioned polarized target thicknesses of around 10^{14} atoms/cm^2.

Conclusion. Optimizing polarized internal target experiments for maximum luminosity should not focus solely on the target thickness, but also requires consideration of machine properties such as the ring vacuum, the transverse acceptance and the radial aperture functions at the target. Decoupling the gas load of a polarized target from the ring vacuum typically should be possible with only two differential pumping stages.

I would like to thank H.O. Meyer and R.E. Pollock for many useful discussions of the luminosity problem. This work is supported by the U.S. National Science Foundation.

REFERENCES

1. R.E. Pollock, Proc. IEEE Particle Accelerator Conf., Chicago 1989, ed. F. Bennet and J. Kopta, IEEE Conf. 1989, p. 17.
2. F. Sperisen et al., IUCF Scientific and Technical Report, May 1989 - April 1990, p. 114.
3. R.E. Pollock, private communication.
4. V.V. Parkhomchuck and A.N. Skrinsky, in "Heavy Ion Storage Rings with Electron Cooling," I.V. Kurchatov, Inst. of Atomic Energy, Moscow, 1990.
5. CE-08/17 collaboration: V. Derenchuk, W. Haeberli, L.D. Knutson, H.O. Meyer, P.V. Pancella, S.F. Pate, A. Pei, W.K. Pitts, R.E. Pollock, S. Price, B. von Przewoski, T. Rinckel, A. Ross, T. Sloan, J. Sowinski, and F.Sperisen.

Protecting the HERMES Experiment from Synchrotron Radiation

M. Düren

Max-Planck-Institut für Kernphysik, Postfach 10 39 80,
W-6900 Heidelberg, Fed. Rep. of Germany

1. Introduction

HERMES is a new proposed experiment at DESY to measure the spin dependent structure functions of proton and neutron in the HERA electron ring [1]. High statistical precision is obtained by using a storage cell that increases the density of the internal polarized gas target to a value of 10^{14} atoms/cm^2. Synchrotron radiation has two effects that may harm this experiment: The storage cell is irradiated by synchrotron photons and radiation scattered from the target gas and from the cell walls enters the front chambers of the detector. In this report the amount of synchrotron radiation is calculated and means are proposed to protect an experiment like the HERMES experiment from synchroton radiation. An efficient collimator system is necessary for the operation of the experiment. Narrow collimators have secondary effects on the beam. One aspect is the additional background production at the edges of the collimators. The Monte-Carlo program SYNTRACK [2] was written which allows a detailed simulation of synchrotron radiation, of scattered radiation, of beam tail generation by rest-gas bremsstrahlung and of the scraping of the beam tails by the collimators.

2. Generation of Synchrotron Radiation

The electrons in the HERA ring radiate synchrotron radiation in the bending and the focussing magnets. In high energy units, the radiation of a dipole magnet with a bending radius ρ is given by the formula [3]:

$$\frac{P}{\Delta\theta}\left[\frac{kW}{mrad}\right] = 0.0141\,\frac{E^4 I}{\rho}\left[\frac{GeV^4 A}{m}\right].$$

The characteristic energy of the synchrotron light is [3]:

$$E_c[keV] = 2.22\frac{E^3}{\rho}\left[\frac{GeV^3}{m}\right].$$

The number of emitted photons is:

$$\frac{\dot{N}}{\Delta\theta}\left[\frac{Hz}{mrad}\right] = 1.3\cdot 10^{17}\,E\,I\,[GeV\,A].$$

50% of the energy is radiated by photons with an energy above E_c. The synchrotron radiation is emitted in the direction of the electron in a cone with an opening angle $\Theta \approx m_e c^2/E = 0.015$ mrad which can be neglected for electrons at HERA energies.

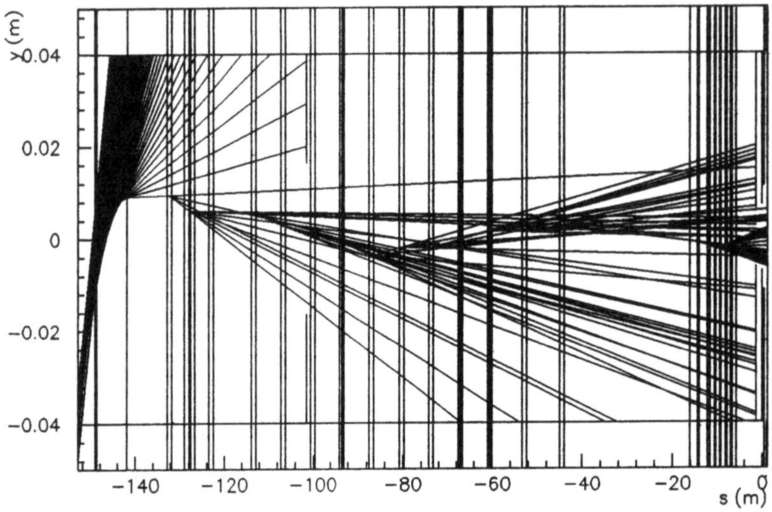

Fig. 1: *Monte-Carlo simulation of synchrotron emission: The plot shows a top view of the straight section at HERA in an expanded scale. A single electron enters at the lower left corner and is bended into the horizontal direction and deflected by several focussing quadrupoles. The target is at s=0. The vertical lines mark the positions of dipoles, quadrupoles and collimators. At each tracking step inside a magnet the electron emits a synchrotron photon that is tracked to the next collimator or to the beam pipe. The collimator system is described in section 4. The program repeats the simulation for a large sample of electrons distributed according to the beam emittance.*

A HERA magnet at the interaction region with a bending radius of $\rho = 1375$ m produces at $E = 35$ GeV and $I = 60$ mA a synchrotron radiation power of $P/\Delta\theta = 0.9$ kW per mrad bending. The radiation is emitted tangentially to the electron track in a horizontal band. At the interaction point at a distance of $l = 6$ m the band has a power density of $P/\Delta x = 150$ W/mm and a photon density of $\dot{N}/\Delta x = 4.5 \cdot 10^{16}$ photons per second and mm. The characteristic energy is $E_c = 69$ keV, the mean photon energy is $\langle E \rangle = 22$ keV.

The electron beam is focussed and defocussed by quadrupole magnets. The magnetic field is proportional to the distance of the electrons from the magnet axis. Therefore the synchrotron radiation in the quadrupoles is described by the relations

$$P \sim r^2 \quad \text{and} \quad E_c \sim r,$$

where r is the offset of the individual electrons from the beam axis. The typical energy spectrum of the quadrupole radiation is softer than that of the dipole radiation. The quadrupole radiation is peaked in beam direction. The total radiation of a single HERA quadrupole in the East Area is typically $P = 18$ W.

Figure 1 shows a Monte-Carlo simulation of the synchrotron radiation emission of a single electron trajectory in the straight section of the East Area.

3. Effects of Synchrotron Radiation

Synchrotron radiation has effects on the operation of the target cell and of the detector:

- The target cell is heated by synchrotron radiation. The heating leads to a reduction of the target density by temperature rise and in extreme cases to a destruction of the thin cell walls. The HERMES cell with no shielding is heated by about 100 W.
- The wall-coating of the cell may be destroyed by photon irradiation, causing a stronger wall depolarization of the polarized target gas. The unshielded HERMES cell is irradiated by about 10^{16} photons per second.
- Synchrotron radiation is scattered off the target gas, the walls of the storage cell, the collimator edges and the vacuum pipe. The main scattering processes are Compton scattering, Rayleigh scattering and the photoelectric effect with subsequent fluorescence light emission. The scattered radiation enters the silicon strip detector and the front drift chambers of the HERMES detector and prevents their operation unless counter-measures are taken.

4. Protection from Synchrotron Radiation

In order to reduce the radiation from the dipole magnets, the straight section in front of the target is enlarged from 5.8 m to 140 m. Secondly the last bending section in front of the straight section is split into a strong bending part and an additional weak magnet with $\rho = 6.3$ km. By this means the power density at the target is reduced from $P/\Delta x = 150$ W/mm to $P/\Delta x = 1.5$ W/mm and the characteristic energy is reduced by a factor of 5 to $E_c = 15$ keV and thus the dipole radiation can be shielded more easily.

A method to reduce quadrupole radiation is to weaken the quadrupoles in the straight section and to place them at a larger distance from the target. The β-functions in the quadrupoles should also be decreased. These requirements are in conflict with the necessary focussing of the beam at the target and with the global HERA beam requirements. Therefore this method is not applicable for the HERMES experiment.

A very efficient way of reducing the remaining synchrotron radiation is a movable collimator system as described in the following (see figure 2):

- A collimator C_0 between the target cell and the last dipole upstream the target shields most of the dipole radiation 100 m away from the target. It prevents scattering of synchrotron radiation at the beam pipe.
- A collimator C_1 at $s = -1.50$ m shields the subsequent collimator and the target cell from direct synchrotron radiation.
- A collimator C_2 at $s = -0.05$ m in the shadow of C_1 shields the target cell from radiation that is edge-scattered at C_1.

With this arrangement only double scattered radiation may hit the detector and the target cell. An exception is the radiation scattered at the target gas which cannot be shielded. The size and the positions of the collimators and of the target cell were chosen to maximize the target density. Table 1 shows the results.

A detailed simulation of the synchrotron radiation was performed [2, 4]. The energy spectrum of the radiation hitting C_1 is plotted in figure 3. C_1 intercepts $8 \cdot 10^{16}$ photons/s with an energy $E_\gamma > 3$ keV, corresponding to 325 W. $15 \cdot 10^{16}$ photons/s, corresponding

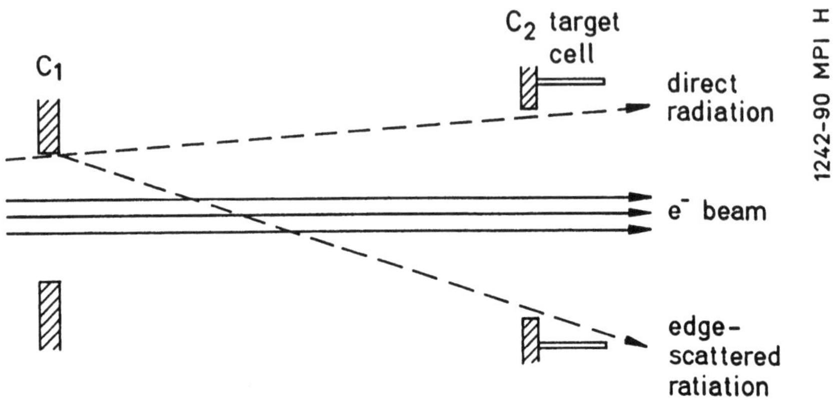

Fig. 2: *Sketch of the collimator system. C_2 is in the shadow of direct synchrotron radiation. The target cell and the detector are in the shadow of direct radiation and of radiation that is edge scattered at C_1.*

Table 1: *The size of the collimators and of the target cell.*

	position	**horiz. opening**	**vert. opening**
C_0:	−101.50 m	±16.3 mm	±2.9 mm
C_1:	−1.50 m	±6.3 mm	±2.3 mm
C_2:	−0.05 m	±7.1 mm	±2.6 mm
Target cell:	0 – 0.40 m	±11.3 mm	±4.2 mm

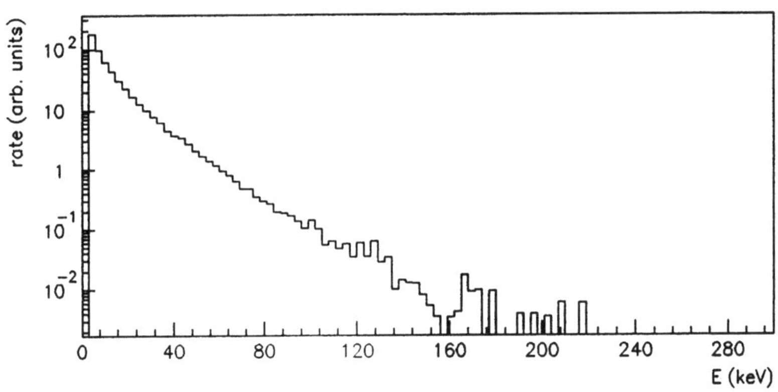

Fig. 3: *Energy spectrum of the synchrotron radiation hitting the collimator C_1.*

to 375 W, go through the hole of the collimators and the target cell. $2 \cdot 10^{16}$ photons/s hit the edge of C_1 within 2 mm distance.

A small fraction of the edge-scattered radiation, $1.5 \cdot 10^9$ photons/s, hits the collimator C_2, and one can estimate that the amount of double scattered radiation entering the detector is reduced to about 25 kHz. This is about one order of magnitude smaller than the direct radiation scattered from the target atoms. Therefore we are confident that the synchrotron radiation can be shielded without causing any problems to the HERMES detector.

5. Secondary Effects of the Collimator System

A positive secondary effect of the collimator C_2 in front of the target is that it is conductance limiter for the gas leaking out of the storage cell. But there are two negative aspects: The electron bunches induce wake-fields and resonant modes. Those modes have to be damped. Secondly the tails of the electron beam are scraped by the narrow collimators. In order to reduce both effects the collimator opening should be larger than 1 mm + 15 σ of the beam size.

The scraped electrons generate electromagnetic showers and also deep-inelastic events in the collimators. This background is reduced by adding a pair of anti-scraping collimators that scrape the beam in larger distance from the target. They are placed half a betatron oscillation upstream the collimator C_1 at $s = -50$ m (horizontal) resp. at $s = -36$ m (vertical). They should have an opening that is (in relative units of the beam size) a little bit smaller than that of C_1. Furthermore, the smallest aperture of the whole ring should be far away from the detector, e.g. in the West Area.

The tails of the electron beam are partially due to bremsstrahlung in the rest gas. This part was simulated by Monte-Carlo [2, 4]. The amount of bremsstrahlung electrons that are scraped at the collimators close to the target cell is of the order of 8 kHz. These electrons produce showers and deep inelastic events in the tungsten collimators. The particle rate that penetrates the collimators and enters the detector was found to be a negligible background for HERMES.

6. Conclusion

A storage cell experiment in an electron storage ring like HERMES has to be protected from synchrotron radiation. A long straight section in front of the target and an additional weak bending magnet reduces the dipole radiation. A well adjusted collimator system shields efficiently from the quadrupole radiation. The remaining indirect radiation does not harm the target cell and its photon background is acceptable for the detector.

References

[1] J. Amann et al., HERMES proposal, DESY, PRC-90/01 (1990)

[2] M. Düren, SYNTRACK program, MPI-Heidelberg.
The program was developed using parts of SYNCHRO2 developed by Pitzl at DESY and of the program DECAY TURTLE by Brown and Iselin at CERN.

[3] H. Winick, S. Doniach, *Synchrotron Radiation Research*, Plenum Press, New York, 1980

[4] M. Düren, HERMES report 1/89 (1989) and HERMES report 1/90 (1990)

A Storage Cell for the Filter Target Test Experiment

K. Zapfe

Max-Planck-Institut für Kernphysik, Postfach 103980,
W-6900 Heidelberg, Fed. Rep. of Germany

To polarize circulating antiprotons in a storage ring a gas target of polarized hydrogen will be used. The gas will be injected into a T-shaped storage cell to increase the target thickness. The status of the storage cell for a test experiment with protons instead of antiprotons in the Heidelberg Test Storage Ring TSR is reported.

The Target Chamber

To demonstrate the successful application of a storage cell in a storage ring, a test experiment will be performed at the Heidelberg Test Storage Ring TSR [1] to polarize protons of 30 MeV by an internal polarized gas target. For this experiment a T-shaped, thin walled storage cell has been built. The cell will be installed in the target chamber, which is shown in figure 1. The chamber is pumped by a cryo pump

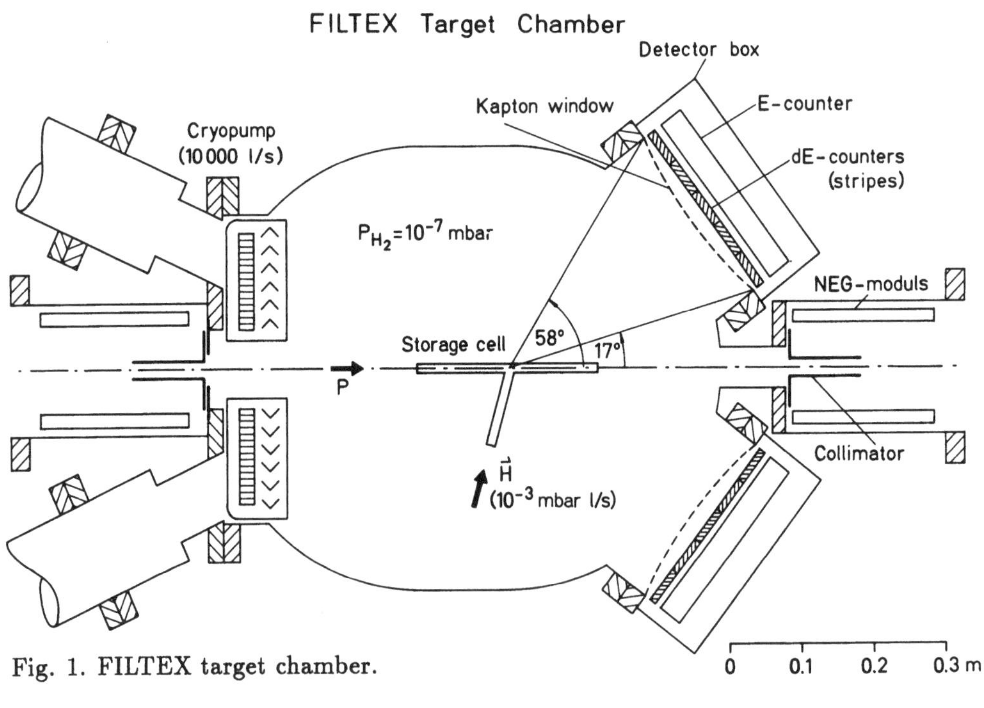

Fig. 1. FILTEX target chamber.

with 10.000 l/s. The resulting pressure in operation will be 10^{-7} mbar. To reduce the pressure to the normal ring vacuum of 10^{-11} mbar, three differential pumping stages of Non-Evaporating-Getter (NEG) modules and collimators will be installed on each side of the target.

The circulating protons pass through the target cell in the center of the chamber. The scattered protons enter through circular, 125 μm thick kapton windows into the detection system consisting of two detectors in the vertical ($\phi = 90^0, 270^0$) and two in the horizontal plane ($\phi = 0^0, 180^0$). The storage cell is filled with polarized hydrogen gas by a high intensity atomic beam source [2]. The polarization of the target will be measured by means of a Balmer polarimeter [3].

The Storage Cell

The storage cell shown in figure 2 is made of 0.5 mm thick aluminum. This causes an energy loss of several MeV to the scattered protons depending on the scattering angle θ. The energy spectrum for 30 MeV protons and the spectrum of the residual energy after passing the storage cell walls is shown in figure 3. The residual energy for protons up to $\theta = 55^0$ enables their detection with the scintillation detectors.

A schematical drawing of the cell is shown in figure 4. It consists of two 250 mm long half tubes. The two sealing surfaces of each half tube are enlarged by 10 mm wide lips to reduce the leakage rate out of the cell. In the closed position inner diameter of the cell tubes is 11 mm. The feed tube for the polarized hydrogen gas from the atomic beam source is 100 mm long with a diameter of 10 mm. It is glued under an angle of 75^0 on one of the half tubes.

The small opening opposite to the feed tube allows for diagnostics with a quadrupole mass spectrometer. The injected atoms diffuse slowly out of the cell performing about 500 wall collisions. The resulting depolarisation is minimized by coating the cell with Teflon [4]. To allign the cell with the beam and source axes there are adjustment

Fig. 2. Storage cell with clam shell mechanism.

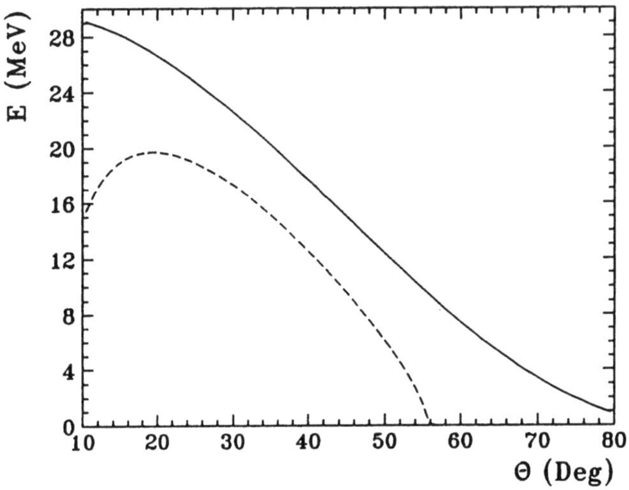

Fig. 3. (—) Energy spectrum of scattered protons for E = 30 MeV.
(- - -) Spectrum of the residual energy after passing the storage cell walls.

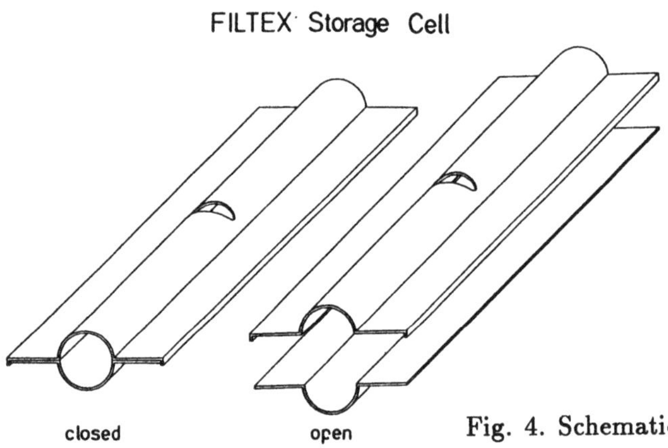

Fig. 4. Schematical drawing of the storage cell.

mechanism inside the target chamber for the x-, y- and z-direction and around three rotation axes.

The required target thickness for FILTEX is $\bar{n} = 1 \cdot 10^{14}$ polarized hydrogen atoms per cm^2 [5]. If an atomic flux of intensity I is injected into a storage cell of length l and conductance C_{tot}, the density $\bar{n}$ is given by

$$\bar{n} = \frac{I}{C_{tot}} \cdot \frac{l}{2} \tag{1}$$

assuming a triangular density distribution along the target. To get a high density, the conductance must be minimized. For a short tube, C is determined by the molecular weight M, the temperature T, the diameter d and the length l according the following formular:

$$C = 3.8 \cdot \sqrt{\frac{T_{[^\circ K]}}{M}} \cdot \frac{d^3_{[cm]}}{l_{[cm]} + \frac{4}{3} d_{[cm]}} \cdot \frac{l}{s} \tag{2}$$

Because of the strong dependence of C on the cell diameter a small beam size at the target is of greatest importance. With a β-function of $\beta_x = 0.314$ m, $\beta_y = 0.345$ m and an acceptance of $A = 66\ \pi$ mm mrad, the beam dimensions in the center of the target are given by $9.2 \cdot 9.6 \mathrm{mm}^2$. Thus a cell diameter of d = 11 mm was chosen. The cell will be cooled down to 100 K to further reduce the conductance. For this temperature the condactance is $C_{tot} = 10.22$ l/s. With an expected intensity of $I = 7 \cdot 10^{16}$ atoms/s injected into the storage cell, a density of $\bar{n} = 8.6 \cdot 10^{16}$ atoms/cm^2 seems attainable.

The acceptance of the storage ring should not be limited by the cell. As the cell diameter is matched to the beam envelope, the two half tubes can be separated to a distance of 40 mm during filling of the storage ring. For this purpose, each half tube is mounted on a linearly moving sledge sliding on a ball bearing from stainless steel. To enable the rolling of the balls in the vacuum at low temperatures, all components have been carefully cleaned to remove all grease. Moreover, the original balls of the ball bearing have been exchanged be balls with a 1/100 mm smaller diameter in order to allow for sufficient play. By means of two racks and a pinion the two bearings are connected via two knuckle joints to a rotary motion drive.

First tests have shown, that the opening mechanism works well under vacuum conditions. Cooling down the cell to 100 K didn't impair its functioning either.

Conclusions

A storage cell for the FILTEX test experiment has been built. First tests in vaccum indicate that the cell with a mechanism for taking its half tubes apart can be operated in a storage ring.

References

[1] E. Jaeschke, this workshop
[2] W. Korsch, Ph. D. Thesis, Universität Marburg, unpublished (1990)
W. Korsch, this workshop
[3] G. Graw et al., Intense polarized hydrogen source for the FILTEX target, in: *Proc. IV LEAR Workshop*, Villars 1987, eds.: C. Amsler, G. Backenstoss, R. Klapisch, C. Leluc, D. Simon and L. Tauscher, (Harwood, Chur, 1988) 221
W. Luck, Ph. D. Thesis, Universität Marburg, unpublished (1990)
[4] W. Haeberli, this workshop
W. Luck, this workshop
[5] H. Döbbeling, Proposal CERN/PSSC/85-80 (1985), and Addendum (1986)

Mechanical Design of Storage Cells

J.F.J. van den Brandt

Physics Department, University of Wisconsin, Madison, WI 53706, USA

1. INTRODUCTION

In this contribution I will describe the mechanical design aspects of storage cells for polarized internal gas targets. Such targets consist of a flux of polarized atoms directed into a thin-walled storage cell. I will limit myself to applications involving medium- and high-energy electron rings, such as the MIT-Bates and NIKHEF pulse-stretcher rings, and the HERA electron storage ring at DESY. As target gas I will only consider polarized hydrogen, deuterium and ^{3}He. Valuable practical experience with polarized internal gas targets has been obtained by the Argonne-Novosibirsk collaboration[1] at the VEPP-3 electron machine. Tests[2] with a storage cell have also been conducted at the DORIS machine at DESY. In the former experiment a 94 cm long elliptical tube with a 46 (24) mm horizontal (vertical) aperture has been used. In the latter case a 50 cm long fixed cylindrical stainless steel tube with a diameter of 15 mm has been employed. Furthermore, experience has been gained during the design and construction of the MIT polarized ^{3}He internal target[3], while valuable in-beam experience was obtained in experiments using the Caltech polarized ^{3}He external target[4].

First, I will discuss the considerations that govern the geometry of a storage cell. The second part concerns the discussion of various possible cell materials. Finally, I will show some design concepts of the target system for the HERMES experiment[5] at DESY.

2. CELL GEOMETRY

In general an internal target system has to be designed in such a way as to guarantee that effects on the quality of the stored beam are minimized. The most important function of a storage cell is to increase the luminosity of the experiment. The parameters that govern the luminosity are shown in equation (1):

$$\text{Luminosity} \sim \frac{L^2}{D^3}\sqrt{\frac{M}{T}}, \tag{1}$$

where L is the length of the target cell, D the diameter, M is the target mass and T is the temperature. From this expression it is clear that one will try to minimize the diameter

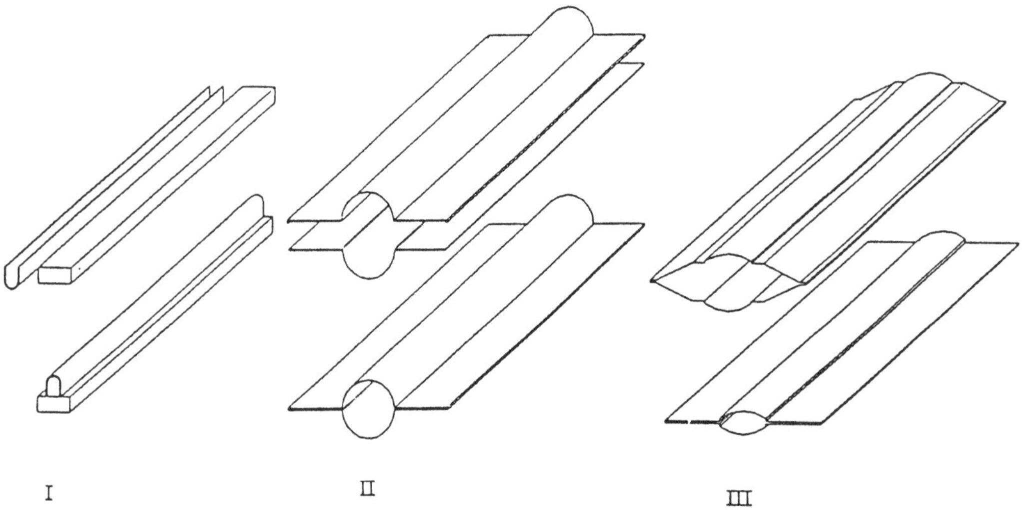

Figure 1. Diagrams showing the open and close position of a storage cell: I the Argonne-Novosibirsk clam-shell target, II the MPI target cell at Heidelberg, and III a proposal for the HERMES storage cell.

of the cell. For this reason the storage cell should be installed in a region with low beta functions, using quadrupoles to strongly focus the electrons into the cell.

The size of the electron beam determines the dimensions of fixed and retractable apertures. For a fixed storage cell one should use diameters that exceed the scaled value for the apertures of the vacuum chambers in the region with largest beta functions.

For the MIT-Bates and the HERMES designs we used $D > \pm 15\sigma + 1$ mm, where $\sigma_{x(y)} = \sqrt{\beta_{x(y)}\epsilon_{x(y)}}$. Here $\beta_{x(y)}$ is the beta function and $\epsilon_{x(y)}$ the emittance of the electron beam in the horizontal (vertical) direction. For movable apertures it has been shown[2] that losses of the electron beam occur at 7 - 8 σ apertures. It is clear that this discussion is extremely machine, tune and experiment dependent, as for example beam halos and emittance ratios are at present poorly understood quantities. In order to ensure flexibility in operating storage cells with agressive diameters, they should be designed in such a way that they can be opened up while the electron beam is tuned. The target cell can then be positioned precisely with respect to the electron beam and closed around it.

Figure 1 schematically shows the possibilities for the construction of a target cell that can be opened and closed. The principle of the Argonne-Novosibirsk target[6] (labeled I) leads to tight constraints on the flatness of the cell halves in order to seal the target over it's entire length. In order to minimize warping of the cell the walls need to have sufficient thickness. These problems have been overcome to some extend in the design developed at the Max Planck Institute[7] at Heidelberg (labeled II). Rather than a line seal one now extends the seal over a large area. A third interesting possibility is the scheme[5] labeled

III. Here the problem with the gas confinement is completely eliminated through the use of foil hinges and by welding the target sides together.
For polarized hydrogen and deuterium another constraint on the dimensions of the storage cell is given by the allowable number of wall bounces. In order to prevent depolarization effects, and given the present status of wall coatings[8], the cell should be designed such that the average number of wall bounces is less than 500. Additional limits on the geometry are placed by the constraint that for certain applications no synchrotron radiation is allowed to hit the cell, as this could result in degradation and/or destruction of the wall coating. Furthermore, these photons can scatter off the cell wall into the detector system giving rise to unwanted background. Elaborate collimation schemes have been designed[9] in order to prevent synchrotron radiation from hitting the cell walls, resulting in tight constraints on the cell geometry.
With respect to the length of the cell one should consider the detector acceptance as well as the variation of the beta function in the field-free region[10] near the interaction point. This dependence is given in equation (2):

$$\beta = \beta_0(1 + \frac{(s - s_0)^2}{\beta_0}). \tag{2}$$

Here β_0 is the value of the beta function at the interaction region and $(s - s_0)$ is the distance from the interaction region.
The thickness of the storage cell is dictated by the contributions to the resolutions due to energy-straggling and multiple-scattering processes. For these reasons one often considers cells with a thickness between 25 and 100 μm. It is clear that various ports have to be connected to the storage cell. One needs an entrance tube for feeding the polarized gas into the storage cell, and for polarized hydrogen and deuterium often exit ports will be needed in order to sample the polarization of the target gas. For certain applications, such as recoil-particle detection (e.g. ^{2}H and ^{3}He particles), one has to consider thin exit windows connected to the sides of the storage cell.
For application in electron rings one favors a straight tube for the shape of the storage cell. This in order to minimize RF heating of the cell. These processes[11] involve single- and multi-bunch effects of the electron beam and in general scale with the bunch charge squared ($\sim q_b^2$). For this reason the magnitude of these RF effects varies strongly depending on the specific machine (and tune of the machine). In order to prevent trapping of electromagnetic fields inside the storage cell one should avoid shapes that resemble cavity-like structures (e.g. pill-box cells).

3. STORAGE CELL MATERIALS

Storage cells are used in ultra-high vacuum environment and therefore only materials having low outgassing properties should be employed. For the same reason it may be

necessary to bake the storage cell. Application in electron rings requires the use of radiation resistant materials, which in general excludes many non-metals. In order to reduce energy-straggling and multiple-scattering processes one favors low-Z materials. Clearly, beryllium and aluminum are very good from this point of view. Because our application involves polarized gases we cannot employ magnetic materials.
An important constraint on the choice of material is the mechanical strength of the structure. Considering all demands given above, I think that probably the best choices are stainless steel, aluminum, titanium and beryllium.
For stainless steel there are several non-magnetic types (such as S.St. 304L and 316L), which have excellent welding properties: structures can be made as thin as 25 μm. An obvious concern are the effects of wall depolarization due to para-magnetic centers in the cell walls. The performance of coated S. St. cells should be investigated.
Aluminum is a good material from the point of view of resolution (low Z); however welding thin aluminum objects is not straightforward. An alternative is to use etching techniques in order to obtain the desired thickness. An important application for pure aluminum is in cryogenic storage cells. As shown in expression (1) the luminosity can be increased up to a factor of four, without increasing the gas load on the machine vacuum, when cooling the cell walls to 20 K. Pure aluminum is excellent for this application as it has an extremely high thermal conductivity (λ = 40 W/cm K) at these temperature. Note that cooling to these low temperatures is only an option for polarized ^{3}He, as for hydrogen and deuterium wall depolarization effects become inhibitive for temperatures below 120 K.
Titanium appears to be a good material because it is strong and non-magnetic. Welding tests have been performed and we found that titanium foils have good welding properties for thicknesses as low as 12 μm.
To conclude this part let me remark that when straggling and multiple-scattering processes are a concern then beryllium may be a good material. It's use however places tight constraints on machining of the structure. Other possibilities for cell materials include kapton and Al_2O_3. Up to now I do not know of any detailed experience with these materials with respect to the fabrication of storage cells.

4. PRACTICAL EXAMPLES

Next a possible realization of a storage cell target will be discussed. In order to allow for maximum flexibility I discuss both the design of an open/close mechanism and of a positioning device.
Figure 2 shows the basic principle of the open/close mechanism of the storage cell. This cell was designed with the properties of the HERA electron beam in mind and has a length of 40 cm, a wall thickness of 0.1 mm and an elliptical cross section with diameters 25.2 $\times$ 7.6 mm. These dimensions are determined mainly by a $\pm 20\sigma$ beam clearance and the constraint that no synchrotron radiation is allowed to hit the cell walls.

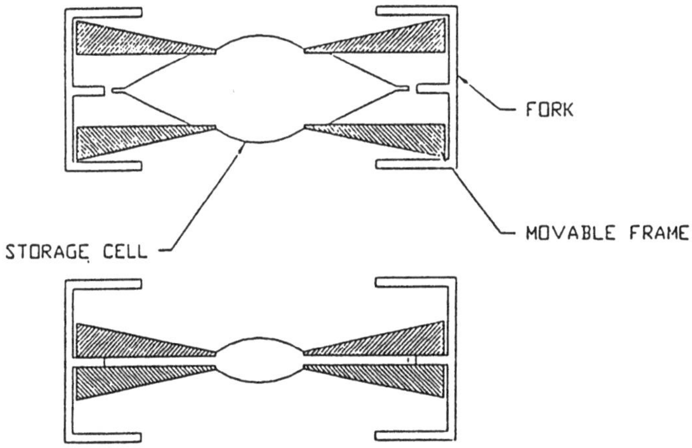

Figure 2. A possible mechanism for a storage cell target showing a cross section in both the open and close position.

The cell structure could be etched out of one sheet of (hard) aluminum with a thickness of 0.5 mm. The hinges are 1 mm wide and have a thickness of 40 μm. The cell walls are 0.1 mm thick, enforced by two 0.5 mm thick ribs. The feed tube and the sampling port are both 10 cm long with a diameter of 10 mm and are connected to the center rib.
The cell is mounted in a fork with an E-profile like cross section through the use of two movable frames. These frames are connected to the fork through stainless steel edge-welded bellows which allows for opening and closing of the cell by pressurizing and evacuating the appropriate bellows. As pressurizing medium one can use a gas (N_2) or a liquid such as alcohol.
The precise and reproducible positioning of the two cell halves with respect to each other and the fork is obtained through strategically placed pins that are fixed in the fork. This precise and reproducible orientation of the cell with respect to the fork, enables the cell to be positioned freely by moving the fork. The positioning device for the cell is shown in figure 3.
The aluminum cell construction is connected rigidly to a stainless steel tube outside the vacuum using two identical edge-welded stainless steel bellows. The tube is connected in the middle to a flange which in turn is mounted to an XY positioning table. This table in turn is mounted on a $\theta\phi$ positioning table. The bellows should be flexible enough to allow for the desired $XY\theta\phi$ positioning of the target cell. The positioning tables can be motorized using appropriate stepping motors. Absolute linear and angular (optical) encoders can be employed for measuring the position of the flange.
A slit system in front of the target cell shields the detector and target cell against secondary scattered synchrotron radiation. In addition these slits act as a conductance limiter and they can be used to locate the electron beam. The slits could be manufactured from thin

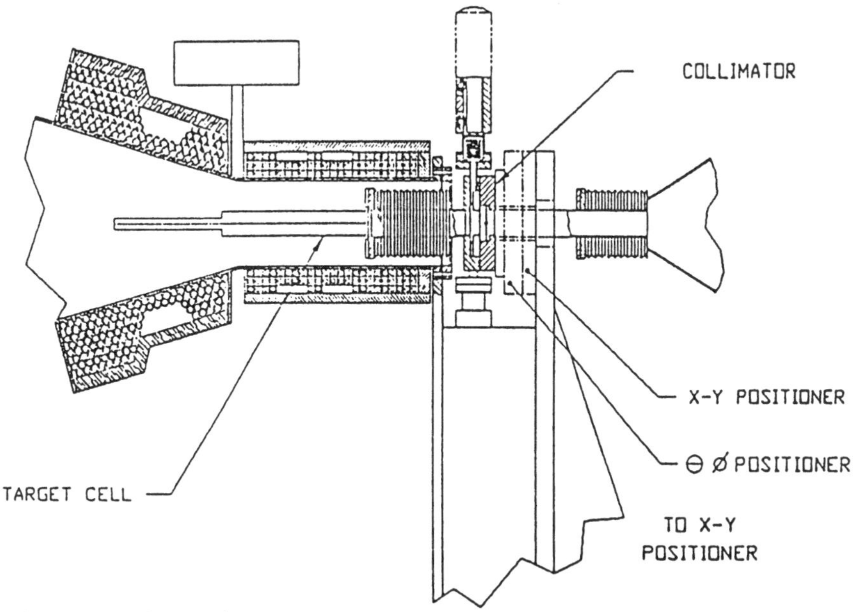

Figure 3. The mechanics for positioning the storage cell.

tungsten plates which move in a stainless steel housing. The slits should be precisely movable using stepping motors. Their position can be determined by linear encoders. During the experiment the operation of the eight motors and the acquisition of the linear and angular encoder data can be performed by computer. First the position of the electron beam is determined by sequentially moving the collimator slits in and out. By monitoring the singles rates in the electron spectrometer the beam position is obtained. Then the cell is moved to the correct position using the $XY\theta\phi$ positioning tables. After this is completed the cell is closed. Closing of the cell can be performed in a controlled manner by exploiting the spring rate of the bellows. It is possible to fine tune the cell position by minimizing the singles rates in the spectrometer.

In summary, I want to point out that the application of storage-cell targets is still in it's infancy. At present only the Argonne-Novosibirsk collaboration is performing actual measurements using the above described methods. More experiments are being planned and we foresee a vivid future for the application of the storage-cell technique.

ACKNOWLEDGEMENTS

I would like to acknowledge discussions with M. Doets, W. Haeberli, R. Holt, K. Lee, R. Milner, B. Kaan, B. McKeown E. Steffens, C. Woodward and K. Zapfe. The author's research is supported in part by the National Science Foundation under Contract No. PHY-8717764.

REFERENCES:

[1] S.G. Popov, Proc. of the Workshop on Targets in Electron Storage Rings, SLAC, January 1989.

[2] R. Brinkman, R.D. Kohaupt, Neeseman and Kose, DESY, private communications.

[3] K. Lee and S. Ottoway, private communications.

[4] C.E. Woodward *et al.*, Phys. Rev. Lett. **60**, 698 (1990).

[5] J.F.J. van den Brand *et al.*, Mechanical Design of the HERMES Polarized Gas Internal Target System, HERMES Report (1989).

[6] R. Holt, private communications.

[7] K. Zapfe, private communications.

[8] W. Haeberli, "Nuclear Physics with Stored, Cooled Beams", AIP Conf. Proc. 128, p. 251 (1985); W. Haeberli, Proceedings of the Workshop on Polarized Beams and Polarized Gas Targets, February 1990, to be published; S. Price, private communications.

[9] M. Düren, HERMES report, 1989.

[10] M. Sands, The physics of electron storage rings, SLAC-121, UC-28, 1970.

[11] J.F.J. van den Brand *et al.* and H.G. Gaul *et al.*, wake field calculations. HERMES reports (1989).

List of Participants

Avakian,R.	Yerevan /USSR
Brand,J.F.J.van den	Wisconsin /U S A
Chupp,T.	Harvard /U S A
Conzett,H.E.	Berkeley /U S A
Dalpiaz,P.F.	Ferrara /Italy
Delheij,P.P.J.	TRIUMF /Canada
Dick,L.	Milano /Italy
Dueren,M.	Heidelberg /Germany
Eversheim,P.D.	Bonn /Germany
Fick,D.	Marburg /Germany
Finger,M.	Prague /Czechoslovakia
Gaul,H.G.	Heidelberg /Germany
Gebel,R.	Bonn /Germany
Gruebler,W.	Zuerich /Switzerland
Haeberli,W.	Wisconsin /U S A
Heil,W.	Mainz /Germany
Hinterberger,F.	Bonn /Germany
Holt,R.J.	Argonne /U S A

Kinney,E.R.	Argonne /U S A
Korsch,W.	Marburg /Germany
Krisch,A.D.	Michigan /U S A
Kuzik,V.E.	Moscow /USSR
Mertig,M.	Dresden /Germany
Meyer,H.-O.	Indiana /U S A
Mori,Y.	KEK /Japan
Nann,H.	Indiana /U S A
Nies,I.	Bonn /Germany
Penttila,S.	Los Alamos /U S A
Pilipenko,Yu.K.	Dubna /USSR
Phelps,R.A.	Michigan /U S A
Prescott,C.Y.	SLAC /U S A
Przevoski,B.von	Indiana /U S A
Raymond,R.S.	Michigan /U S A
Roser,T.	Michigan /U S A
Schueler,P.	Yale /U S A
Sperisen,F.	Indiana /U S A
Steffens,E.	Heidelberg /Germany
Toporkov,D.	Novosibirsk /USSR
Tran,Minh-Tam	Lausanne /Switzerland
Widdra,W.	Marburg /Germany
Young,L.	Argonne /U S A
Zapfe,K.	Heidelberg /Germany

Part IV

Polarized Solid Targets

First Use of the Polarized Target Facility at Fermilab

P. Chaumette[1], *J. Derégel*[1], *H. Desportes, G. Durand*[1], *J. Fabre*[1], *D. Hill*[2], *T. Kasprzyk*[2], *L. van Rossum*[2], *and J. Sheppard*[2]

[1]DPhPE/STIPE/STCM, C.E.N.-Saclay, F-91191 Gif-sur-Yvette Cedex, France
[2]Argonne National Laboratory, Argonne, IL 60439, USA

The first run of the E704 experiment (measurement of $\Delta\sigma_L$) was performed at Fermilab on a 180 GeV polarized proton-antiproton beam on a longitudinally polarized proton target.The dilution refrigerator for the target and the polarizing magnet could be serviced and translated in running conditions in or out of the beam line, thus giving good flexibility in the time sharing with alternating operation as hydrogen target or beam polarimeters. The average polarization values were around 75 % and the polarization was reversed once a day. Excluding the polarization reversal periods, the readiness of the target was around 95 % for the whole run.

The polarizing solenoid and the dilution refrigerator for this target (Fig.1) have been built at Saclay[1,2], the Argonne National laboratory being in charge of the whole equipment around this system and of its installation and maintenance[3]. The principle of operation of this polarized target is based on the long nuclear spin relaxation time at very low temperature.

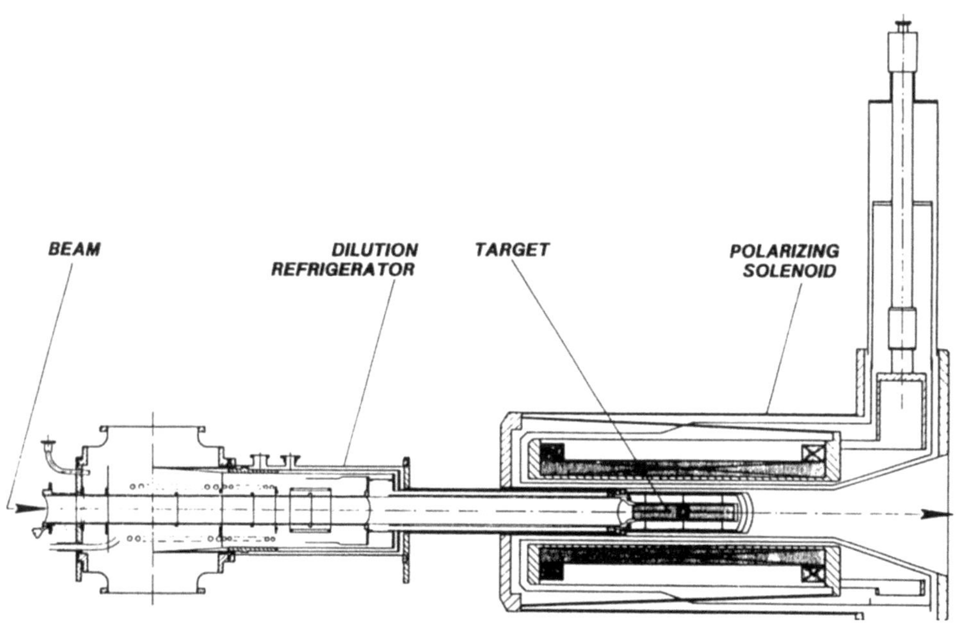

Fig.1 Dilution refrigerator and polarizing solenoid

The dimensions of the target, 200 mm in length by 30 mm in dia., require a ^3He-^4He dilution refrigerator with an horizontal axis. In the polarization mode the refrigerator has to provide a cooling power of 150-200 mW and the temperature must stay below 500 mK inside the target. The frozen spin mode requires a temperature below 50 mK in order to get depolarization time constants of more than 10 days in a low magnetic field of 0.3 T. This refrigerator has been designed for a quick loading of the target.

The polarizing superconducting solenoid axis is concentric to the dilution refrigerator. It provides a maximum field of 6.5 T with an homogeneity better than $\pm$ 5×10^{-5} in the target volume. A central warm bore of 95 mm dia. receives the nose of the dilution refrigerator containing the target. This solenoid also serves to provide the holding field in frozen spin mode for the first physics experiments, with small scattering angles. A pair of large aperture superconducting holding coils might be constructed and installed in the future, allowing the various spin orientations and large scattering angles.

The tests of the refrigerator were performed at Saclay, with a pumping device at our disposal of 3000 m^3/h, instead of the 5500 m^3/h pumping system installed at FNAL. During these tests, a temperature of 47 mK was obtained in the mixing chamber, with a ^3He flow of 2,5 mmole/sec, and with 10 mmole/sec for ^4He in the frozen spin mode and with no power dissipation in the mixing chamber. With a 150 mW power dissipation in the mixing chamber the temperature is stabilized at 470 mK with 35 mmole/sec flow rates for both ^3He and ^4He - The 5500 m^3/h pumping system available at Fermilab did not improve the temperature nor the cooling power between 0 and 150 mW.

Fig.2 gives the temperature versus the dissipated power in the mixing chamber. In frozen spin mode the molecular gas ratio running through the pumps is 88% ^3He - 12% ^4He.

After completion of the tests at Saclay, the DR and solenoid were shipped in 1988 to Fermilab. [1] The Polarizing solenoid system and the dilution refrigerator were mounted on separate concrete blocks using a precision rail system (Fig.3). The cooling down of the dilution refrigerator in 1989 and the completion of the entire polarized target facility permitted to achieve a two months run in may and june 1990. The best values of polarization were around -83% and +79%. With a field ranging between 1.8 T and 2.5 T, the polarization

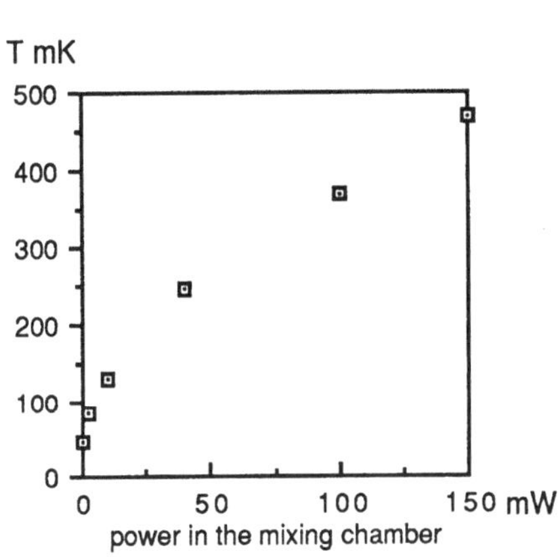

Fig.2. Temperature in the mixing chamber versus power

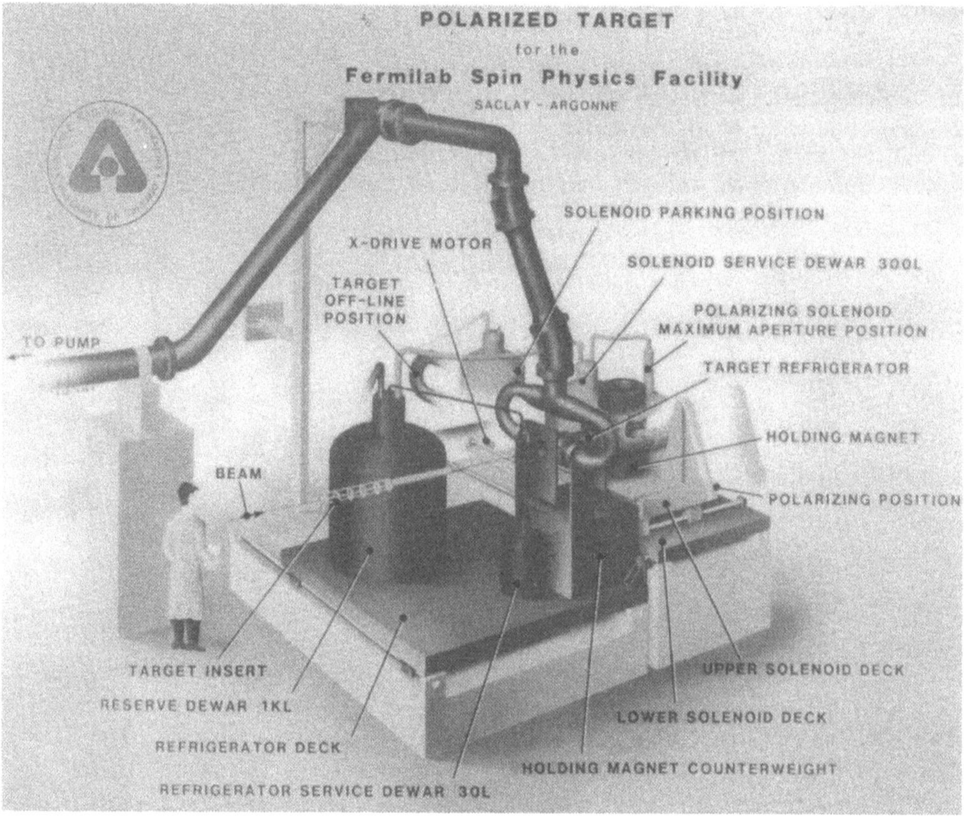

Fig.3 Central target assembly

lost in the frozen spin mode was around 1% per day. The average polarization for running was around 75 %. The time for the daily polarization reversal was around 2 hours (no plug), or as much as 5 hours (severe plug).

Our major problem during the installation came from the use of an industrial low price pumping device consisting of a 5500 m3/h roots pump in serial with two roughing pumps. Those 3 pumps are built with an external motor and a rotary ceramic-metal seal. Due to the large amount of oil (60 lit.), to the oil stress due to the seal friction, and to the leaks developing at each stop of the pumps a large amount of unwanted gas was generated. We have reduced the speed of the roots and installed an helium trap in serial with the nitrogen trap usable in the frozen spin conditions. However the pumping system was still generating 0.5 lit/day hydrogen, 0.5 lit/day air and 0.5 lit/day methane and other hydrocarbons. It was necessary to partially warm up the dilution refrigerator once a week to remove the hydrogen collected during the polarizing periods. For the next run it is planned to replace the pumping system by an hermetic pumping device, and it is considered to add an holding coil system for the wider scattering angles.

REFERENCES

1. P. Chaumette et al. AIP conf.proc.187 ,1331 (1989)
2. P. Chaumette et al. Advances in Cryogenic Engineering, Vol 35, 1067 (1990)
3. D. Hill et al. AIP conf.proc.187 ,1334 (1989)
4. R. Bernard et al. Nuclear Instruments & Methods 134, 176 (1986)

The Bonn Frozen Spin Target for Experiments with Real Photons

H. Dutz, S. Goertz, D. Krämer, W. Meyer, G. Reicherz, L. Reichertz, W. Thiel, A. Thomas, and J. Weber

Physikalisches Institut der Universität Bonn,
Nußallee 12, W-5300 Bonn 1, Fed. Rep. of Germany

At the new Bonn Electron Stretcher Accelerator **ELSA** *scattering experiments with real photons in the single pion photoproduction $\gamma N \rightarrow \pi N$ and in the photodisintegration of the deuteron $\gamma N \rightarrow pn$ are prepared. To determine these processes completely, it is nessesary to measure single and double polarisation observables. For these experiments a new 'bonn frozen spin target'* **BOFROST** *was developed. The target with a maximum volume of* 30 cm^3 *is cooled in a vertical* $^3He/^4He$ *- dilution refrigerator. In the 'frozen spin mode' a lowest temperature of 64 mK was reached.*

The magnetic system consists of two superconducting solenoids: A polarisation magnet with a maximum field of 7 T, a homogenity of 10^{-5} *over the defined target volume and a 'vertical holding magnet' with a maximum field in the target region of 0.57 T.*

At a field of 3.5 T a maximum polarization of + 94% and − 100% in Ammonia NH_3 *and + 92% and − 96% in Butanol was obtained. In the following the 'frozen spin target' and the first polarization measurements are described.*

1 Introduction

The measurements of spindependent observables are indispensable for a complete experimental determination of the reactions with real photons. Since 1970 experiments with polarized solid state targets were a substantial part of the experimental program at the Bonn 2.5 GeV electron synchrotron. The first experiments with the polarized target deal with target asymmetry measurements in the single pion photoproduction $\gamma N \uparrow \rightarrow \pi N$ [1] [2]. Based on the developements of the new targetmaterials NH_3 and ND_3 [3] and with the construction of a more powerful polarization refrigerator target asymmetry measurements in the photodisintegration of the deuteron $\gamma d \uparrow \rightarrow pn$ [4] were performed. Due to the exellent polarization resistance against radiation damage a tensor polarized ND_3 target was used in a first electron scattering experiment, to get information about the formfactors of the deuteron [5][6].

At ELSA double polarization experiments are prepared at the PHOENICS-detector. In combination with the new frozen spin target and the tagged linear polarized photon beam, it will be possible to measure not only the single polarization observables Σ, T and P, but also the double polarization observables G and H in an engergy range from 0.2 - 0.95 GeV [7].

2 The Bonn Frozen Spin Target (BoFroST)

The 'bonn frozen spin target' is based on the 'saclay frozen spin target' design [8]. The main components of the frozen spin target are a $^3He/^4He-dilution\ refrigerator$ and a superconducting magnet system [9]. The magnet system consists of three superconducting solenoids, a polarization magnet and two 'holding' magnets. The arrangement of the two 'holding' coils is given by the

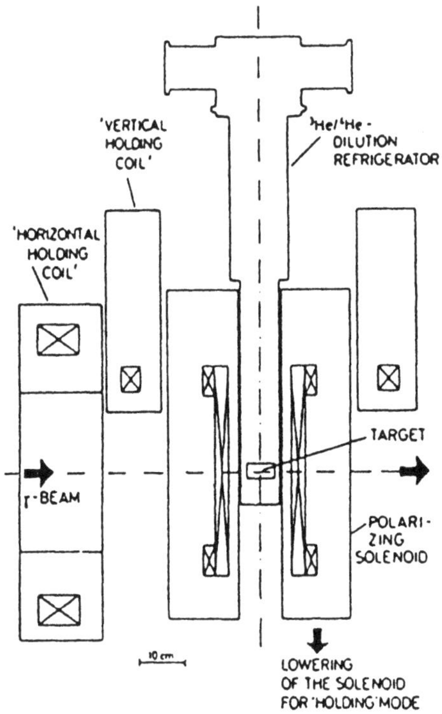

Fig.1 The Bonn Frozen Spin Target

polarization directions which are necessary for the measurement of the polarization observables. For the double polarisation observables G and H in the single pion photoproduction, the polarisation of the target nucleons must be orientated in the scattering plane and for the single polarization observables Σ, T and P vertical to the scattering plane. The schematically arrangement of the bonn frozen spin target is shown in fig.1.

At the moment, the target material will be polarized at a field of 3.5 Tesla produced by the polarization magnet (max. field 7 Tesla). After reaching the maximum value of polarization by the DNP, the microwaves will be switched off. In this case the temperature in the refrigerator will be lowered on values of $T \leq 80\ mK$. The nucleon relaxation time τ of the target nucleons is a steep function of the temperature :

$$\tau \simeq exp^{-T}.$$

By this, it is possible to reduce the field in the polarization magnet onto values of 0.5 Tesla. The field in one of the 'holding' magnets then will set on 0.5 Tesla at the target area. In the next step the polarization magnet can be lowered and the target area will be open for the γ-beam. To operate the target in the frozen spin mode, the extrem low temperatures must be maintained. Therefore the thermal heat input into the refrigerator by the γ-beam is limited onto few μW. This restricts the γ-beam intensity onto max. $10^7 - 10^8\ \gamma's/sec$.

2.1 Magnet System

Up to now the magnet system of the frozen spin target consists of a superconducting polarization magnet and a 'vertical holding' magnet. The magnets were build at CEN Stipe Saclay. The polarization magnet has a maximum field of $B_{max} = 7T$ at a nominal current of $I_{max} = 183\ A$ and a field homogeneity over the defined target volume of $\Delta B/B \leq 10^{-5}$.

The 'vertical holding' magnet has in the center of the coil a maximum field of $B_{max} = 1.03\ T$ at a nominal current of $I_{max} = 200\ A$. The target area is placed 200 mm below the center of the coil, where under these conditions the maximum value of the field is 0.57 T.

The full acceptance of the PHOENICS detector (1.2 sr) is accessible after lowering the polarization magnet. At 0.35T scattered particles with a momentum larger than $50 MeV/c$ can be detected.

The 'horizontal holding' magnet, built at Saclay, has a warm bore of 450 mm with a maximum field in the target area of 0.5 T. The distance between the target and the center of coil will be 350 mm.

2.2 The ^{3}He/^{4}He Dilution Refrigerator

The vertical $^3He/^4He$ dilution refrigerator is based on developements of the vertical Saclay refrigerator. However, the 'Bonn Cryostat' includes a ^4He separator and a different dilution unit. This refrigerator is built for a maximum target volume of 30 cm^3. The pumping system consists of a roots pumping system, 1000 m^3/h and a 350 m^3/h roots blower (Alcatel MIV), and a 60 m^3/h rotary blade pump (Alcatel 2060H). The minimum temperature with this performance is $T_{min} = 64\ mK$ at a ^3He flow rate of $\dot{n}_3 \approx 3\ mMol/sec$ and a ^4He-consumption of approximately $2l\ liq.\ He/h$. The maximum cooling power of this system at $T_{Qmax} = 280\ mK$ is $Q_{max} = 40\ mW$ (fig. 2) with a ^4He-consumption of $4l/h$.

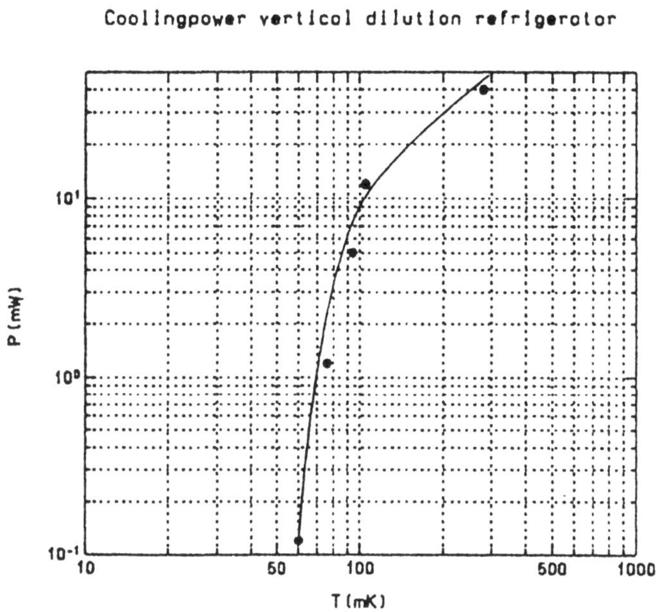

Fig.2 The coolingpower of the ^{3}He/^{4}He dilution refrigerator

3 First Polarization Measurements With The Frozen Spin Target

During the last months we tested the frozen spin target in the laboratory. We measured the polarization of different target materials: radiation doped Ammonia $^{14}NH_3$ and chemical doped (Porphyrexide) Butanol C_4H_9OH at a polarization field of 3.5 T. This magnetic field requires a microwave frequency for DNP of about 98 GHz which is generated by a phase lock looped IMPATT-diode with a stability of $\Delta f/f = 10^{-8}$ [10]. The polarization was measured with a NMR-system based on a computer controlled 'Liverpool-Box' at a frequency of 150 MHz [11]. The accurancy in the polarization measurement with this system is ± 2.5 %. The temperature during the DNP in the refrigerator can be specified between 150 - 200 mK.

Under this conditions we measured for NH_3 maximum polarization values of + 94% and − 100%. In Butanol the max. polarization values are + 90% and − 96%. The typical build-up times $\tau_{0.7}$ were in Ammonia about 100 min for both polarization directions [12] and in Butanol 22 min (positive polarization) and 12 min (negative polarization) respectively [13].

For the operation of the frozen spin target, the relaxation times of the target materials are a very important value. At a temperature of $T_{min} = 80\ mK$ and at a 'holding field' of $B_H = 0.33\ T$ the following relaxation times were measured NH_3 : $\tau = 30\ days$ and Butanol : $\tau = 5\ days$ (see [12] and [13]). From this point of view, both target materials at these temperatures are useful for polarization experiments with real photons.

4 Conclusions

The frozen spin target and the tagged (linear polarized) photon beam at the PHOENICS-detector is an ideal combination to measure single and double polarization observables in the single pion photoproduction and the photodisintegration of the deuteron. First measurements of the target asymmetry in the single pion photoproduction with Butanol as the target material will be started in the middle of 1991 at the PHOENICS-detector.

References

1. K.H. Althoff et al., Nucl. Phys. **B 53** (1973), 9
2. K.H. Althoff et al., Nucl. Phys. **B 137** (1978), 269
3. T.O. Niinikoski, J.M. Rieubland, Phys. Lett. **72 A** (1979), 141
   U. Härtel et al.,Proc. of the Int. Sym. on High Energy Physics with Pol. Beams and Pol. Targets, eds. C. Joseph and J. Soffer, Birkhäuser Basel, (1980), 451
4. K.H. Althoff et al., Z. Phys. **C 26** (1984), 175
5. B. Boden et al., Bonn-ME-90-06
6. H.D. Schablitzky, Bonn-IR-89-32
7. W. Meyer, Bonn-IR-88-60
8. R. Bernard et al., Saclay DPhPE 86-04
9. H. Dutz, Bonn-IR-89-15
10. J. Weber, Bonn-IR-90-19
11. W. Thiel, Bonn-IR-91-05
12. W. Thiel, in this proceedings
13. D. Krämer, in this proceedings

The PSI Frozen-Spin Target

B. van den Brandt, J.A. Konter, S. Mango, and M. Weßler
Paul Scherrer Institute, CH-5232 Villigen PSI, Switzerland

For measurements of the 2-spin and 3-spin transfer parameters in the n-p system in the 300 - 600 MeV range, a new large (100 cc) frozen spin polarized target is under construction at PSI. The holding coil system allows a virtually arbitrary polarization direction, quick polarization reversal and large opening angles.

Introduction.

A vertical dilution refrigerator will be incorporated in a cryostat in which a twofold superconducting holding coil system for vertical and horizontal direction of polarization will be integrated. Rotatability of the holding coil cryostat around the vertical axis will provide the third degree of freedom in polarization direction. This cryostat will be positioned on the beam, and a high homogeneity solenoid, mounted in a room temperature bore cryostat, will be brought into "polarizing" position, or down into "data taking" position, sliding it along a common supporting column (see fig. 1).

The dilution refrigerator.

The ^{4}He reservoir, supplying the holding coil system, will at the same time serve as a buffer volume for the two precooling loops of the dilution refrigerator insert, thus avoiding thermal instabilities during dewar changes.
Precooling of the incoming ^{3}He will be accomplished in a triple heat exchanger between two streams of ^{4}He from the 4.2 K and sub-λ bath, with maximal recovery of the enthalpy of the outcoming ^{3}He-gas from the inside pumping tube. Calculations show that the Roubeau design [1], with the gas cooler inside its own pumping tube (a design favourable for its high degree of vacuum-integrity), becomes less efficient at high flow rates, at least for the length of cryostat chosen.
The target loading will be performed with an up-scaled version of the present PSI design [2], in which the target cell is an integral part of the target holder. This insert rod essentially consists of a μwave tube plus baffle system and a Vespel block with a 5° conical section, mating the central acces hole in the Vespel supporting structure of the heat exchanger.

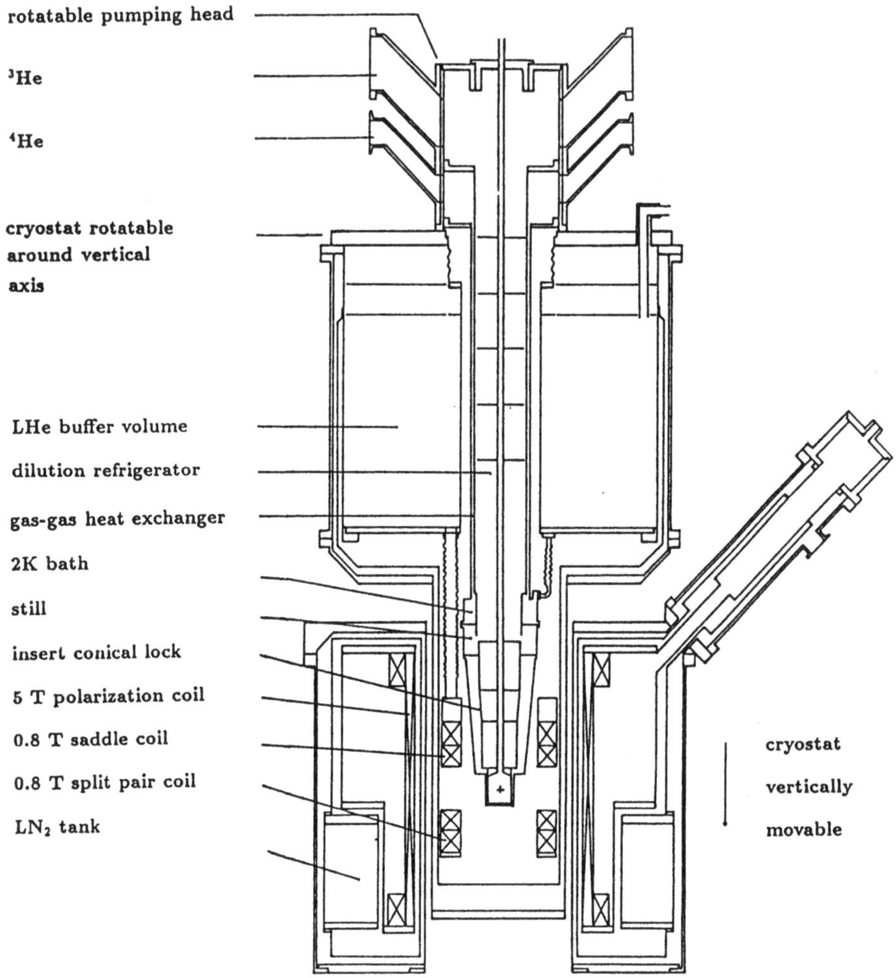

Fig. 1: *Dilution refrigerator cryostat with integrated holding coil system and 5 Tesla polarization coil cryostat shown in polarization position.*

This loading system enables us, if required, to load a rectangular target-cell into an also rectangular mixing chamber, thus minimizing the amount of background material in the target region.

The design goals are:

- $\dot{Q}(T = 200mK) \cong 100mW$
- $\dot{Q}(T = 100mK) \cong 30mW$
- $T_{min} \leq 50mK$
- target size 100 cc
- flexibility in target geometry
- possibility of dummy target

The magnet system.

The magnet system consists of 3 superconducting magnets, a high homogeneity 5 T polarizing coil in a room temperature bore cryostat and a twofold holding coil system, integrated in the dilution refrigerator cryostat. The holding coil system consists of a split pair magnet and a saddle coil magnet, wound on a common former and contained in a vacuum tight can, providing resp. the vertical and horizontal holding fields.

A linear combination of these fields together with the rotatability of the cryostat around the vertical axis allows virtually any quantization direction in space, only limited by the 2 pillars of the magnet support.

Moreover, the twofold coil system allows a quick polarization reversal. Factory tests have shown a reversal time of 12 min.

The characteristics of the magnet system are:

- Polarizing Coil
  - 5 Tesla coil in room temperature bore cryostat
  - $B \parallel$ z-axis
  - $\Delta B/B \simeq 10^{-4}$ over 100 cc cube
  - $\dot{B}/B \simeq 10^{-6}$ persistent mode
- Vertical Holding Coil
  - 1.2 Tesla superconducting split pair coil
  - B_{design}=0.8 T $B\parallel$ z-axis
  - field constant = 1.105 mT/A
  - $\Delta B/B \simeq 15.5$ % over target volume
  - L= 3.87 H
- Horizontal Holding Coil
  - 1.1 Tesla superconducting saddle coil
  - B_{design}=0.8 T $B\perp$ z-axis
  - $\Delta B/B \simeq$15 % over target volume (computed)
  - L= 0.8 H
- Opening angles of the holding coil system (see fig. 2)
  - $2\alpha = 145°$

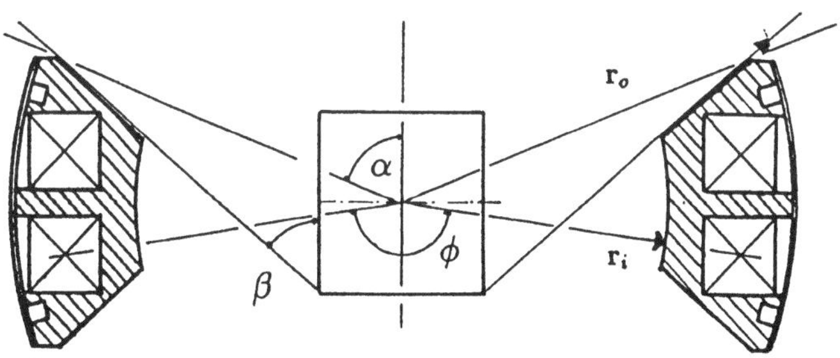

Fig. 2: *Horizontal cut through pillars of holding coil system showing opening angles.*

- $2\beta = 100°$
- $\phi = 165°$

- Dimensions of the holding coil system
  - r_i= 73 mm
  - r_o=110 mm

The pumping system.

The pumping and gas-handling system are designed for a circulation rate of $\dot{n}_3$=30 mmole sec^{-1}
The pumping system comprises of a series combination of Roots blowers Balzers WKP 4000, Alcatel RSV 1000, RSV 300B and ADP 80, all equipped with canned motors and compressing completely oil-free.

References

[1] P. Roubeau, J. Vermeulen, Cryogenics **11** (1971), 478-482.

[2] B. van den Brandt, J.A. Konter, S. Mango, NIM **A 289** (1990), 526-531.

Spin Physics in CHAOS

E.L. Mathie[1], *B. van den Brandt*[2], *J.A. Konter*[2], *and S. Mango*[2]

[1]University of Regina, Regina, Saskatchewan, Canada
[2]Paul Scherrer Institute, CH-5232 Villigen PSI, Switzerland

This paper is not really about the chaotic state of spin physics, but about our efforts to extend our studies of pion physics with an emphasis on applications of spin through a new detector facility at TRIUMF, which we call CHAOS (*Canadian High Acceptance Orbit Spectrometer*). To date much of the experimental program in pion spin physics at TRIUMF has utilized arrays of counters together with polarized protons, deuterons and ^{13}C targets.

Many experiments require a larger solid angle and higher resolution than what is typically permitted with arrays of scintillation counters. The TRIUMF QQD pion spectrometer is only suitable for low energy pions and not suitable for any low rate experiments.

The solution to these problems at TRIUMF is CHAOS, which is a magnetic detector with vertical cylindrical geometry, enabling the simultaneous measurement of nearly whole angular distributions. The detector will have an horizontal acceptance of 260° and a vertical acceptance of 18°. The magnetic field orientation is vertical and will have a strength of up to 1.4 T which leads to a typical dP/P of 1% FWHM. Particle trajectories will be determined with four planes of wire chambers and particle time of flight and energy loss information will be available from plastic scintillators. The device is shown schematically in Figure 1. CHAOS is being constructed by a collaboration which includes TRIUMF; the Universities of British Columbia, Victoria, Regina, Colorado, Trieste, and Karlsruhe.

SPIN PHYSICS of the NUCLEON with PIONS

Measurements of πp elastic and inelastic scattering with a polarized proton target will lead to improvements in the πN phase shifts, contribute to the πN Σ term debate and thus to the strange quark content of the proton and the proton spin.

In their recent papers, Locher[1], Sainio[2] and Gasser[3] have discussed how low energy pion scattering can be used as a testing ground for chiral perturbation theory. In particular the σ term, which can be related to the πN S-wave scattering length, is a measure of the size of explicit chiral symmetry breaking of QCD due to the quark mass term in the QCD Lagrangian. The Σ term can be related to the strange quark content of the proton in the framework of chiral perturbation theory and is linked to σ by dispersion relations: $\Sigma = \sigma + 5MeV$, where $\sigma = 35MeV/(1-y)$ and, $y = 2 < p \mid s\bar{s} \mid p > / < p \mid \bar{u}u + \bar{d}d \mid p >$

From low energy pion scattering data one presently finds $\Sigma = 65MeV$, which leads to the conclusion that y is about 0.4. Calculations of the effect of different values of Σ on the analyzing power for π^-p are shown in figure 2 [4]. Clearly this πN experiment addresses very important fundamental aspects of the structure of the nucleon.

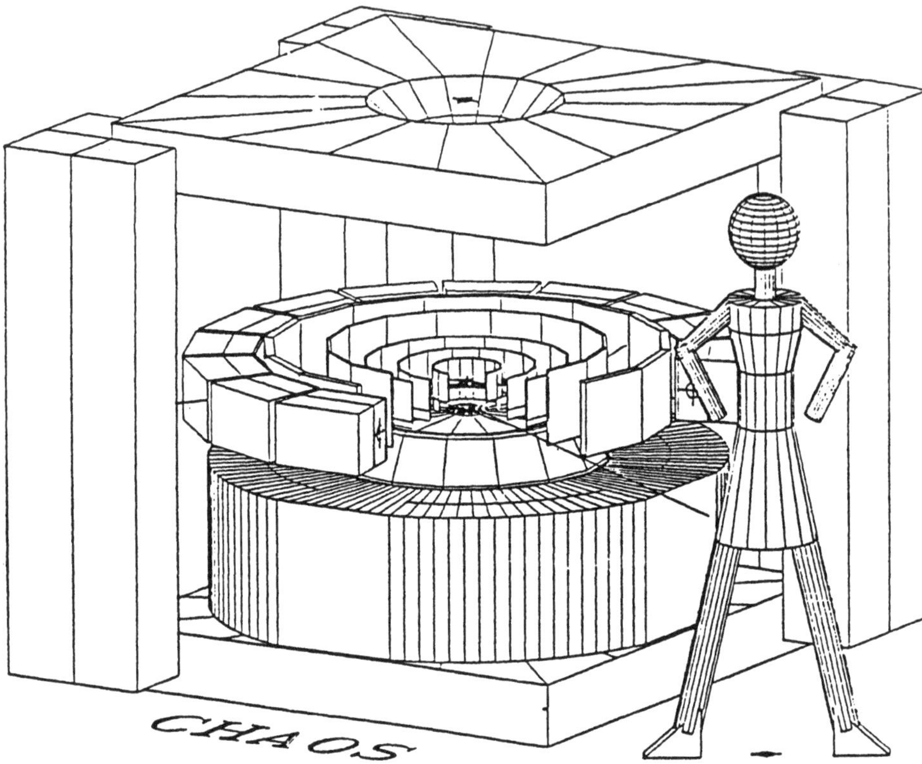

Figure 1: Schematic View of CHAOS

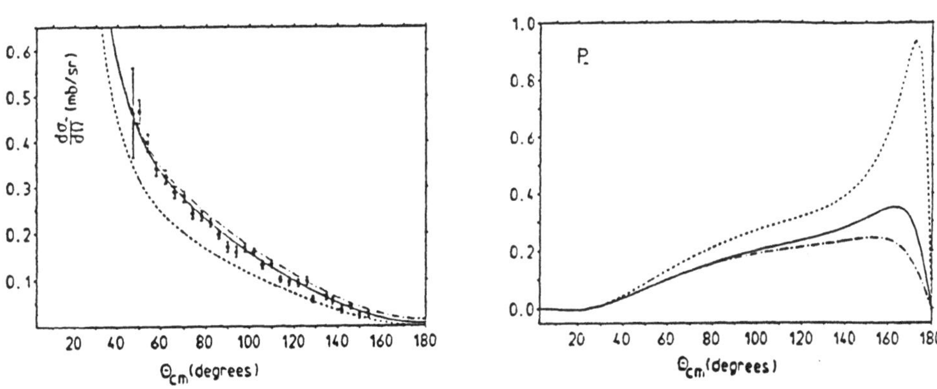

Figure 2: Cross sections and analyzing power for negative pion elastic scattering at 51 MeV, solid line corresponds to Σ=59MeV and dotted line corresponds to Σ=39 MeV [4].

SPIN PHYSICS of the FEW BODY πNN SYSTEM

The πNN system remains one of (if not the) optimum testing grounds for conventional theories (without explicit quark dependence). The theoretical results are challenged by more complete empirical descriptions of πd elastic and the πd to pp reaction (via partial wave expansions); and with more precise analyzing power results from πd breakup.

Many who work with the πNN system believe that lurking in the shadow of the conventional understanding are indications of quark degrees of freedom, however the absence of calculations with explicit quark dependences means one must be content to very carefully evaluate the theoretical results based upon only meson and nucleon degrees of freedom.

The modern theoretical descriptions reflect the coupling nature provides between the various reaction channels (as they should) but even so do not agree at a precision level with our empirical understanding based on experiments.

The two body reaction channels πd elastic and πd absorption are becoming much better represented by partial wave analyses etc. In one CHAOS experiment with a polarized deuteron target it is hoped to dramatically improve the partial wave solutions for the elastic reaction.

The three body final state of the πd breakup reaction is particularly well suited to experiments with CHAOS, where the large phase space is much better covered than in earlier experiments, polarized and unpolarized [4] alike.

SPIN PHYSICS of NUCLEI with PIONS

Polarized Nuclei have been used in only a few pion experiments very recently. One expects to determine subtle aspects of the pion nuclear interaction, and more specifically investigate the polarized Δ-core spin-orbit interaction.

POLARIZED TARGET FOR CHAOS

The magnetic field in CHAOS is not sufficiently high or uniform to enable dynamically polarizing a target but is well suited to act as a holding field for a frozen spin target.

CHAOS is based upon an existing magnet which has a 16cm bore through yoke and pole pieces, while a large polarizing solenoid available at TRIUMF has a 12 cm room temperature bore. A polarizing solenoid cannot be positioned inside CHAOS for polarizing and then removed because of the geometry constraints imposed by proportional chambers etc., hence the target must be polarized outside of CHAOS and moved into position in frozen spin mode which means one must supply a holding field which goes with the target dilution refrigerator.

In the current plan we envisage placing the solenoid above the yoke hole of the CHAOS magnet, polarize the target in it, then pull down and slide away the solenoid and lower the target in frozen spin mode into its operational position between the pole pieces of CHAOS. A new target projected for use at PSI for University of Karlsruhe experiments there will be made compatable with the CHAOS geometry and used at TRIUMF as well. This cryostat, based on the PSI dilution refrigerator-compact cryostat design [6], will incorporate a split pair superconducting holding magnet (see figure 3). The problems of target transport into CHAOS, the interaction of target with the CHAOS magnet environment, and modifications to pumping equipment are being addressed at TRIUMF.

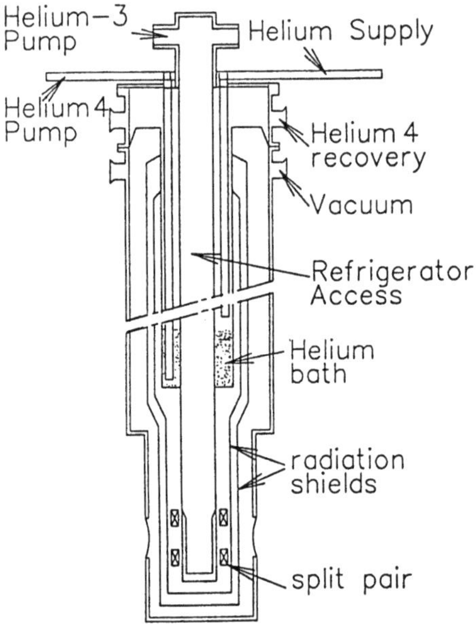

Figure 3: Modified PSI cryostat for use in CHAOS, showing location of superconducting split pair magnet

SUMMARY

A variety of interesting spin physics will be pursued at TRIUMF making use of pion beams, a new polarized target and the CHAOS facility.

Eventually we hope to make use of both the target and CHAOS at KAON.

Finally the authors would like to acknowledge discussions on this topic with D.Healey, P.Delheij, R.Tacik, G.R.Smith and M.P.Locher.

References:

[1]M.P.Locher, "Chiral symmetry, the elementary πN interaction and the πNN system", PSI preprint PR-90-19

[2]M.Sainio,Proc.Symp. on πN and NN Physics, Leningrad,(1989)219.

[3]J.Gasser,Proc.Symp. on πN and NN Physics, Leningrad,(1989)179.

[4]M.P.Locher and M.Sainio, Czech.J.Phys.B39(1989)943.

[5]E.L.Mathie, et al., Phys.Rev.C41(1990)193.

[6]B.van den Brandt, J.A.Konter, S.Mango, Nucl.Inst.Meth.A289(1990)526-531.

The New Polarized Target Facility at the Kharkov Institute of Physics and Technology

V.P. Androsov, A.A. Belyaev, V.A. Get'man, A.P. Dzyubak, I.M. Karnaukhov, A.A. Lukhanin, Yu.P. Popkov, P.V. Sorokin, E.A. Sporov, Yu.N. Telegin, and I.A. Tolmachev

Kharkov Institute of Physics and Technology,
Ukrainian SSR Academy of Sciences, SU-310108 Kharkov, USSR

In Kharkov a new polarized target facility has been constructed to carry out nuclear physics experiments and do development work on polarized target materials.

The parameters of the target shown in Fig. 1 are:

Magnetic field	2.7 T
Magnetic field homogeneity in the volume $30 \times 30 \times 30 \times mm^3$	2.0×10^{-4}
Polarizable sample size	$20 \times 20 \times 20 \times mm^3$
Resonator volume	$40 cm^3$
SHF frequency	72 GHz

With the present facility polarization values obtained for free protons P_p and deuterons P_d were: in 1,2-propanediol $P_p \simeq 76 - 80\%, P_d \simeq 30\%$ and in irradiated ammonia $P_p \simeq 70\%, P_d \simeq 37\%$.

In order to cool down samples of target materials a combined cryostat has been built providing the following operating regimes:

- ^{3}He - ^{4}He dilution in the temperature range 0.05 - 0.5 K with the cooling capacity up to 30 mW at T = 0.3 K;
- ^{3}He pumping out (W $\simeq$ 250 mW, T = 0.5 - 1 K);
- ^{4}He pumping out (W $\simeq$ 1.5 W, T = 0.8 - 4 K).

The cryostat can be used as a ^{3}He or ^{4}He evaporation cryostat or as a ^{3}He - ^{4}He dilution refrigerator. It consist of the continuous and discrete heat exchangers for ^{3}He - ^{4}He cooling and condensation, the run valve, the ^{3}He evaporation chamber and the dilution chamber. The discrete heat exchanger has a larger surface and is made in the form of the rings of porous copper sintered on a thin-walled stainless steel tube. The tests showed that the given construction provides the steady temperatures in the region 0.05 - 4 K with stability of 0.001 - 0.02 K/hour as well as a quick sample replacing (less than half an hour).

To study the spectral characteristics of target materials an insertion has been made comprising the measuring SHF-modulus of electron-spin-resonance (ESR) and nuclear-magnetic-resonance (NMR) spectrometers. The resonant cavity (Fig. 2) consists of the open reflecting resonator and the sample container placed inside of the resonator. The resonator is tuned by means of the moving spherical mirror. The container is made in the form of a rectangular coil a double diffraction lattice, and operates as a RLC-circuit at nuclear transition frequency. Depending on the experimental conditions the diffraction lattices can be either deposited on dielectric plates (fluids being investigated) or winded on a nonconducting frame (the material under investigation is frozen onto the diffraction structure, e.g. ammonia). The resonant cavity is constructed so that it can be used to observe the spectra of both the ESR and NMR, to study their relaxation characteristics and to investigate the processes of dynamic nuclear polarization.

The samples of alcohols doped with paramagnetic complex Cr(V)(2-hydroxil-2-methyl-butyric-acid and 1-hydroxil-cyclohexil-carboxylic-acid) as well as irradiated ammonia have been investi-

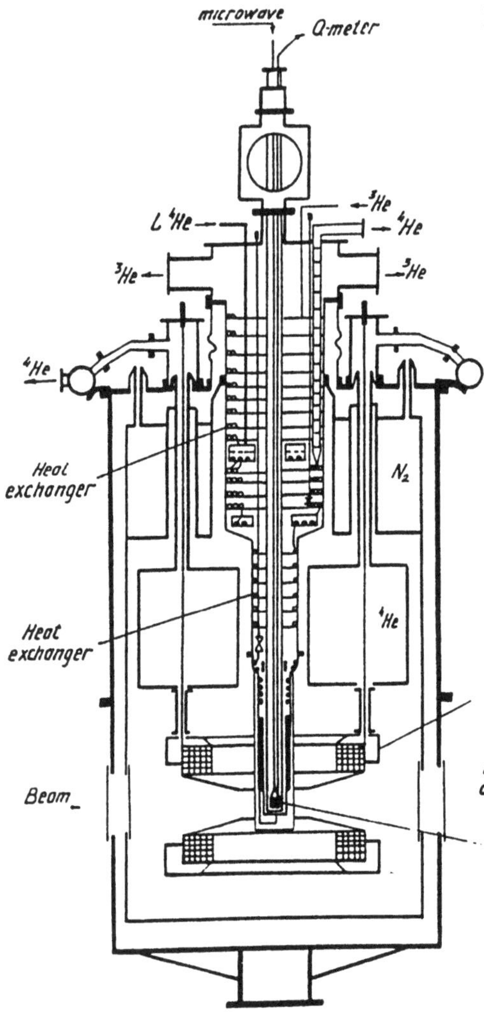

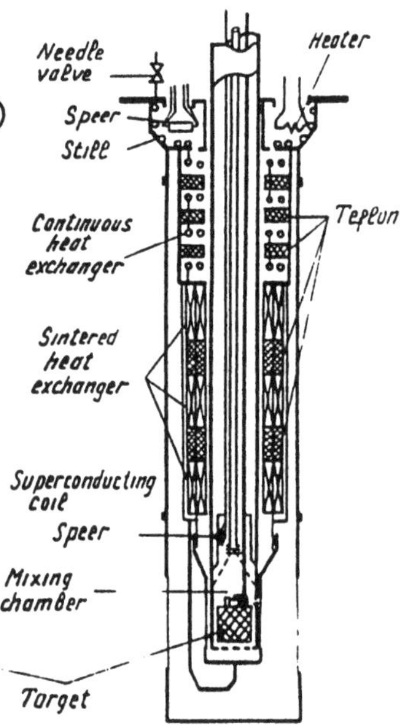

Fig. 1 Polarized target with ^3He-^4He dilution cryostat

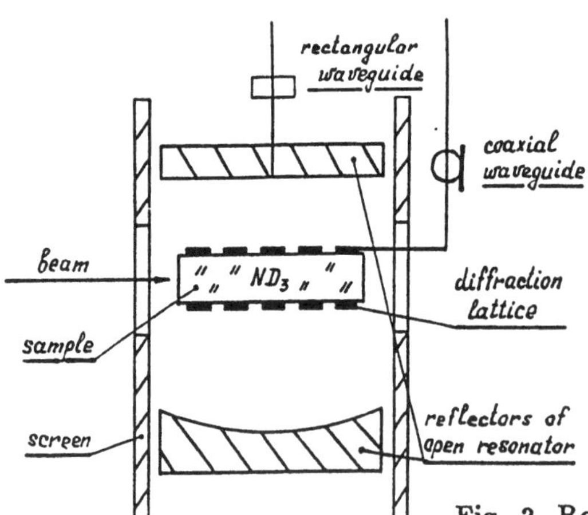

Fig. 2 Resonant cavity of the ESR-spectrometer

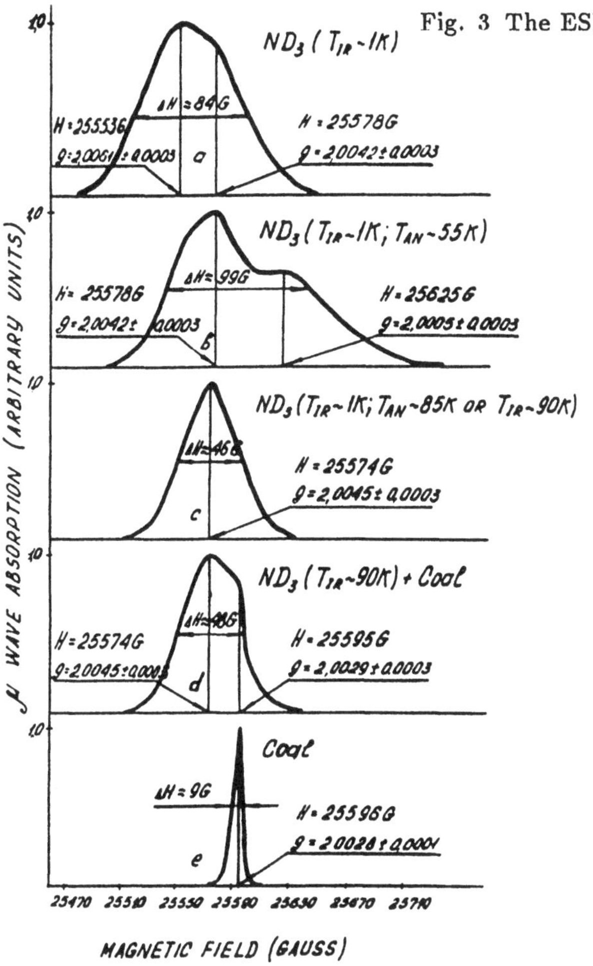

Fig. 3 The ESR-spectra in irradiated ammonia.

gated. It has been found that all the materials with the complex Cr(V) being studied possess a high stability, similar characteristics and give high polarization values. By incorporating the crown-ether into these complexes one can increases their stability by a factor of 2-3 [1]. The disadvantages of the chemically doped materials are their relatively low radiation resistance. Detailed investigations have shown that irradiated ammonia has no disadvantages compared to materials with complexes Cr(V) [2].

The spectral characteristics of deuterated ammonia irradiated at low temperatures have been studied to understood the high radiation resistance of the polarization [3]. Three types of paramagnetic centres were observed: $\dot{N}D_3^+, \dot{N}D_2$ and e-traps with corresponding g-factors 2.0061 ± 0.0003, 2.0042 ± 0.0003 and 2.0005 ± 0.0003 (Fig. 3). During successive annealing the radical $\dot{N}D_3^+$ is the first annealed, then e^--traps and at T ~ 130 K the radical $\dot{N}D_2$. Under high temperature irradiation (T ~ 90 K) only the radical ND_2 accumulates.

The obtained target materials have been employed in pion photoproduction experiments [4,5]. In the following table the measured values of T-asymmetry for π^--meson production on polarized deuterons of irradiated ammonia at an average polarization value $P_d^{\pm} \simeq 35\%$ are given. In the nearest future we intend to investigate systematically the polarization observables in pion photoproduction on nucleons and to study ^6LiD as a possible target material.

θ_π^*	E_γ, MeV	T
	300	0.601 ± 0.102
	320	0.695 ± 0.115
90°		
	350	0.848 ± 0.038
	400	0.760 ± 0.060
	300	0.480 ± 0.060
120°	350	0.750 ± 0.090
	400	0.440 ± 0.067

References

1. Vertiy A.A., Shestopalov V.P., Popkov Yu.P., et al. Investigations of the dynamic polarization of protons in 1,2-propylenglycol with a stable radical Cr(V). DAN, USSR, 1981, v. 260 (5), p. 1093-1095.

2. Proc. 4th Int. Workshop on Polarized Target Materials and Techniques, Bad Honnef (FRG), 1984.

3. Belyaev A.A., Get'man V.A., Dzyubak A.P. et al. Polarisation evolution in ND_3. 8th Int. Symp. on High Energy Spin Physics, Minneapolis (USA), 1988, v. 2, p. 1346-1351.

4. Get'man V.A., Gorbenko V.G., Derkach A.Ya. et al. Positive pions production from polarized protons by linearly polarized photons in the energy range 280-420 MeV. Nucl.Phys., 1981, v. B188, p. 397-415.

5. Belyaev A.A., Get'man V.A., Gorbenko V.G. et al. Experimental studies of the Σ, T, P polarization parameters and the $\gamma p \to p\pi^o$ reaction multipole analysis in the first resonance region. Nucl. Phys., 1983, v. B213(2), p. 201-222.

A Cryostat for Sample Irradiations in Liquid Argon

V.A. Get'man, A.A. Belyaev, A.P. Dzyubak, I.M. Karnaukhov, A.A. Lukhanin, N.V. Sorokin, E.A. Sporov, and I.A. Tolmachev

Kharkov Institute of Physics and Technology,
Ukrainian SSR Academy of Sciences, SU-310108 Kharkov, USSR

Irradiated ammonia having relatively high polarizable contents and high resistivity to radiation damage is one of the preferable materials for polarized targets [1]. The irradiation can be performed either in a cryostat of a polarized target ("low temperature" irradiation) at $T_{irr} \approx 1K$, or in a medium of liquid argon ("high temperature" irradiation) at $T_{irr} = 90K$.

Previously we have published the results of studies on the nuclear polarization in samples obtained by the "low temperature" irradiation method [2-3]. In the given paper we describe the "high temperature" irradiation technique and present the results of the polarization measurements in irradiated deuterated ammonia samples. In Fig.1 shown is an argon refrigerator. It is a unit consisting of 2 parts, an argon liquifier and an irradiation cell. The refrigerator operates as follows. The dewar (5) is filled with a liquid nitrogen. The argon is supplied through the tube (11) from the buffer vessel ($V = 0.5\ m^3$) under pressure 1-1.2atm. The cooling and subsequent condensation take place in the heat exchangers (6). The condensed argon flows down to the hotwell(4), then through the tube (3) enters into the irradiation cell (1). The evaporated argon enters to the gas collector (7) and condensates again in the heat exchangers. By this means the steady circulation is assured. The starting cooling and condensation stage take half an hour. To the end of this stage the pressure in the cryostate reaches 0.72 atm.

For the sample loading one uses a unit which consists of the lock chamber (9) and the valve (8). The loading procedure is performed as follows. The container ($V = 30\ cm^3$) preliminarly filled with solid ammonia and placed in the bath with N_2 is inserted into the lock chamber and is fixed to the movable rod (10). After the lock chamber has been pumped out to pressure 10^{-1}mm Hg

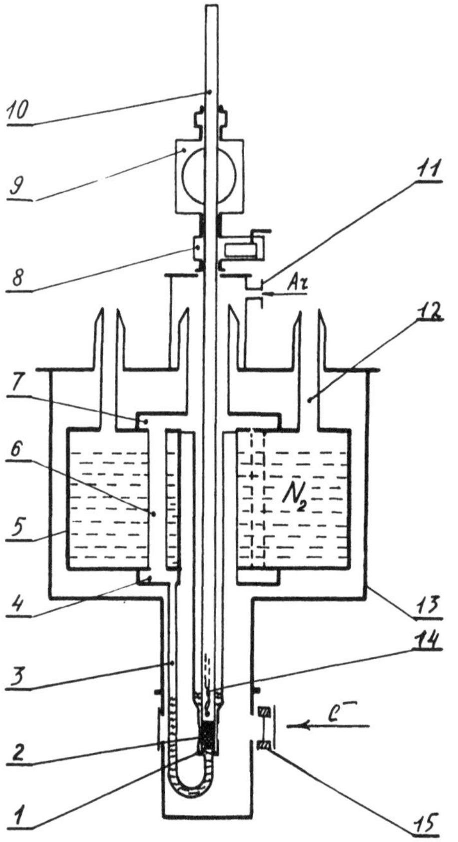

Fig.1. Argon refrigerator. 1 - appendix; 2 - container; 3 - liquid argon feeding tube; 4 - hotwell, 5 - N_2 dewar; 6 - heat exchangers; 7 - gas collector; 8 - valve; 9 - lock chamber; 10 - rod; 11 - argon gas feeding; 12 - liquid nitrogen feeding tube; 13 - vacuum chamber, 14 - thermocouple; 15 - beam monitor

it is displaced in the irradiation cell. The temperature of liquid argon is controlled with the thermocouple (14) being in immediate contact with the sample. A total number of incident electrons through the target is measured with the monitor (15) mounted in the vacuum jacket at 5 mm from the container. Irradiations are performed with beads of slowly frozen ammonia of a size 1-3 mm.[4] At $T \approx 130K$ the full annealing of the paramagnetic centers formed in ammonia takes place, that presents rigid requirements to the procedure of sample reloading. The measurements showed that up to the level of the gas collector the temperature is lower than 11°K. Hence in the cryostat the annealing of para-

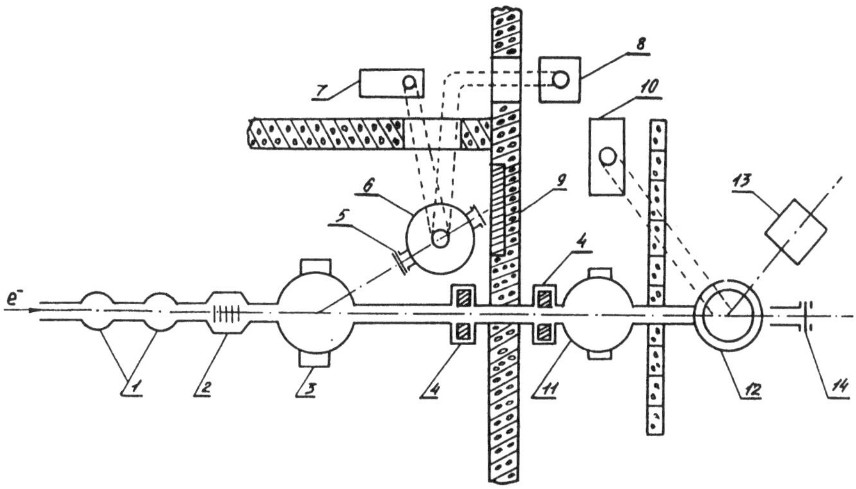

Fig.2. Experimental technique for irradiations of materials in liquid argon. 1 - correctors; 2,5,14 - beam monitors; 3 - deflecting magnet; 4 - photon collimators; 6 - argon cryostat, 7,10 - control panels; 8 - automatic liquid N_2 filling; 9 - radiation shielding; 11 - clearing magnet; 12 - polarized target cryostat; 13 - Faraday cup

magnetic centers can occur only in the lock chamber where the container should be situated not longer than 2-3 min. The anneling of paramagnetic centers during loading stage was not observed experimentally.

Irradiations with 500 MeV electrons were performed at the Kharkov 2GeV linac. The experimental technique is shown in Fig.2. The argon cryostat (6) is mounted after the deflecting magnet (3). The panel with the remote control and monitor units (7) and the unit of the automatic filling (8) of the nitrogen bath are placed behind the radiation shielding (9).

The uniform irradiation of the sample was provided by the electron beam scanning on the target. The scanning have been performed by varying the current in the beam position corrector (1) (vertically) and in the magnet (3) (horizontally). A sufficient irradiation dose $\Phi = 10^{17}\ e^-/cm^2$ at electron current $\sim 2\,\mu A$ was reached after 26-30 hours. After irradiation the ammonia crystals were dark-violet that was indicative of a sufficient number of paramagnetic centers [1]. The samples were stored during some months in the dewar with liquid argon. The color of the samples did not change in this time.

For samples of deuterated ammonia irradiated in liquid argon the deuteron polarization values up to $P^{\pm} = 36\,\%$ were reached. The polarization build-up time $T_{0.7}$ (the time required for the polarization to reach 70% of the peak value) was 60 min that exceeded essentially the value obtained for the samples irradiated under "low-temperature" conditions. After additional irradiation of the target material ($T_{irr} = 0.2K$) to a dose $\Phi = 2 \times 10^{13}$ eq. quanta/cm^2 the initial polarization value increased (from 36% to 38%) and the polarization build-up time $T_{0.7}$ decreased (from 60 to 15 min).

Experience on the irradiated ammonia allows us to conclude that the "low temperature" irradiation ($T_{irr} \approx 1K$) with a subsequent annealing at $T_{ann} = 85K$ and the irradiation in a liquid argon medium ($T_{irr} = 90K$) provide the production of the target material with similar properties.

References

1. Proc. of the 4-th Intern.Workshop on Polarized Target Materials and Technique. (Bad Honnef, september 3-6, 1984), PIUB-1984, 210 p.
2. Belyaev A.A., Get'man V.A., Dzyubak A.P. et al. Dynamic proton and deuteron polarization in irradiated NH_3 and ND_3 at 1K and 0.5 K. VII Intern.Symp. on High Energy Spin Physics.(Protvino, september 22-27, 1986), Serpukhov, 1987, v.2, p.221-225.
3. Belyaev A.A., Get'man V.A., Dzyubak A.P. et al. High deuteron polarization in ammonia (ND_3). Voprosy atomnoj nauki i tekhniki. Ser. Common and nuclear physics, 1988, Issue 1(41), p.105-107. (In Russian).
4. Belyaev A.A., Dzyubak A.P., Lukhanin A.A. et al. Production of samples from ammonia for polarized targets. Problemy yadernoj fiziki i kosmicheskikh luchej. 1987, N 28, p.45-47.(In Russian)

A Multipurpose Q-Meter with Switching Phase Shift Detector

N.A. Bazhanov and A.I. Kovaljov

Leningrad Institute of Nuclear Physics,
Gatchina, SU-188350 Leningrad Distr., USSR

The use of Q-meters with amplitude sensitive detectors leads to the following problems:
1) The NMR signal is overlayed by a large background
2) The NMR signal is a mixture of absorption and small fraction of dispersion

To solve this problems, we have designed a Q-meter with a serial resonance circuit and a phase shift detector. If the cable length l (fig.1a) is zero we have:

$$U_{psd} = (ReZ(\omega, X)) \cos \psi + (ImZ(\omega, X)) \sin \psi$$

where the input impedance Z of the circuit is

$$Z(\omega, X) = i\omega L_0(1 + 4\pi\eta(\chi' - i\chi'')) + r + 1/i\omega C \simeq$$
$$\simeq r[(1 + X''(\omega)) + i(X'(\omega) + 2Q\Delta\omega/\omega_0)],$$

ψ is the phase shift between the reference and the input voltage of the PSD. $\chi' - i\chi''$ the nuclear susceptibility, $X'' = 4\pi\eta\chi''$ and $X' = 4\pi\eta\chi'$ where η is the filling factor. For $\psi = 0$ the NMR signal is

$$S(\omega) = X''(\omega).$$

and the base line becomes a horizontal straight line. If $\psi \neq 0$ the base line will have a saw tooth shape like it is shown in Fig. 1c. The voltage difference between $U1$ and $U2$ is used as input for the Automatic Phase Tuning system (APT) (see fig.1c) designed by Dr.V.Poljakov from LNPI [1] which keeps $\psi = 0$.

Note that for a linear amplitude detector under the same condition we have

$$S(\omega) = X'' + X'^2/2 - X'/Q$$

The cable connecting the Q-meter with the coil in the cryostat complicates this simple picture. To understand the properties of the Q-meter with cable, we have to carry out computer simulations. The Q-meter diagram ($l \neq 0$) which is used for the calculations is shown in fig.1a, the Q-meter parameters are $L_0 = 0.14\,\mu H$, $r_c = 2\,\Omega$, $R_a = 5(or 1.5)\,k\Omega$, $C_a = 0.5\,pF$, $l \cdot \sqrt{\varepsilon} = 2.9\,m$, $\rho = 50(or 150)\,\Omega$, $\omega/2\pi = 103\,MHz$ for the proton mode and $\omega/2\pi = 16\,MHz$ for the deuteron one. The width of frequency sweep is 400 kHz for protons and 600 kHz for deuterons.

The dependence of phase from frequency is assumed to be linear

$$\psi = \psi_0 + \psi_1 \Delta\omega/\omega_0,$$

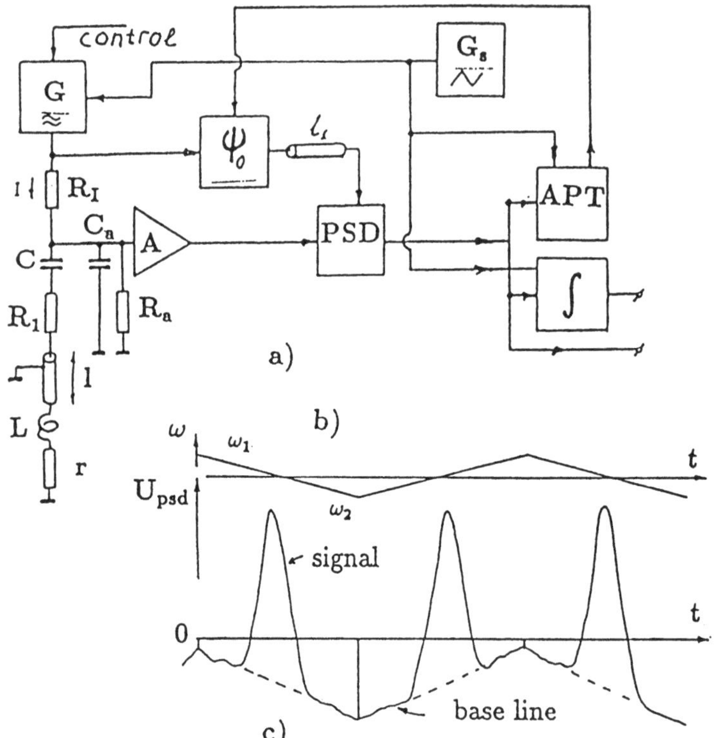

Fig.1.a - block diagram of the Q-meter for mathematical simulation: G - RF generator, G_s - sweep voltage generator, Ψ_0 - controled phase rotator, A - RF amplifier with input resistance R_a and capacitance C_a, PSD - phase shift detector, APT - automatical phase tuning block, $\int$ - integrator, L - MNR coil $L = L_0(1 + 4\pi\eta(\chi' - \chi"))$, R_1 - resistor for Q factor tuning. b - frequency sweep time diagram. c - NMR signal + base line (PSD output), u_1 and u_2 voltage dated for APT.

ψ_0 is the phase for the center frequency, $\psi_1 = \partial\psi\partial\omega$ is a constant connected with the cable length l_1 as $\psi_1 = \omega_0 l_1/c$. The main results of the calculations are

1)The frequency dependence of the circuit resistance $r_e = ReZ(\omega, 0)$ is linear for the first order,

2)The APT system will produce $\psi \neq 0$ to get the voltage difference $U1 - U2$ to zero value. This gives some admixture of dispersion to the NMR signal.

3)By choosing l_1 properly the second order contribution to the background can be compensated.

4)In our present setup the APT leads (in the case of protons) to distortions of the NMR line shape caused by admixture of dispersion. To reduce them it is necessary to increase the cable wave resistance to ($\rho \sim 130 - 150\,\Omega$).

5)The distortions mentioned under 4) do not influence the square under the signal. So the corretions for the polarization measurement of protons are in the order of 0.1%.

6)For deuterons $\tan\beta l \neq 0$ and $\psi \neq 0$. Here it is desirable to increase the wave resistence too. Corrections for determining polarization from the line shape are in the order of 2% for $\rho \simeq 150\,\Omega$.

Phase shift detector Q-meters have been proposed earlier, there is a publication in 1980 [2]. The development of fast electronic components gave us the possibility to return to this device. We have built a PSD Q-meter with following features:

1)By using a switch based PSD we have free choice of frequency.
2)The frequency dependences of the phase shifts are the same for the NMR signal and the reference chanel.
3)The RF amplifier has a frequency band width from 3 to 200 MHz.
4)The NMR coil is the same for all possible nuclei.
5)The Q-meter has two oscillators, so it is possible to work with two different kinds of nuclei at the same time.

In a detector setup like we described it above the phase noise from the oscillator is cancelled out [3].

Fig. 2 shows some results of measurements with the new Q-meter. A proton TE signal with base line (background) is shown in fig.2a. Fig. 2b,c show two deuteron spectra of propandiol which have base lines with different order of compensation.

The PSD diagram is shown in fig.3a. It has been proposed in [4], we were successful to increase the operation frequency to 130 MHz. PSD consist of four independent switches. They require two opposite phase rectangular (meander)

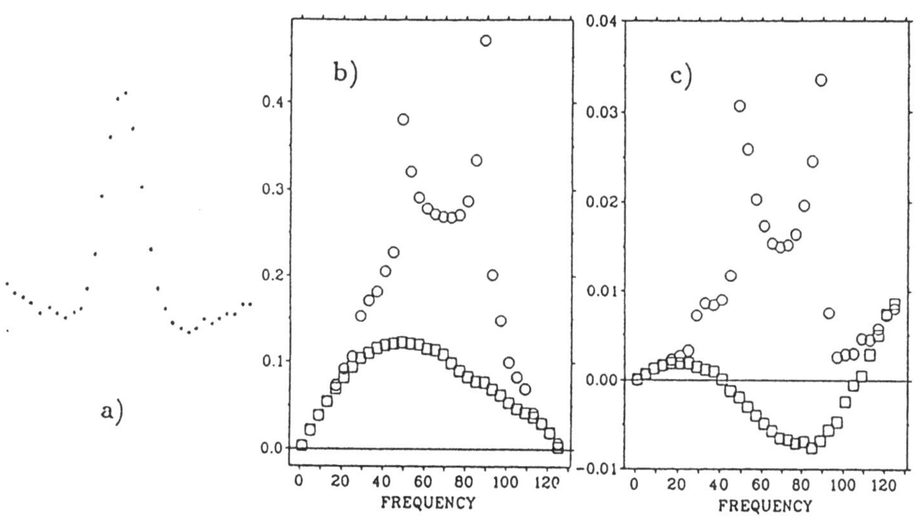

Fig.2.a - TE proton spectrum with base line. b - Deuteron spectrum with base line (background) that has no compensation of the second order terms. $P_D = -40\%$, propandiol fully deuterated. c - The base line is compensated by choosing l_1 properly. $P_D = -20\%$, the sample is 50 % deuterated propandiol.(The signal is four times smaller than in the figure b).

waveform reference voltages (are located at input1 and input2) and two NMR signals inversed on phase (inputs H.RF). The switches produce two symmetrical (opposite sign) low frequency voltages (outputs 1 and 2) which then are subtracted. Such way allows to improve the stability of PSD operation. The reference voltages have a value of $4.5\,V$ peak to peak and are produced from sinusoidal generator one. The maximum input signal is $0.3\,V$.

The follower with high input resistence and low noise is on fig 3b. We used a Darlington transistor and following feed-back. Operate frequency $f_{op} \leq 200\,MHz$. $Z_{input} \geq 5\,k\Omega$.

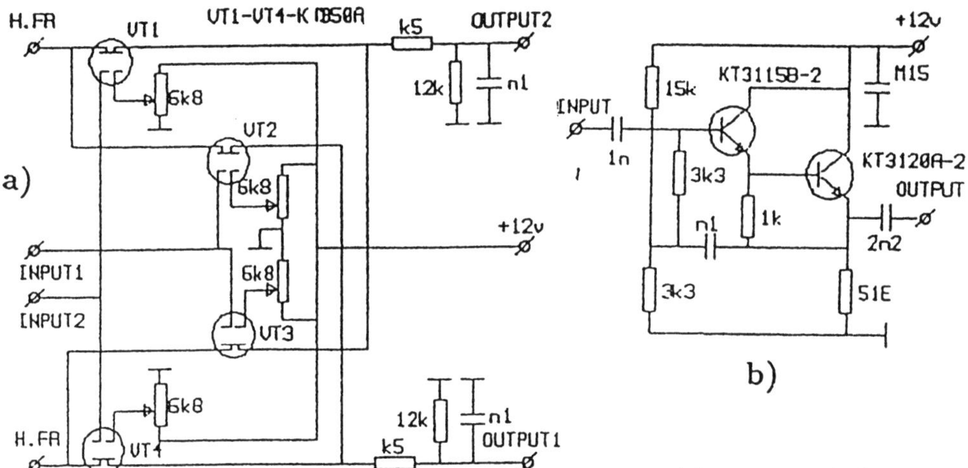

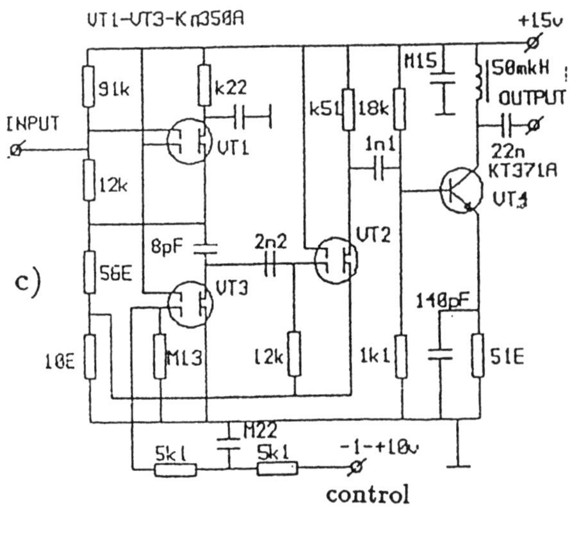

Fig.3. a - switch phase shift detector, b - input high resistence follower, c - controled phase rotator.

The controled phase rotator [5] is shown on fig.3c. We increased its operating frequency more then 150 MHz. The range of change of the phase is140°. Input signal is $0.3\,V$.

REFERENCE

1.E.I.Bunjatova,Yu.F.Kiseljov,N.A.Kozlenko et al. Polarized proton target with horisontal spin. PTE^1, 1988, #6, pp.36-39

2.D.Gifford. A Q-meter with RF Phase Sensitive Detector. Proc. the 2-nd Workshop on Polarized Tardet. RL-80-080. 1980.

3.J.A.Pierce. Itercontinental Frequency Comparison by Very Low-Frequency Radio Transmission. Proc. IRE. # 6, pp794-803, 1957.

4.Ch.O.Kadmar,S.A.Musaev,E.Yu.Salaev. High Presision Phase Sensitive Detector PTE^1, 1982, #5, p.103.

5.V.I.Kostenko,O.N.Demjanenko. Controled Phase Rotator. PTE^1, 1975, #5, p.138.

[1] - Soviet Journal:"Pribory i Tekhnika Experimenta"

Design Considerations for Variable-Frequency Pulsed NMR Deuteron Polarization Measurement

T.O. Niinikoski[1], *H. Garchette*[1], *H. Johansson*[1], *A. Rijllart*[1], *L. Klostermann*[2], *and S. Dhawan*[3]

[1]CERN, CH-1211 Geneva 23, Switzerland
[2]NIKHEF, NL-Amsterdam, The Netherlands
[3]Yale University, New Haven, CT 06511, USA

INTRODUCTION

The power of Variable Frequency Pulsed NMR (VFP NMR) [1] has been demonstrated in the measurement of wide-line NMR spectra, specifically those of deuterons in solid materials. These demonstration experiments use about 90° pulses, but they may not enable simple extrapolations to be made for signals obtainable with small tipping angles, compatible with low saturation as is required in the calibration and measurement of nuclear polarization in polarized targets.

In this paper we shall discuss briefly the principles of polarization measurement, and some of its circuit theoretical and solid-state physics aspects. Comparing with the CW-NMR techniques, the pulsed NMR methods are seen to have the advantage that the Thermal Equilibrium (TE) calibration signal measurement is free of oscillator noise. Furthermore, the TE measurement would be to a larger extent free of circuit drift as well. The desirable technical specifications of the RF part will be outlined, together with considerations of the digital treatment of the converted free precession signals. It is proposed to analyse the free precession signals using fast Digital Signal Processors (DSP), in order to obtain the real and imaginary parts of the RF susceptibility of the target material.

POLARIZATION MEASUREMENT

The accurate measurement of nuclear spin polarization in polarized solid targets is based on the Curie law, which enables one to calculate the polarization of a spin system in thermal equilibrium with the lattice of the material. Once having measured the spin-lattice relaxation time (also called longitudinal relaxation time) and ascertained the absence of effects related with molecular spin conversion (such as are well known in H_2 and CH_4, for example), one can accurately determine the errors due to deviations from thermal equilibrium (TE), mainly caused by the saturation by the usual CW NMR measurement. The discussion of the further details of the calibration error analysis is beyond the scope of this paper.

Knowing the target polarization at a given stable temperature, usually around 0.5 - 1.2 K, allows us to calibrate the sensitivity of the CW Q-meter circuit or any other potential circuit, which is capable of measuring the imaginary (absorption) part of the RF susceptibility of the target due to the given nuclear spin species. This calibration is reliable only if the the circuit is linear, i.e. its output responds linearly to the variation of the susceptibility of the target material.

The CW Q-meter circuit has two main sources of non-linearity. The first one is purely circuit-theoretical, and it is easy to show that best linearity is obtained using the series-tuned circuit very loosely coupled to the oscillator and the amplifier [2]. The cost of this is a low level of the signal, and a good coupling of the oscillator noise and parasites into the amplifier.

Compared with protons, the measurement of wide-line NMR signals, such as those of deuterons or other nuclei with spin 1 or higher having an electric quadrupole moment, is much more difficult under the TE calibration conditions. The static nuclear spin susceptibility of the deuterated material is 42 times lower than that of is undeuterated counterpart, assuming equal spin densities, fields and temperatures. The absolute width of the deuteron absorption line, which is quadrupole broadened due to the electric field characteristic of the C-D bond, is almost ten times larger that the width of the proton line. Because of this, the signal-to-noise ratio of the deuteron TE absorption signal is several hundred times smaller than that of the corresponding proton TE signal; the TE signal is in fact completely buried in the noise of the standard CW Q-meters.

The leading source of noise in the best Q-meters is that of the oscillator. The noise, however, can be averaged out by accumulating the signal over a long period of time, so that the characteristic features of the deuteron resonance spectrum can be easily observed in less than one hour. The accurate measurement of the integrated absorption signal, in fact, is mainly limited by the stability of the circuitry over such extended periods, and by the systematic variations of the Q-value of the tank circuit due to the temperature and magnetoresistance effects during field shifts required for the baseline measurements.

Pulsed NMR measurements, in practice, are much more sensitive than the CW NMR measurements, because the oscillator is disconnected during the acquisition of the NMR free precession signals and therefore does not contribute to the system noise. Furthermore, the pulsed NMR principle is based on the measurement of the voltage due to the free precession signal, in contrast to the CW Q-meter which is based on the measurement of the circuit impedance modulated by the RF susceptibility of the target material. In pulsed NMR variations of the circuit impedance thus influence the signal only in a multiplicative way, whereas in the CW method the circuit drift is additive with a large numeric coefficient. It has thus been possible to measure accurately the deuteron NMR signals in a solid material, in TE at room temperature, using the VFP NMR technique [1].

SATURATION

The VFP NMR tests were carried out using large flip (or tipping) angles of spins at resonance, up to 90° or even higher [1,3]. Such a signal measurement would destroy the polarization and necessitate the repolarization of the target. The dynamic nuclear polarization of the target could therefore not be followed by live NMR polarization measurement, and on-line optimization of the microwave parameters would be impossible. Working under such conditions would resemble steering a car by looking only into the reversing mirror.

This is clearly undesirable for polarized targets, and the VFP NMR method must therefore be modified for small flip angles at resonance

$$\theta = \int_0^{t_p} dt\, \gamma_d H_1(t)\ , \tag{1}$$

where t_p is the pulse duration, γ_d the gyromagnetic ratio of the deuteron, and $H_1(t)$ the RF field which may be shaped to optimize the spectral width of the excitation. A flip angle $\theta \approx 0.5°$ would give an adequate saturation time constant of order 30 h under measurement repeated once per second. Assuming a rectangular pulse of 1 ms duration, coil inductance of 1 μH and Q-factor 5, frequency 16.35 MHz (for $B_0 = 2.5$ T operation), and requiring the tipping angle of $\theta \approx 0.5°$ at a radius r = 2 mm from the coil wire buried in the target material, the current

$$I = \frac{2\pi\, r\, B_0\, \theta}{\mu_0\, \omega_0\, t_p} \tag{2}$$

in the wire should be $I \approx 2.1$ mA. Making reasonable assumptions on the circuits coupling the coil to the oscillator and the preamplifier, the oscillator power must be -6 to 0 dBm. The power can be substantially scaled down if a lower inductance coil can be used, and if the coupling to the oscillator can be made strong; playing with these parameters might also help optimising the signal-to-noise ratio of the receiver front-end.

The advantage of the small flip angle is that the response of the spin system is linear, i.e. all components of the magnetization can be considered as additive, and perturbation expansions can be used for analytic calculations.

EXCITATION SPECTRUM

From the point of view of simplicity in its mathematical treatment, the easiest pulse shape is the rectangular one, i.e. the amplitude of the $H_1(t)$ field is constant during the pulse, and zero before and after. Such a pulse is non-adiabatic (diabatic) in the whole spectrum of the spin frequencies, which necessitates the individual tracking of all spins during and after the pulse. For a train of pulses with stepped frequency, interleaved with data acquisition of the free precession signals, the response of the spin system to such pulses depends on the phase coherence of the successive pulses, and on the spin-spin interactions which result in the spectral diffusion of the transverse magnetization, which has a strong variation in the frequency spectrum due to the sudden switching of the RF field.

Preliminary results from our numeric simulation of rectangular pulse VFP NMR would indicate that the free precession signal is very strongly damped, because the excitation spectrum has the shape of $\mathrm{sinc}(\omega t_p)$, where ω is the offset from the pulse frequency. This has strong sidelobes extending far through the wings of the deuteron resonance line. The resulting response to a series of pulses, although based on such a simple model, are in a clear disagreement with the good experimental results of Sindorf and Bartuska[1]. In particular, it would seem that the suggested harmonic relationship[1]

$$t_p = n\left(\Delta f\right)^{-1} , \tag{3}$$

cannot suppress the spectral density due to the sidelobes, although the signals resulting from earlier pulses may be reduced. Here $\Delta f = \Delta\omega/2\pi$, the size of the frequency step between two successive pulses.

Wemmer, Wolff and Mehring[4] point out that the Fourier transform of the free-induction decay (FID) signal following a rectangular pulse, with H_1 applied in the x-direction, is

$$G(\omega) = F(\omega)[X(\omega) - i\,Y(\omega)] , \tag{4}$$

where $F(\omega)$ is the lineshape function, and

$$Y(\omega) = \frac{\sin t_p\omega_e}{\omega_e} ; \qquad X(\omega) = \frac{\omega}{\omega_e^2}(1 - \cos t_p\omega_e) ; \tag{5}$$

and

$$\omega_e^2 = \omega_1^2 + \omega^2 .$$

The Eq. (4) immediately shows how the sidelobes of the Fourier transform of the excitation pulse will sample the whole resonance lineshape, resulting in a rapid decay of the FID signal. The signal decay time from (4) would be close to the inverse width of the resonance line; preliminary results from our numeric simulations confirm this. In general, symmetric wide-pulse wide-line NMR has the problem that although large transverse magnetizations can be generated, the frequency components of M_x and M_y stay coherent only during the symmetric pulse itself, and the FID signal disappears almost as soon as the driving RF field disappears [5].

Many schemes have been proposed to improve the performance of rectangular pulses in NMR. These may involve a sequence of many pulses, with or without amplitude and/or phase modulation. In our case of interest, the accurate and linear measurement of the wide quadrupole broadened deuteron resonance line, we shall consider the single pulse excitation of the spin system using amplitude modulation, and possibly phase modulation of the RF source.

The design of shaped pulses for selective excitation NMR has received much attention in the fields of both high-resolution spectroscopy and magnetic resonance imaging (MRI). Shaping improves the frequency domain response by localizing the excitation to a selected range around the oscillator frequency; obvious shaping functions are the Gaussian [6]

$$G(t) = C\, e^{-a(t - t_0)^2}\,, \tag{6}$$

Hermite [7]

$$G(t) = C\left[1.0 - 0.957\, a\, (t - t_0)^2\right] e^{-a(t - t_0)^2}\,, \tag{7}$$

and Sinc [8,9]

$$G(t) = C\,\frac{\sin a(t - t_0)}{t - t_0}\,. \tag{8}$$

Here C and a are selected so that the desired flip angle and excitation width are obtained. Although all these pulses eliminate the side-lobes in the excitation frequency spectrum, their principal basic problem is that the transverse magnetization components exhibit a large frequency-dependent linear phase shift [10]. Correction of the phase error is possible by software routines, but this leads to a "rolling baseline" in the case of wide-line NMR [10]. Also, as was said above, the FID signal overlaps the symmetric pulse, and unless the trailing edge of the pulse is truncated, little of the FID can actually be observed in the case of a wide NMR line, unless refocusing is used. Refocusing (or spin echo) is obtained by applying a 180° pulse or a more complex sequence of pulses; these are excluded in our application because the polarization is lost in such a measurement.

The many solutions to these problems involve shaping of the pulse. The half-Gaussian pulse [5] rises with a Gaussian curve but is truncated at the peak. No sidelobes are introduced in the excitation spectrum due to this sharp falling edge, and the M_y has a positive peak with no negative lobe on either side, unlike the Gaussian excitation. This can be understood by following the magnetization trajectories in the rotating frame: as the B_1 rises slowly, off-resonant spins follow the the effective field adiabatically without any significant M_x component. Spins close to the RF frequency rotate slightly about the effective field, but because this grows slowly, they will not develope M_y -components in the negative direction. When the B_1 is turned off rapidly, the FID has a large absorption (y-) component with slow decay, and the dispersion (x-) component begins at zero and exhibits oscillations about zero with relatively small amplitude.

Other obvious asymmetric shapes are half-Hermite and truncated-Sinc. However, it is not possible to compare the shapes accurately neither analytically nor numerically, because of circuit imperfections and other artifacts which finally determine the performance of the system. There are, however, numeric optimization routines, which can be used to improve a particular parameter such as the excitation spectrum shape. Such routines give interesting results for the pulse shape; for example requiring a "top-hat" excitation profile results in an oscillatory shape close to a truncated-Sinc [10].

The simplest pulse shape is perhaps the asymmetric sawtooth, which has been already studied by Kennedy, Vold and Vold [3]. Their one-point data acquisition at 800 µs delay demostrates beautifully that the tail of the FID signal is near zero for a near-uniform distribution of Larmor frequencies, for both rectangular and sawtooth pulse shapes using a small flip angle. Only the abrupt edges of the deuteron signal are visible as sharp peaks. The VFP NMR with single point FID acquisition thus requires a large flip angle, which results in a non-linear transverse magnetization and therefore a larger and longer FID signal, approximately proportional to the magnetization of the spins.

Other shaping techniques can be deviced to improve upon the circuit imperfections such as mismatches and non-linearities, and on necessary artifacts such as filters and the probe ringing. This leads to pre-distortion shaping [11] which aims at correcting the pulse shape prior to the probe coil in such a way that a wanted amplitude and phase evolution is obtained in the RF field.

As our main aim is to develope the technique so that very small flip angles are feasible, we are thus led to propose the acquisition of the full FID signals in quadrature, starting promptly after the extinction of the excitation pulse.

CIRCUIT DESIGN

The proposed circuit block diagram is shown in Fig. 1. The precise choice of the frequencies other than the f_d = 16.35 MHz DMR, fixed by the 2.5 T field, is somewhat arbitrary. We feel, however, that upconversion after the preamplifier to a constant frequency of about $f_1 - f_0 = 30$ MHz may be preferable to a downconversion, because of easier control of sidebands and spurious signals, and because the filters may be easier to design. The fixed-frequency receiver downconverts the FID signal to a frequency around $\Delta f \approx 30$ kHz in a quadrature mixer, giving out the signals proportional to $M_x = S_{FID} \cos(\Delta f\, t)$ and $M_y = S_{FID} \sin(\Delta f\, t)$ which can be digitized and stored in a two-port memory. Two digital signal processors (DSP) then analyse the signals giving out the $M_y(t_p)$ and possibly some other useful parameters related with the signal shapes and phasing.

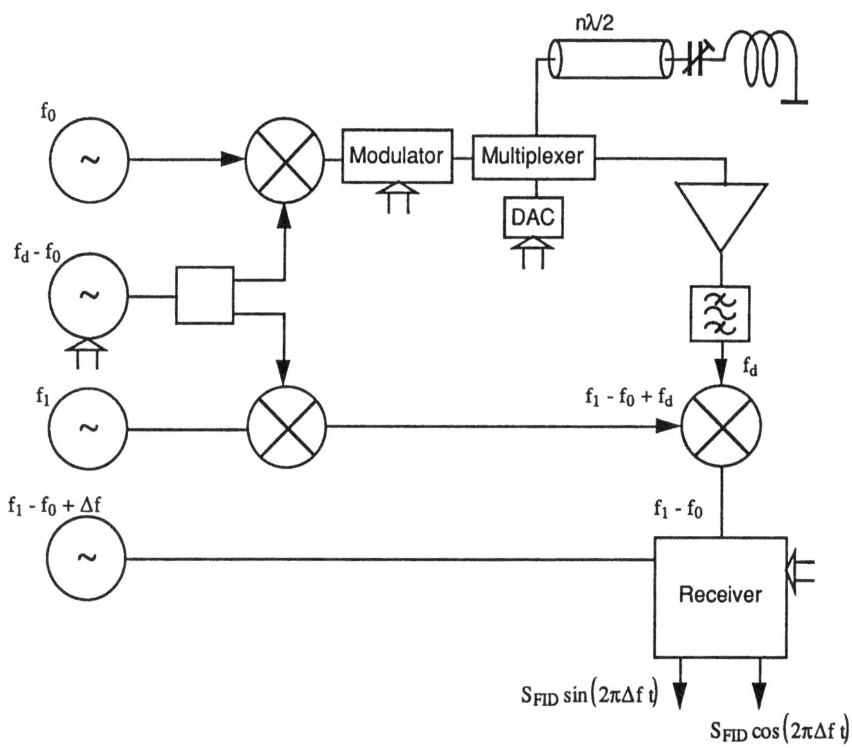

Fig. 1: Schematic diagram of the proposed Variable Frequency Pulsed NMR circuit for deuteron polarization measurement. Operation at 2.5 T requires a center frequency of f_d = 16.35 MHz, with 0.5 MHz scan around it, generated with a scanned synthesizer having a center frequency $f_d - f_0$. The two local oscillator frequencies can be chosen at convenience or on the basis of optimization. Convenient values are, for example, f_0 = 10 MHz and f_1 = 40 MHz. A processor will control the timing of the pulse modulator and input multiplexer signals; the same timing circuits will also control the scanned synthesizer and the possible muting of the receiver circuits during the pulse and subsequent spurious transients.

The timing of the pulse sequence could involve 1 - 2 ms pulse lengths, 0.5 - 1 ms acquisition times and appropriate delays for amplifier and filter recovery. This would require that the FID's must be analysed in some 2 ms time. Basing on execution times reached in a PC running a suitable similar code [12], it would seem that modern DSP's with 30 M flop/s throughput can handle the task.

CONCLUSIONS

Although simulations and demonstration experiments so far fail to show the feasibility of accurate measurement of the deuteron NMR signal shape and size using very small flip angles, there seem to be ways to improve the performance so that pulsed NMR could become the standard way of polarization measurement for nuclei with quadrupole moment in solids. We may list the following specifications and requirements, which are being studied and evaluated:

- Pulse flip angle about 0.5° at resonance;
- The RF non-uniformity is not a problem in FID signal measurement. Buried coils therefore can be used as is customary in polarized targets;
- Asymmetric pulse shape needs to be optimized for this specific application;
- Pre-distortion compensation of circuit mismatches, filters, probe Q-factor etc. may be required;
- Relatively broad filters with sharp fall-off are required, because of the fast decay of the FID signals and the requirement that extrapolation to the end of the pulse is desired;
- Probe tuning to each pulse frequency needs to be made;
- Problems related with the 3-level system of spin-1 need to be studied in order to understand possible distortions, due to pulse sequencing, in the center of the resonance line.

REFERENCES

1. D. W. Sindorf and V. J. Bartuska, J. Mag. Res., **85** (1989) 581.
2. T. O. Niinikoski, Second Workshop on Polarized Target Materials, Cosener's House, Abingdon, Oxford, U.K., Report RL-80-080, Ed. G. R. Court, S. F. J. Cox, D. A. Cragg and T. O. Niinikoski (Chilton, Didcot, SRC, Rutherford Laboratory, 1980) p. 80.
3. M. A. Kennedy, R. L. Vold and R. R. Vold, Submitted to J. Magn. Reson., July 1990.
4. D. E. Wemmer, E. K. Wolff and M. Mehring, J. Mag. Res., **42** (1981) 460.
5. J. Friedrich, S. Davies and R. Freeman, J. Magn. Reson., **75** (1987) 390.
6. C. Bauer, R. Freeman, T. Frenkiel, J. Keeler and A. J. Shaka, J. Magn. Reson., **58** (1984) 442.
7. W. S. Warren, J. Chem. Phys., **81** (1984) 5437.
8. A. J. Temps and C. F. Brewer, J. Magn. Reson., **56** (1984) 355.
9. M. S. Silver, R. I. Joseph and D. I. Hoult, J. Magn. Reson., **59** (1984) 347.
10. H. Geen, S. Wimperis and R. Freeman, J. Magn. Reson., **85** (1989) 620.
11. W. S. Warren and M. S. Silver, *The Art of Pulse Crafting: Applications to Magnetic Resonance and Laser Spectroscopy*, in **Advances in Magnetic Resonance**, (Academic Press, New York 1988) p. 248.
12. O. Hartmann, private communication, .

Instrumentation for the Two-Litre Polarized Target of the SMC NA47 Experiment

A. Rijllart[1], *T.O. Niinikoski*[1], *N. Hayashi*[2], *and R. Ent*[3]
[1]CERN, CH-1211 Geneva 23, Switzerland
[2]Nagoya University, Nagoya-shi, 464, Japan
[3]University of Virginia, Charlottesville, VA 22901, USA

The conceptual design of the instrumentation for a large polarized target is discussed. The target consists of two halves with opposite polarizations, and is designed for the measurement of inclusive spin asymmetries in deep inelastic muon scattering in the SPS muon beam at CERN. The target features deuterated and normal hydrocarbon materials, field rotation for spin reversal and computer optimization of microwave parameters during dynamic nuclear polarization.

1. Introduction

The Spin Muon Collaboration (SMC) experiment NA47 at the CERN SPS is designed to measure the spin structure functions of the proton and the neutron to a high enough accuracy to test the Bjorken sum rule at the level of 10% precision. This sets severe criteria on the accuracy of the target polarization measurements, to be discussed below. The experiment uses a longitudinally polarized positive muon beam and a longitudinally polarized proton or deuteron target, and aims at statistics about 20 times better than what was obtained in the predecessor experiment NA2 of the European Muon Collaboration [1].

The polarized target volume is about 2 litres and consists of two halves, polarized in opposite direction, shown in Fig.1. Microwave reversal of polarization and magnetic field rotation create 4 different configurations of magnetic field and polarization orientation for both target halves.

Dynamic nuclear polarization will be done at a magnetic field of 2.5 T and data will be taken in frozen spin mode at a holding field of 0.5 T. A total of 10 Continuous Wave (CW) NMR channels is foreseen to obtain an accurate value of the polarization over the target volume. In each half there will be 4 NMR channels tuned for 2.5 T and 1 channel for 0.5 T operation.

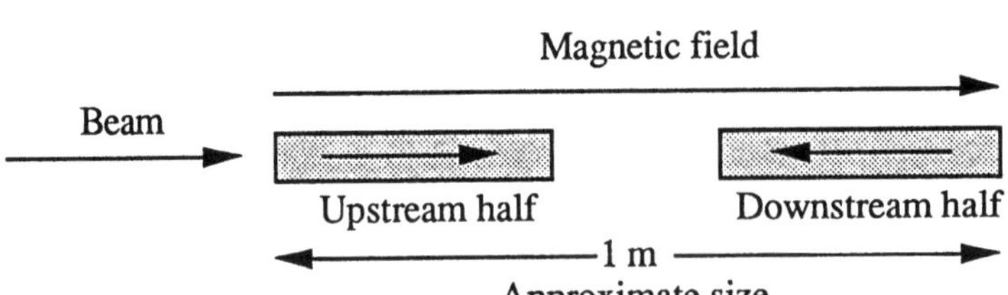

Fig.1. The SMC polarized target with polarization directions.

An on-line computer system consisting of a μVAX-3 with Pilot control of tasks written in FORTRAN has been chosen to control all target systems, magnet, microwaves, NMR and dilution refrigerator, and interface to the users.

2. NMR system requirements

The minimum requirements for the NMR measurement of polarization are:

- Absolute statistical accuracy < 0.2 % in a few seconds on proton or deuteron integrated signals during dynamic polarization (for fast feedback purpose);
- Relative systematic accuracy < 2 % on proton and deuteron polarization measurement;
- Good long term stability, variation < 0.5 %/day, for both high field and frozen spin modes.

The relative systematic accuracy below 2% requires ultimate care in the calibration procedure for the integrated NMR signals. The main problems here are the determination of the tiny deuteron NMR signal at the 1 K calibration temperature sufficiently accurately, and the absolute measurement and homogeneity of the target temperature during calibration. For the deuteron signal measurements the CW-NMR techniques are unsuitable because of circuit drifts and oscillator noise; a variable frequency pulsed NMR technique is being designed with the appropriate specifications [2]. The temperature measurement problems will be covered by filling the target with pure ^{4}He during calibration, and by using a ^{3}He vapour pressure thermometer for accurate absolute temperature measurement in the 2.5 T field.

The proton NMR system concepts adopted to meet the requirements are:

- Continuous-wave NMR using 10 Q-meters.
- All channels averaged simultaneously i.e., 8 channels at 2.5 T and 2 channels at 0.5 T. This is achieved by synchronizing three 4-channel NMR processor units, each containing a 68000 CAMAC processor (STAC) and a 4-channel NMR interface,
- A digital frequency synthesizer is swept in steps by the master STAC,
- Averaged data is transferred through CAMAC to the μVAX, which calculates polarization, time evolution of the polarization and generates graphics outputs,
- The μVAX controls all data acquisition parameters, keeps baselines and calibration constants. It also controls the magnetic field and the microwave sources, directly or using an intermediate Personal Computer (PC).

The new deuteron NMR system is planned with the same number and allocation of channels; initially the CW Q-meters will be used basing the calibration on the lineshape measurement.

3. Signal averaging

One of the key improvements in the CW-NMR system design is that all NMR signals can be simultaneously averaged. A schematic diagram of the NMR system is shown in Fig.

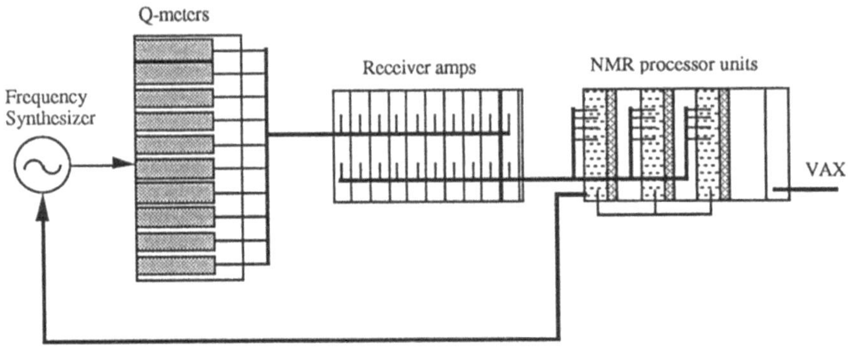

Fig. 2. 10 Channel NMR system layout

2. This is achieved by implementing a control mechanism into the 4-channel NMR processor units which allows to synchronize many of these units to each other. The asynchronous data-transfer-acknowledge signal from an address, reserved for this purpose, is being output through an open-collector gate.

This signal is linked to all units to be synchronized. In the software of the STAC processor this reserved address is being accessed just before the moment when all units will issue their 'hold' commands to the input sample and holds. Due to this wired-or logic the last processor to complete this instruction allows the continuation of all processors. In this way they execute this 'hold' instruction perfectly synchronously at every step. This technique can be expanded to synchronize a large number of inputs, only limited by the loading and time delay on the synchronization signal.

4. The computer system

The μVAX-3 computer has 16 Mbyte RAM, a 800 Mbyte hard disk, and is one of seven μVAXes in the SMC VAX cluster. The PT μVAX is the centre of a distributed computer system and connects through multiple interfaces, local processors and PC's to the 4 main subsystems of the PT, as shown in fig.3. These subsystems are:

- The dilution refrigerator instrumentation
- The magnet system: the 2.5 T solenoid, 12 trim coils, and a 0.2 T dipole
- The NMR system
- The microwave system: 2 EIO sources

Its main on-line tasks are:

- Polarization measurement
- Magnetic field control
- Microwave frequency and power control of both EIO sources

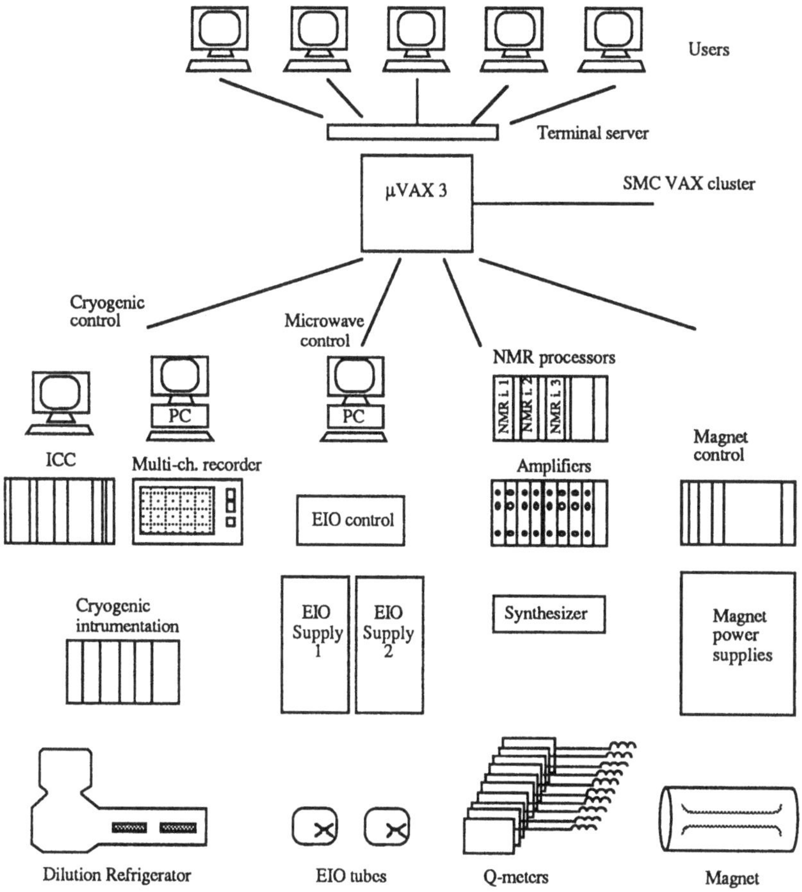

Fig. 3. SMC Polarized Target computer system.

- Combine information for multiple controls, such as using the NMR line centres for magnetic field centering and using polarization growth data for microwave parameter optimization
- Alarms and some control functions for the cryogenic instrumentation.

5. Software concepts

- Interactive control based on status and menus.
  The status display has a general field where target, magnet and polarization configuration status are shown, and a specific field which depends on the menu one has selected. Menus are sub-divided into different levels, where the highest level allows global changes to the system, and the lowest level allows to change the state of one single output.
- Top-down approach
  The definition of all levels has been translated into menus, which together form a skeleton program. Inside this skeleton program variables and procedures have been defined to describe the state of the various sub-systems and how they can be changed.

- Bottom-up approach
  Lower level routines are being written first in the skeleton program where each hardware module can be tested, and each programmed function can be checked to have the correct result. Once the lowest level functions are operational higher level tasks can be programmed and tested.

6. Menu levels

- First (or top) level
  Here the highest level commands are given such as polarize, freeze, repolarize, rotate magnetic field, optimize microwave parameters, etc.
- Second level
  Inside the second level the top-level operation has to be checked and allow for interrupt, cancel or override of specific actions.
- Third level
  The third level menu gives direct control of a sub-system, such as NMR system, magnet system, etc. This is above all needed for setting-up and testing of all functions of the sub-system.
- Fourth level
  This level gives access to all settings in the third level menus. This includes NMR sweep parameters, calibration constants, microwave optimization and modulation control parameters.
- Fifth (or lowest) level
  In this level one can check each input or output individually, have test procedures for electronic modules, and test the communication links with the different processors and PC's.

7. Conclusions

The instrumentation of a large polarized target is so complex that it justifies a distributed computer system. Using dedicated pre-processors for fast incoming data, personal computers to handle slow control of output devices and the μVAX for data processing and display there is an excellent load balance between the various computers. For the user this will show up as fast response to commands.

The modularity of the computer system, in hardware and in software allows easy implementation of other tasks. One of such tasks could be the implementation of variable frequency pulsed NMR, which requires a digital signal processing system based in VMEbus, for example. This technique is being considered because it could enable us to calibrate the deuteron thermal equilibrium NMR signal, due to its stability and low noise.

References

[1] EMC, J. Ashman et al, Nuclear Physics B, **328**, 1-35 (1989).
[2] T.O. Niinikoski, H. Johansson, L. Klostermann and A. Rijllart, in these Proceedings.

The PSI NMR-System for Polarized Targets

M. Weßler, J. Böhler, B. van den Brandt, J.A. Konter, and S. Mango
Paul Scherrer Institute, CH-5232 Villigen PSI, Switzerland

We present the latest version of a NMR system with high sensitivity, developed for experiments with polarized nuclei other than ^1H.

The system is essentially a computer controlled fast Q-meter with straightforward amplitude detection, based on a 1-250 MHz programmable frequency synthesizer.

To achieve the high sensitivity needed to measure the TEP (thermal equilibrium polarization) signals of nuclei like D in deuterated alcohols or ^{15}N in $^{15}NH_3$, we use a high Q parallel resonance circuit, in which the capacitance C is positioned as close as possible to the NMR coil. In our ^3He refrigerators the capacitor is placed just above the liquid ^3He level and connected directly to the NMR coil, while in our dilution refrigerators it is placed just above the still and connected to the NMR coil by a 14 cm long piece of semirigid stainless steel 50 Ω cable. We use a combination of fixed ceramic capacitors and voltage controlled GaAs varactor diodes, mounted on a small circuit board together with a 1 MΩ input impedance GaAs FET source follower (See fig. 1a). With one varactor diode we can tune the circuit to resonance over a range of about 500 kHz, and by combining several diodes in parallel this range can be extended to more than 1 MHz.

When the LC circuit is brought to resonance at the center frequency of a sweep of interest by the use of an appropriate combination of capacitors and varactor diodes, its high Q produces a strongly structured 'baseline', which limits very much the possibility of amplification after the detector and can cause troubles in as much as the linearity of the detector itself is concerned. More importantly, the LC circuit is at resonance strictly speaking only at the center frequency of the sweep, and this will introduce distortions of the NMR signal in dependence of the distance from the center frequency of the sweep. To get rid of these problems we keep the LC circuit constantly at resonance over the sweep, by changing the value of C synchronously with the frequency. We achieve this 'permanent tuning' by modulating the reverse bias voltage U_v applied to the varactor diode(s), so as to fulfil the resonance condition $C(\omega) = \frac{1}{L\omega^2}$ for each frequency of the sweep. Because the dependence $C(U_v)$ is not a simple analytic expression and furthermore varies from diode to diode, we preferred to determine experimentally the voltage function $U_v(\omega)$ required to keep the LC circuit at resonance over a certain frequency range.

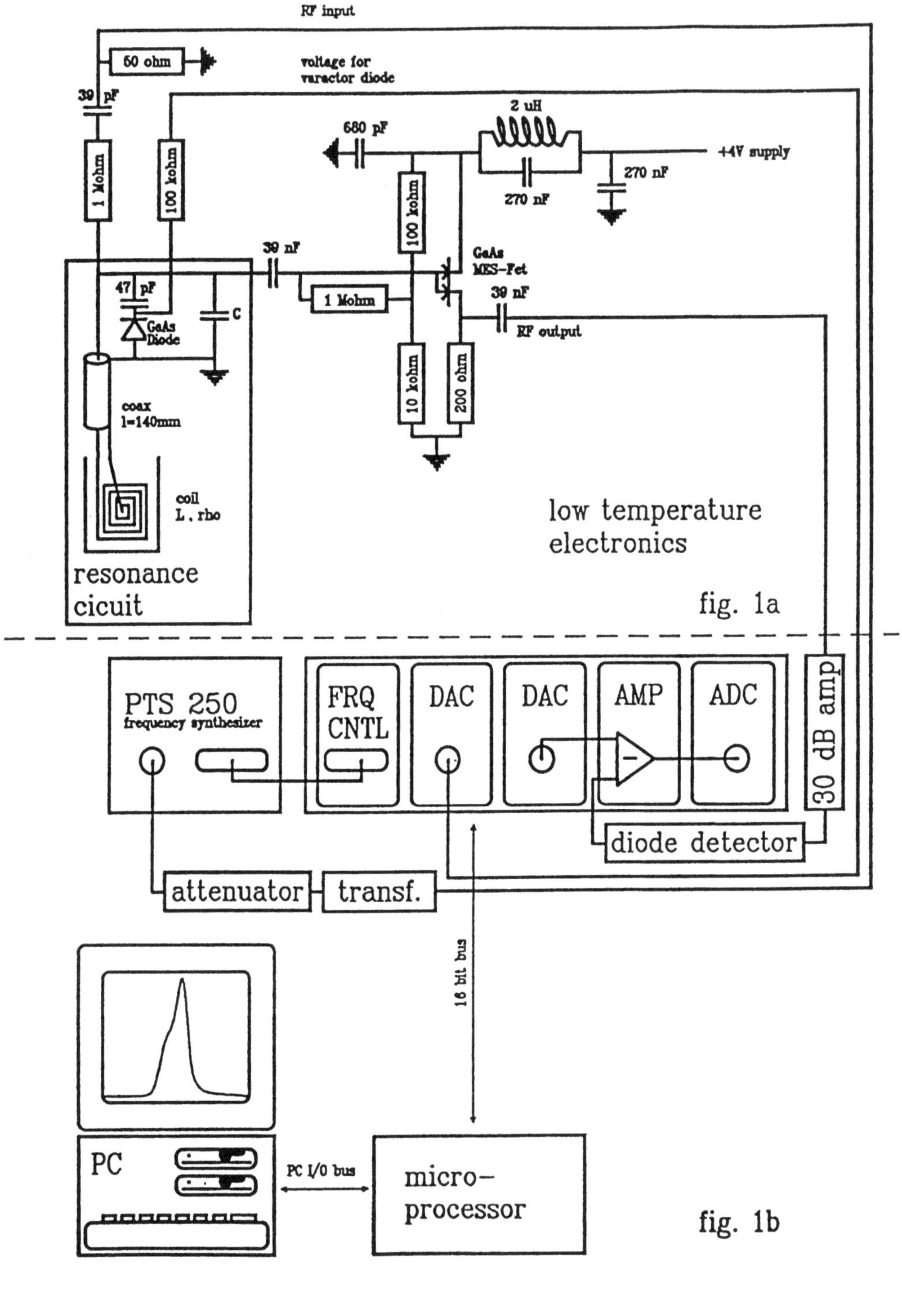

Figure 1: Fig. 1 Scheme of the Q-meter circuit (1a), and the system hardware (1b).

The hardware used to drive the system is shown in fig. 1b. Frequency control, 'permanent tuning' and signal averaging are done by a microprocessor, which is connected to the I/O bus of a personal computer (PC). On that microprocessor we run a highly optimized assembler program which has access to the synthesizer, two digital to analog converters (DAC) and a fast analog to digital converter (ADC) over a bus system with high data transfer rate. The PC is used for downloading the assembler program to the microprocessor and for backloading of the data to the PC, where they are displayed and stored. The DAC's and the ADC have separate power supplies and are galvanically isolated from each other. To protect the system from RF noise produced by the computer, all readouts are optically coupled. Additionally, the synthesizer is isolated by a RF transformer, which allows us to have only one real ground connection, that is chosen to be in the cryostat.

Together with the machine code of the assembly program the PC loads three data arrays into the memory of the microprocessor, i. e. the BCD coded frequency values for the programmable frequency synthesizer and the binary data for the two DACs. In a 4th array, which allocates four bytes per data word, the data read from the ADC are stored. The width of four bytes allows a maximum averaging of one million 12 bit ADC values. The maximum length of each of the four arrays is 4096, which is also the maximum number of frequency steps in a sweep.

When starting a measurement the assembler program transfers the first frequency value from the memory to the synthesizer, which needs some time to settle in frequency and amplitude. This settling time is the main contribution to the overall acquisition time and strongly depends on the synthesizer used. During that time the data from the second and third array are transferred to the DACs, the first DAC generates the voltage U_v for the varactor diode and the second generates a compensation voltage, which is fed to the second input of the differential amplifier (see fig. 1b). When the RF is stable in frequency and amplitude, the first stage of averaging takes place. In a very fast stream mode (1.5 μsec) a selectable number of ADC data is transferred to the first data word of the 4th array. Now the program proceeds to transfer the second frequency value to the synthesizer, and the process just described above is repeated. After the last frequency has been processed the system enters the second stage of averaging, where it switches back to the first frequency. The ADC data from the second and all further sweeps are added up to the values from the first sweep. When the preset number of sweeps is finished, a message is sent to the PC, to tell it that the data are ready. The measured signal is then transferred to the PC, where it is loaded in a graphic environment, which is used to analyse the signal.

To describe the procedure used to determine the voltage function $U_v(\omega)$, we remind that in a constant current Q-meter the voltage drop across the resonance circuit is proportional to its impedance Z.

Fig. 2 shows how the resonance circuit reacts for different varactor voltages U_v. In this figure the leftmost peak belongs to 0V and the rightmost to 10V sup-

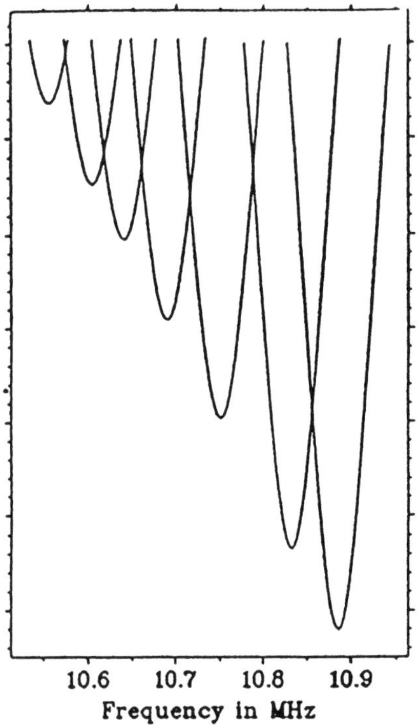

Fig. 2 Resonance curves for different capacity values.

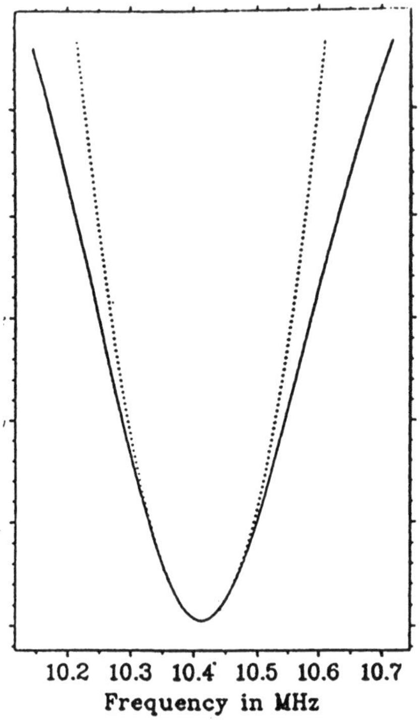

Fig. 3 Determination of the peak position by fitting a parabola (dotted line) to the resonance peak (solid line).

plied to the varactor diode. For technical reasons the peaks are inverted, which means that the minimum of the rightmost peak in fig. 2 has the highest Z value.

The rise of impedance at the different resonance frequencies ω_0 can be explained by circuit analysis. The absolute value of the complex impedance of a LR||C circuit, if R is sufficiently small for the resonance frequency to be $\omega_0 = \sqrt{\frac{1}{LC}}$ and if $C{\cdot}R^2 \ll L$, as in our resonance circuit, can be reduced to

$$|Z(\omega_0)| = \frac{L^2 \cdot \omega_0^2}{R} \tag{1}$$

which shows a quadratic dependence from the resonance frequency ω_0.

To determine ω_0 from the measured curves, a parabola is fitted to the top of the peaks (see fig. 3). We chose this method since such a fit can be done arithmetically and therefore does not need as much calculation time as fitting the real line shape of the curve. The system allows to do the measurement and peak fitting in parallel, which saves us a factor of two in time.

We end up with a table of voltages and corresponding resonance frequencies. Since we need an analytical expression for $U_v(\omega)$, these points are fitted by a 6th order polynomial. Using the results from this fit we are now able to calculate for

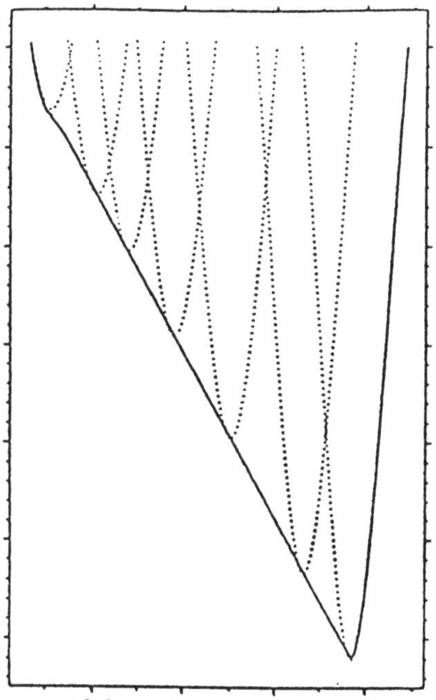

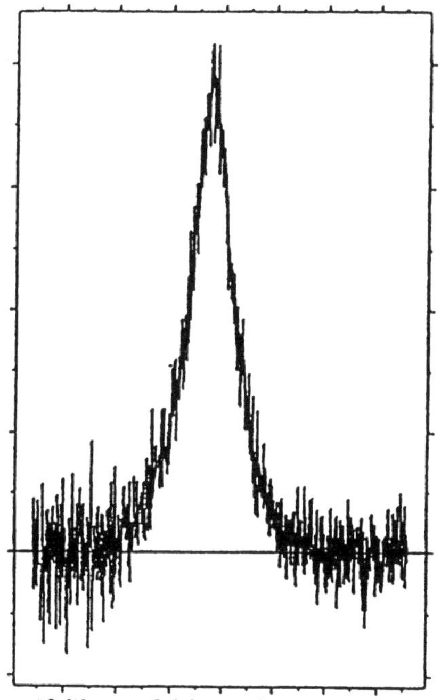

Fig. 4 The solid curve shows the baseline, obtained by varying the capacitance continuously. It is superimposed by resonance curves (dotted lines), taken at fixed capacitance values.

Fig. 5 ^{15}N thermal equilibrium signal at 2.5 tesla and 2.17 K.

each frequency the voltage U_v, which tunes the circuit at resonance. The solid curve in fig. 4 shows the result of this synchronous tuning. It is once more superimposed by resonance curves at constant voltages U_v (dotted lines), to show the composition of the baseline.

Offset and slope of the baseline are compensated at the input of the dc differential amplifier following the detector, by adding a ramp signal generated by a DAC, which is inverse to the incoming signal (see fig. 1b).

The compensation, the low noise and the good stability of the system allow us to use an amplification of 10^4 without exceeding the dynamic range of the 12 bit ADC, and enabled us to measure e.g. the ^{15}N TEP signal at 2.5 T and 2.17 K as shown in fig. 5.

Calculation and Simulation of the Deuteron NMR Spectrum in Propanediol

S. Dhawan[1], *J. Reynolds*[1], *K. Morimoto*[2], *and G. Court*[3]

[1]J.W. Gibbs Laboratory, Yale University,
New Haven, CT 06511, USA
[2]National Laboratory for High Energy Physics, KEK,
Oho-Tsukuba, Ibaraki-ken 305, Japan
[3]University of Liverpool, Department of Physics,
Liverpool L69 3BX, United Kingdom

The Q-meter used in the target polarization measurement is simulated. We calculate the complex magnetic susceptibility of the target material as a function of frequency from the first principles and use this data file as an input to the Q-meter. The output from the Q-meter is studied for parameter changes like $\lambda/2$ cable, phase angle, etc.

Mathcad[1] is used for this simulation work. All the equations, formulas, constants, etc. are defined as part of the document and appear as these would on paper. The results can be easily plotted on the screen and become part of the document.

Calculating $\chi''(\omega)$

We calculated the absorptive part of the complex magnetic susceptibility ($\chi''(\omega)$) of deuterons in propanediol-d8 using the method of calculation and parameters in the paper of O. Hamada et al.[2]

The deuteron nucleus has a nuclear spin of 1. Therefore, in deuteron nuclear magnetic resonance (DNMR), there are two Zeeman transitions. In propanediol, there are two different deuteron bonds: carbon-deuteron (C-D) and oxygen-deuteron (O-D) bonds. The field gradient is not the same for these two bonds, and as a result, the DNMR spectrum for propanediol has two components. As the polarization of the deuterons changes, the relative intensity of the two Zeeman transitions also changes, causing an asymmetry of the peaks in the DNMR spectrum. An internal equilibrium is established in the deuteron spin system, defining a unique "spin temperature." The deuteron polarization is given by:

$$P = (1-R^2)/(1 + R + R^2).$$

R is the Boltzmann factor or the relative intensity of the two Zeeman transitions.

The whole DNMR spectrum is given by:

$$S(x) = [2/(1 + R)]\{C_{OD}[RF_{OD}(x) + F_{OD}(-x)] + C_{CD}[RF_{CD}(x) + F_{CD}(-x)]\},$$

where the functions F(x) and F(-x) are the components of the spectrum which correspond to 0 ↔ -1 and 1 ↔ 0 transitions, respectively, for each bond. C_{OD} and C_{CD} are the relative number of deuterons: $C_{OD} = 0.75$ $C_{OD} = 0.25$.

In the program DNMRPOL.MCD, we first obtain the value of R that corresponds to the specified P and then calculate the four components of the DNMR spectrum, from which the whole spectrum was constructed as described above. The spectrum is normalized such that $|S_{max}| = 1$.

The spectra obtained from these calculations correspond well with experimentally observed data.

Calculating χ'(ω)

The real (dispersive) and imaginary (absorptive) parts of the magnetic susceptibility, χ'(ω) and χ"(ω), are related to each other by the Kramers-Krönig relations. For our purpose the relation is:

$$\chi'(\omega_0) = (2/\pi) \, {}_0\!\int^{\infty} \omega\chi''(\omega)(\omega^2-\omega_0^2)^{-1} \, d\omega .$$

The method of calculation described in the appendix of Sperisen's[4] paper is used.

Circuit Modelling

The circuit diagram is shown in Fig. 1 with the parameters. In this simulation, the frequency is swept over a range of ±250 kHz around 16.3 MHz (B=2.5T). A constant current of 10 mA is fed into the circuit. The variable capacitor is adjusted to tune the circuit near its resonant state.

The Z_{total} is the total impedance of the circuit and includes:

$$L(\omega) = L_0 \, (1 + 4\pi\eta\chi(\omega)) \qquad \eta = \text{Filling factor/scale factor}.$$

η is adjusted for a signal modulation level of approximately 5%.

The output voltage is:

$$V_{out}(\omega) = I \, Z_{total} e^{i\phi} \qquad \phi = \text{phase angle}.$$

The circuit without a sample material is so tuned that (i) the response parabola of $|V_{out}|$ is centered at ω_{res} by C_{TUNE}, and (ii) the response parabola of the real part of V_{out} ($Re[V_{out}]$) is centered at ω_{res} by phase

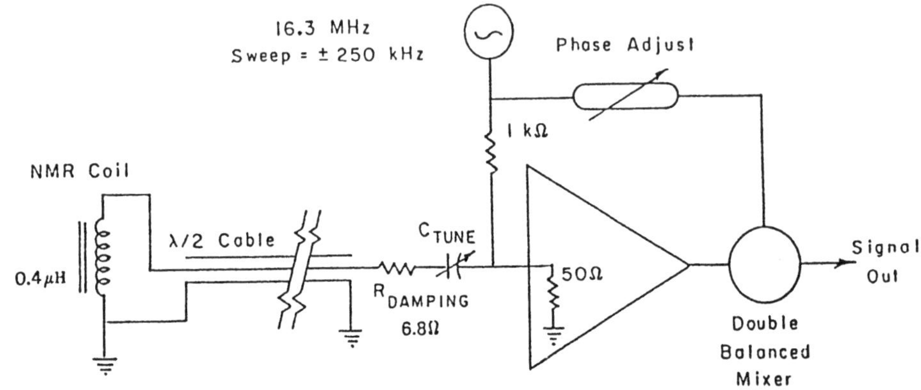

Fig. 1. Schematic of Q-meter for simulations.

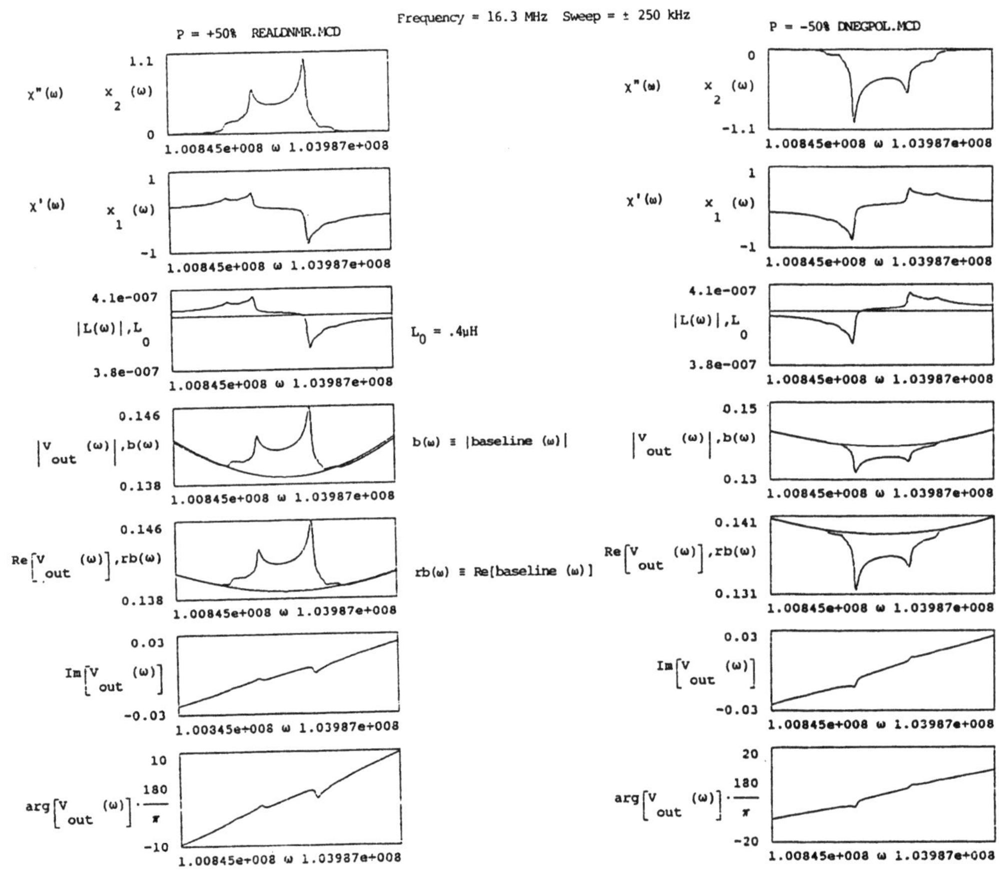

Fig. 2. Plot of complex magnetic susceptibility, inductance of voltage and phase angle for P = +50% and P = -50%.

adjust. From these two steps we determined the phase angle difference "ϕ", which was 6.6° in this circuit.

Program DMAKEBAC.MCD calculates the background (circuit response) without any sample material. Program REALDNMR.MCD uses the $\chi(\omega)$ of the sample material and calculates the response of the circuit and then displays $|V_{out}(\omega)|$, $|Baseline(\omega)|$, $Re[V_{out}(\omega)]$, $Re[Baseline(\omega)]$, $Im[V_{out}(\omega)]$, and $arg[V_{out}(\omega)](180/\pi)$ (the phase of the signal in degrees). Baseline(ω) is the response of the circuit without the sample material in the coil. These are shown in Fig. 2.

Polarization Measurements

We used this program simulation to investigate and compare the two different methods of measurement: the asymmetry ratio and the integral of the signal. In both measurement methods, we worked with the DNMR signal as defined by:

$$sig(\omega) = Re[V_{out}(\omega)] - Re[Baseline(\omega)].$$

To determine the asymmetry ratio R, we took the value of the signal at each peak ($\omega_{res} \pm 2\pi(59$ kHz$)$) and subtracted from it the height of the pedestal resulting from the other Zeeman transition. To make this measurement process efficient, we determined a frequency at which the signal had a value similar to that of the pedestal at the frequency of the peak. We then just used the formula:

$$R = [sig(\omega_{peak1}) - sig(\omega_{ped2})]/[sig(\omega_{peak2}) - sig(\omega_{ped1})].$$

This method produced values that corresponded fairly well with the specified polarization. The integral of the signal was simply evaluated by:

$$I = \int_{\omega low}^{\omega hi} sig(\omega)\, d\omega.$$

In the following pages, we will be evaluating the changes in the polarization measurement as the result of changes in the circuit and other parameters. We looked at two polarizations: P = +50% and P = -50%. We will be presenting two final results: δPR, the change in polarization as measured by the ratio method; and δPI, the change as measured by the integral method.

$$\delta PR = P(R) - P(R_0)$$

$$\delta PI = [I/I_0]\cdot 50,$$

where R_0 and I_0 are the measurements made from the signal of the original tuned circuit. δPR and δPI are given in percent and are absolute changes based on 50% (e.g., a two percent increase means that the measurement went from 50% to 52%.)

Investigating Parameter Changes

1) Dispersive Contribution to Re[Vout(ω)]

(programs REALDNMR.MCD (P = +50%) and DNEGPOL.MCD (P = -50%), fig. 3)

We studied the contribution of the dispersive part of the susceptibility ($\chi'(\omega)$) on the real part of the output voltage and how it affected the polarization measurements.

In fig. 3 sig(ω,1,1) corresponds to the standard signal. sig(ω,1,0) is the signal that would result if $\chi(\omega) = \chi'(\omega)$ (no absorptive component), and sig(ω,0,1) is the signal that would result if $\chi(\omega) = i\chi''(\omega)$ (no dispersive component). The contribution of the cross-modulation ($\chi'(\omega)$) to the signal integral is:

$$\frac{I(sig(\omega,1,0)}{I(sig(\omega,1,1)} = \begin{array}{l} 9.7\%,\ P = +50\% \\ 11.6\%,\ P = -50\%. \end{array}$$

The values δPR and δPI come from the comparison of sig(ω,0,1) (only $\chi''(\omega)$) to the regular signal.

$P = +50\%$: $\delta PR = -0.41\%$ $\delta PI = 4.8\%$

$P = -50\%$: $\delta PR = -0.38\%$ $\delta PI = 5.8\%$

The cross-modulation affects the integral more than the ratio.

2) Effect of Phase Drift

(program DPHASE.MCD fig. 4)

We studied the effect that a drift in phase tuning occurring between the measurements of the baseline and the signal would have on the polarization

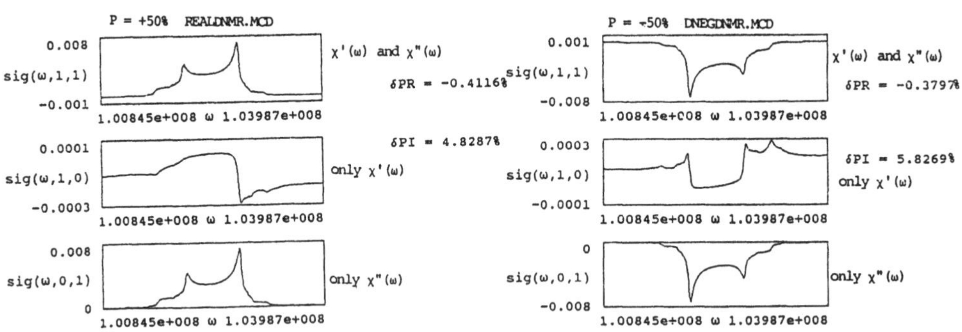

Fig. 3. Signal spectra with and without χ' and χ'' for P =50% and P = -50%.

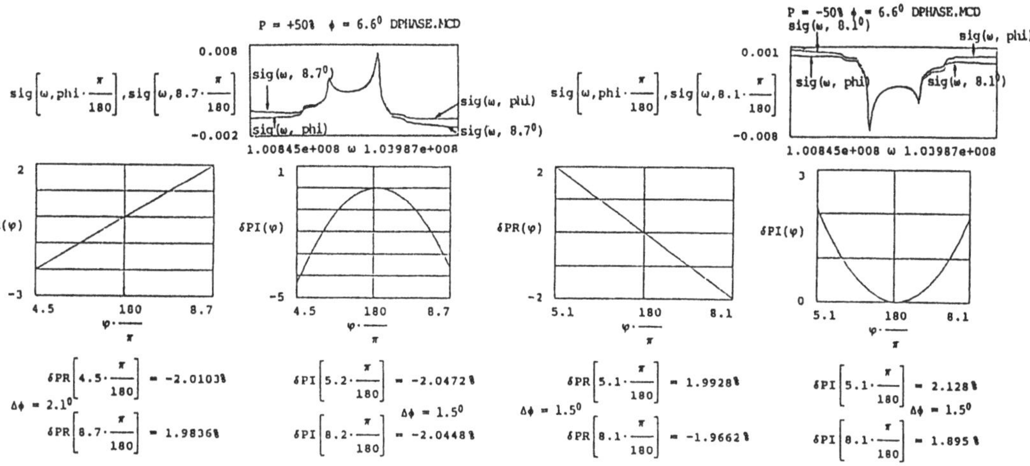

Fig. 4. Changes in signal and δP as a function of phase changes from 6.6°.

measurements. We calculated $\mathrm{Re}[V_{out}(\omega)]$ using a range of values for ϕ, and then determined the signal spectrum using the original baseline (ϕ = 6.6°).

In fig. 4, the values of $\delta PR(\phi)$ and $\delta PI(\phi)$ are plotted for a range of ϕ values. With positive polarization, the ratio method is slightly less sensitive to phase drift than the integral method. A phase drift of 2.1° will produce a change in polarization of ≈ 2% using the ratio method, while a drift of only 1.5° will produce the same change using the integral method. For negative polarization, a phase drift of 1.5° will produce a polarization change of ≈ 2% using both methods.

3) Effect of using $\phi(\omega)$
(program DPHI.MCD fig. 5)

In the original circuit, ϕ is adjusted so that the circuit is in tune at ω_{res}. However, for a constant ϕ, the circuit is out of tune in the entire frequency sweep. This could make a difference in the polarization measurement, especially the ratio method where neither of the peaks are located at ω_{res}.

We constructed a linear function $\phi(\omega)$ that compensated for the phase ($\arg[V_{out}(\omega)]$) of the voltage:

$$\phi(\omega) = \mathrm{phi} - m[\omega - \omega_{res}],$$

where:

phi = 6.6° (resonance tuning)

$$m = (\arg[V_{out}(\omega_{hi})] - \arg[V_{out}(\omega_{low})])/(2\pi(500\ \mathrm{kHz}))$$

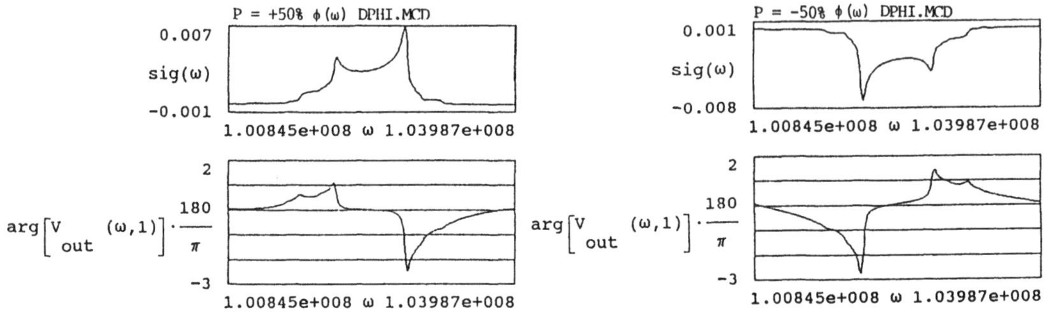

$\delta PR = 0.3373\%$ $\delta PR = -0.1915\%$

$\delta PI = -5.5541\%$ $\delta PI = -5.9938\%$

Fig. 5. Changes in signal and δP with phase = $\phi(\omega)$.

We then calculated the new baseline and output voltage, and obtained the polarization from the resulting signal. The values δPR and δPI are:

P = +50%: $\delta PR = 0.34\%$ $\delta PI = -5.6\%$

P = -50% $\delta PR = -0.19\%$ $\delta PI = -6.0\%$.

Conclusion

Here we have shown the effect of small changes in the parameters of the polarization measurement. The effects are slightly different for positive and negative polarization. We hope to utilize these results to improve on the basic Q-meter circuit.

References

1. MathSoft, Inc., One Kendall Square, Cambridge, MA 02139, USA.
2. O. Hamada et al., NIM 189 (1981) 561-568.
3. A. Abragam, The Principles of Nuclear Magnetism (Clarendon, 1961).
4. F. Sperisen, NIM A260 (1987) 455-462.

Polarization Studies with Radiation Doped Ammonia at 5 T and 1 K*

D.G. Crabb†

Randall Laboratory of Physics, The University of Michigan,
Ann Arbor, MI 48109, USA

The polarization properties of radiation doped ammonia have been studied in a ^{4}He evaporation refrigerator operating at a temperature of 1 K in a magnetic field of 5 T. The effects of an intense proton beam on these properties were also studied.

Introduction

As part of an ongoing study of proton-proton scattering at high $P_\perp^2$ we, at the University of Michigan, have operated polarized targets in intense proton beams over a number of years. In order to study such low cross section processes the cryogenic system must be able to withstand the heat load generated by the proton beam and still maintain an acceptable level of polarization. The polarizable material must be resistant to radiation damage and the polarization be recovered by annealing.

In recent years we have made a number of improvements to our 2.5 T/0.5 K evaporation refrigerator such as changing the fluid path and changing the refrigerant to a mixture of ^{3}He and ^{4}He[1,2]. This has allowed operation in a proton beam of about 2 x 10^{10} protons/sec; above this level the polarization reduces to an unacceptable level.

Over the same period of time radiation doped ammonia (NH_3) has been developed as a target material[3]. It has superior resistance to radiation damage, compared to chemically-doped materials and moreover, the polarization can be recovered completely after repeated anneals, again in contrast to chemically doped materials.

A typical polarizing curve at 0.5 K and 2.5 T for NH_3 is shown in Fig. 1. A polarization of about 70% is reached in about 2 hours. However, with an incident beam of 2 x 10^{10} protons/sec it is possible to maintain a polarization of only about 50%. In order to improve the figure of merit

* Research supported by a grant from the U.S. Department of Energy.

† The collaborators on this work were: C. B. Higley, A. D. Krisch, R. S. Raymond, T. Roser, J. A. Stewart, and G. R. Court.

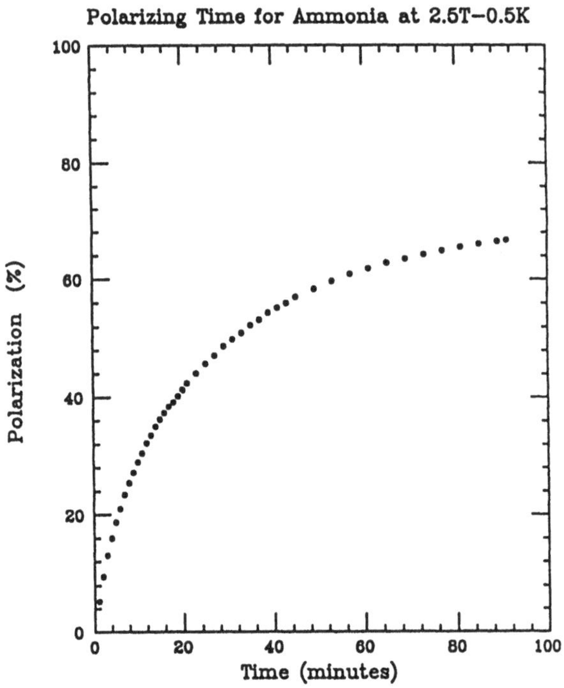

Fig. 1 Polarizing time for ammonia at 2.5 T–0.5 K.

(P^2I, the polarization squared times the beam intensity) for our scattering experiments it was decided to build a ^4He refrigerator operating at 1 K together with a 5 T magnet. The expectation was that the polarization at 1 K and 5 T should be at least the same as at 0.5 K and 2.5 T and that the superior thermal properties of ^4He and operation at 1 K would allow a polarization of 70% to be maintained at high beam intensity. Thus an increase in the figure of merit by at least a factor of two was expected. Data from Seely et al.,[4] using the SLAC-Yale target indicated that we might do even better.

Results from this 5 T/1 K system have already been published[5] and this report summarizes these results and presents new data from operation in an intense proton beam.

Polarized Target System

The refrigerator and magnet system is shown in Fig. 2. The split coil superconducting magnet was supplied by Oxford Instruments and operated at a field of up to 5.1 T with a field uniformity of $\pm 10^{-4}$ over a 4 cm diameter sphere. The ^4He refrigerator was inserted vertically along the bore of the magnet, while the target cylinder, containing NMR coils

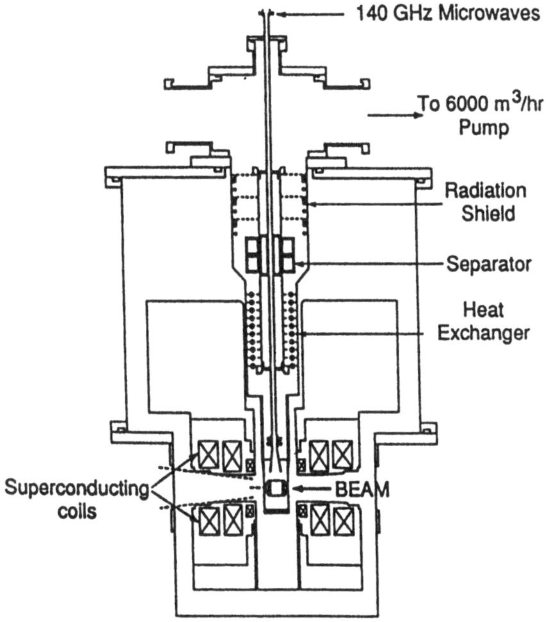

Fig. 2 Diagram of the new polarized-proton-target apparatus. The superconducting magnet produces a highly uniform 5 T field. The ^4He cryostat produces about 0.9 W of cooling power at 1 K. The target material is contained in the small cavity at the bottom of the cryostat. The 140 GHz microwaves are fed into the target cavity via the horn. The proton beam will pass through the target cavity.

with attached cables, microwave guide and liquid level and temperature sensors, was inserted into the refrigerator along the vertical axis of the magnet. Microwaves were supplied by a Varian EIO tube operating at 140 ± 2 GHz with a maximum power output of 21.8 watts. The polarization was measured using an NMR coil as part of a series LCR Q-meter of the Liverpool type. Further operating details of the system can be found in reference[5].

Results

a) Initial Results from NH_3

These were reported in reference[5] but are shown again in Fig. 3. Polarizations of 95% were obtained along with rapid growth (90% in 25 minutes). The polarization character of the same material at 0.5 K and 2.5 T is shown for comparison. The NH_3 had been irradiated at the MIT Bates Linac according to the prescription given in reference[6]. This particular sample was irradiated at an average electron beam current of

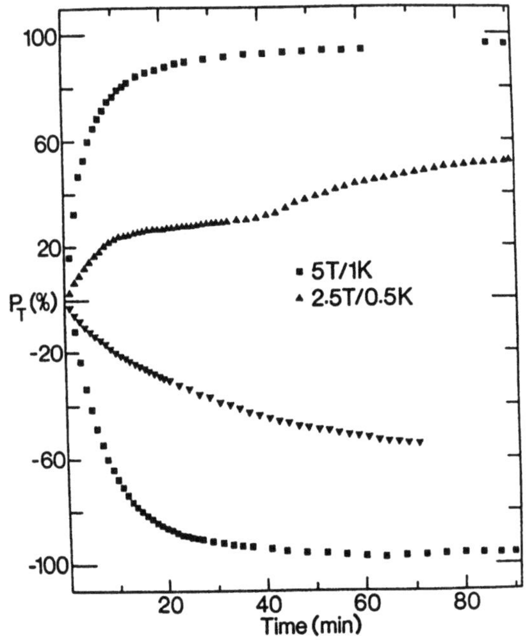

Fig. 3 The spin polarization of the free protons in NH_3 is plotted against the time of microwave irradiation. The new data at 5 T and 1 K are shown as squares; the earlier data at 2.5 T and 0.5 K are shown as triangles.

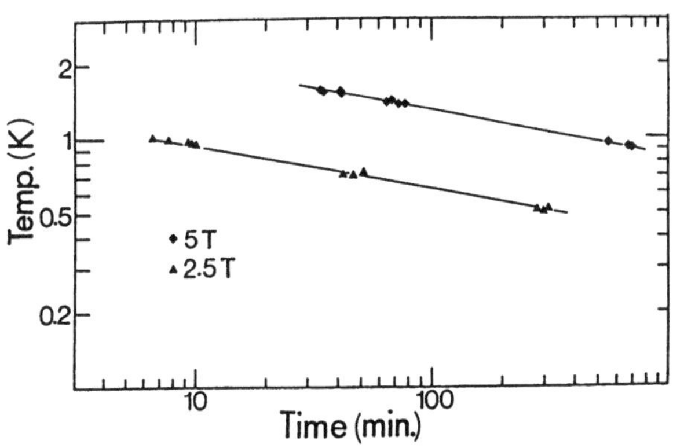

Fig. 4 The spin-relaxation time for the protons in NH_3 is plotted against temperature. The new data at 5 T are shown as diamonds; the earlier data at 2.5 T are shown as triangles.

1.5 μA to a total of 10^{17} electrons incident. The relaxation times for each refrigerator condition are compared in Fig. 4. The cooling power of the refrigerator was 0.9 watts at 1 K and 1/3 watts of microwave power was delivered at the target during normal operation.

b) Results after Installation in Spectrometer

After the initial test period in our laboratory the polarized target assembly was moved and set-up as the target in our spectrometer at the AGS at Brookhaven National Laboratory.

The polarization results from this new environment are shown in Fig. 5. As can be seen the positive enhancement growth and ultimate value are similar to those in Fig. 3, whereas the negative enhancement has a much slower growth. This we attributed to the presence of a large iron spectrometer magnet very close to the polarized target and which was distorting or shifting the uniform field region away from the target volume.

Fig. 6 shows the frequency dependence of the polarization of the protons in NH_3 and is similar to that obtained at 2.5 T and does not show any further separation of the positive and negative enhancement peaks[7]. The negative enhancement shows the same slow rise as in Fig. 5.

c) Results with beam on target

Fig. 7 shows the polarizations obtained over the course of a three month experimental run in a 24 GeV/c proton beam at the AGS. The average beam intensity was about 2 x 10^{11} protons per pulse. The polarization is plotted against the dose received by each of the two target batches. Each point represents the average polarization measured over

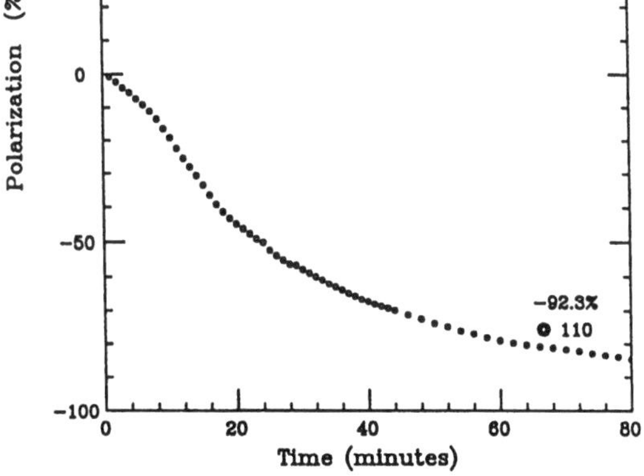

Fig. 5 Polarizing time for ammonia at 5 T–1 K 1st NH_3 target; before beam.

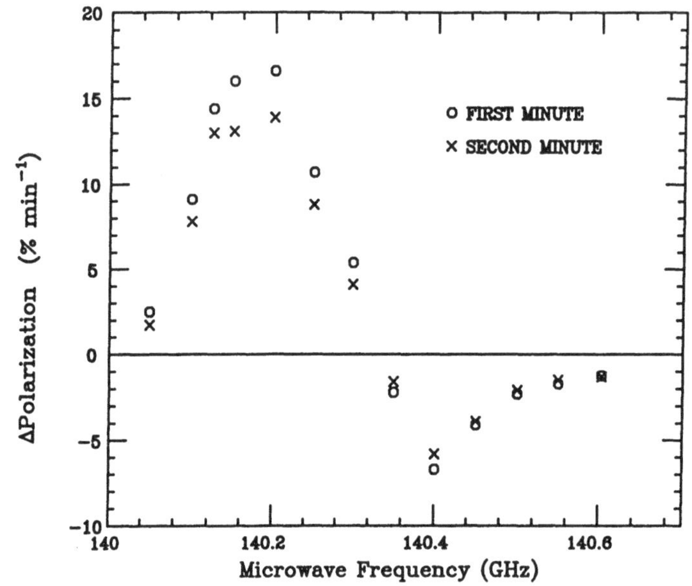

Fig. 6 Frequency sweep for NH_3 at 5 T–1 K.

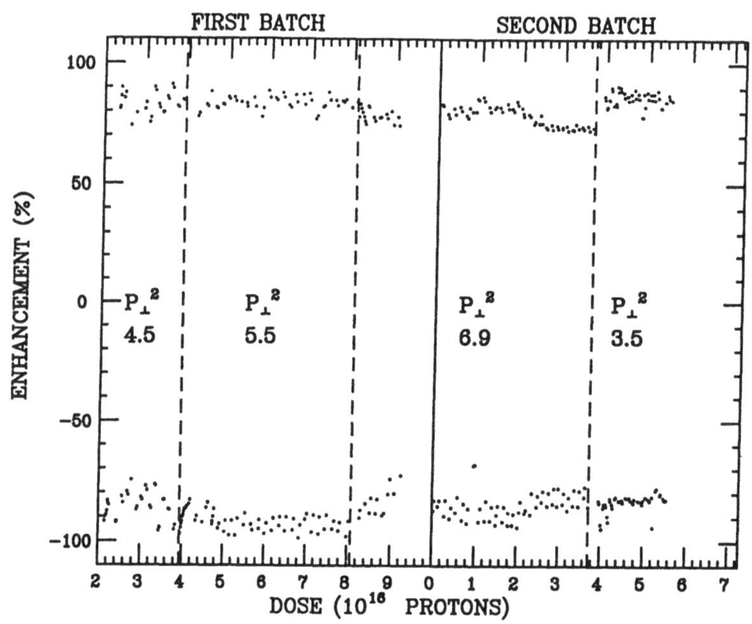

Fig. 7 Average enhancement.

the course of a data run of duration of about one hour. An average polarization of about 85% was maintained over the course of the experiment.

The NH_3 target was usually annealed after a set of four data runs over which time about 10^{15} protons had been incident on the target. The annealing procedure was to maintain the target material at 80 K (as measured by a gold-iron/chromel thermocouple buried in the material) for two minutes. After an anneal the polarization returned to its previous high value. However after about 9 x 10^{16} protons had traversed the target this was no longer true, as annealing at 80 K had no effect and the polarization continued to fall as the target was further irradiated. Finally an anneal at 100 K for fifteen minutes did have an effect, in that the polarization achieved was considerably reduced. We note that 9 x 10^{16} protons on the target is approximately the same as the number of electrons that were delivered in the original irradiation. A simple explanation is that the density of polarizing centers ($\dot{N}H_2$) became too high and that an 80 K anneal was not sufficient to reduce the number significantly. The 100 K anneal was at high enough temperature for significant recombination of $\dot{N}H_2$ to take place thereby affecting the resulting polarization. A possibility for further study would be to anneal at gradually increasing temperatures and see how the polarization is affected.

Another feature of Fig. 7 which should be noted is the change in polarization behavior at changes in the spectrometer set-up for different $P_\perp^2$ data sets. This generally meant a change in the field of the iron magnet next to the target magnet and/or a change in relative field directions for the two magnets. This would tend to exacerbate or compensate the effect on this uniform field region discussed above.

Fig. 8 shows the microwave frequency necessary to optimize the polarization over the course of the experiment. The tuning system is manual

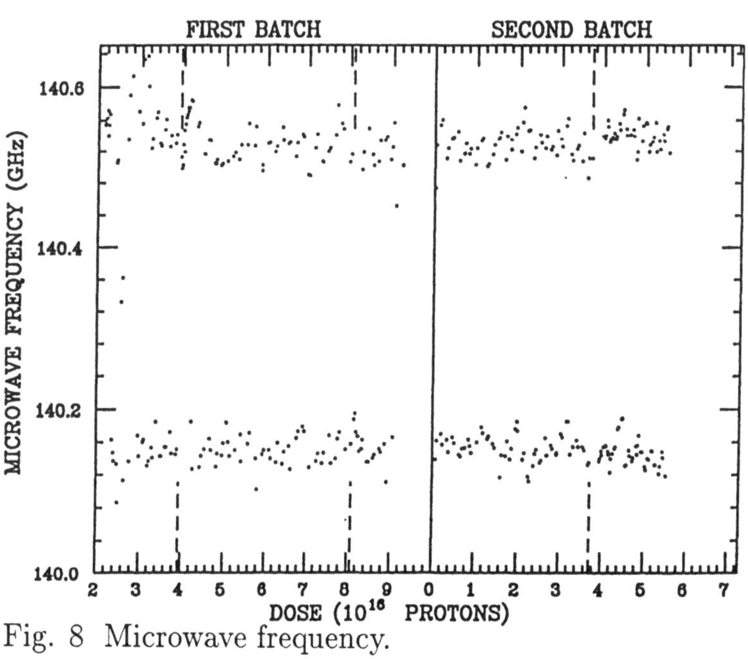

Fig. 8 Microwave frequency.

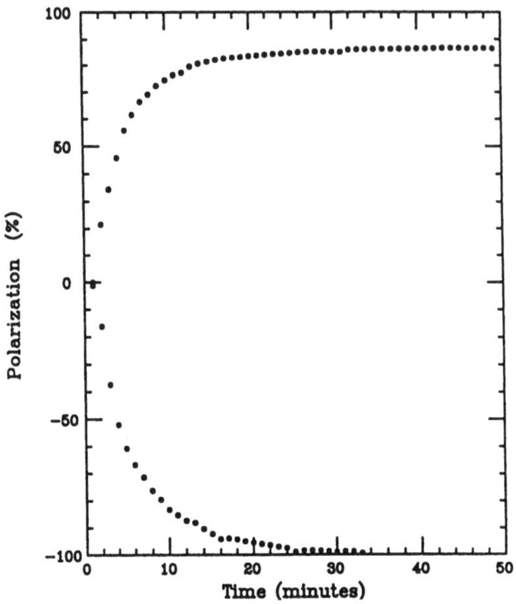

Fig. 9 Polarizing time for ammonia at 5 T–1 K 1st NH_3 target, after 3 x 10^{16} protons.

and so all experimenters were obliged to tune the polarization. The learning curve is indicated in the figure; there is a large scatter in frequency at the start, but settles down as the experiment progressed. The separation between the frequencies for the two enhancements was about 375 MHz. However at the start, at zero radiation damage, the difference was about 250 MHz (see also Fig. 6). This change in frequency difference has been noted before[8] at lower fields.

Other data of interest are shown in the following figures: Fig. 9 shows the polarization build-up after some damage to the material. There is now a rapid build-up of the negative enhancement compared to Fig. 5, while the positive enhancement still rises rapidly. Fig. 10 shows the fall in polarization as a function of radiation dose. There are two different exponential regions with characteristic values 4 x 10^{15} protons/cm^2 and 1.3 x 10^{16} protons/cm^2. These values are in good agreement with the data of Seely et al.[9].

Fig. 11 shows the effect on the polarization of beam heating of the target. A proton beam of ~ 2 x 10^{11} protons/pulse (~ 7 x 10^{10} protons/sec) reduces the polarization from 87% to 85%. Thus the figure of merit has been improved over the previous system by a factor of 10.

d) Other results

At the end of the scattering experiment further tests were carried out but were limited by a curtailment of the liquid helium supply.

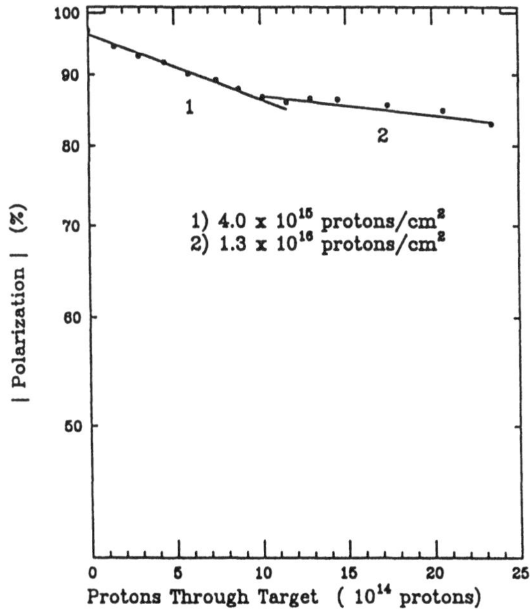

Fig. 10 Polarization decay due to radiation damage of NH_3 at 5 T–1 K, first target of E794.

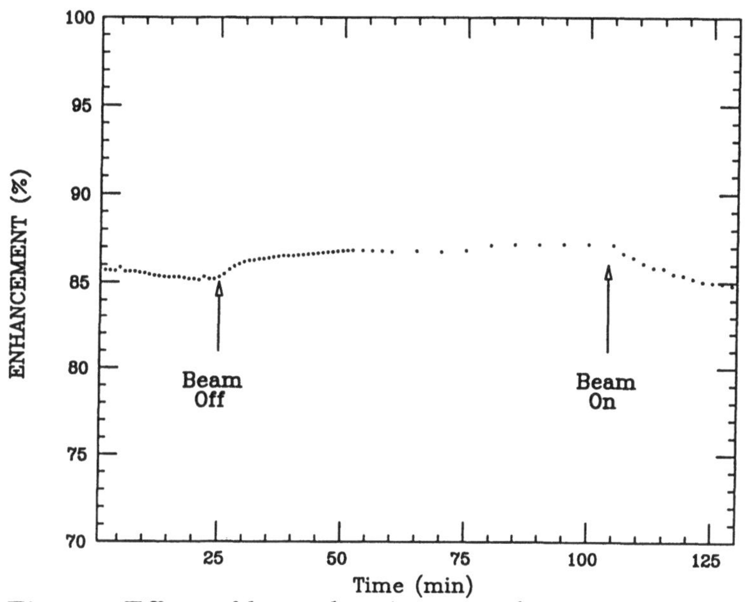

Fig. 11 Effect of beam heating on polarization during run 1778.

Three different samples of NH_3 were tested; they had been irradiated at electron currents of 1 μA, 1.5μA and 2μA each to 10^{17} incident electrons. The polarization results are shown in Fig. 12; these are to be compared with Fig. 13 which shows the same material polarized at

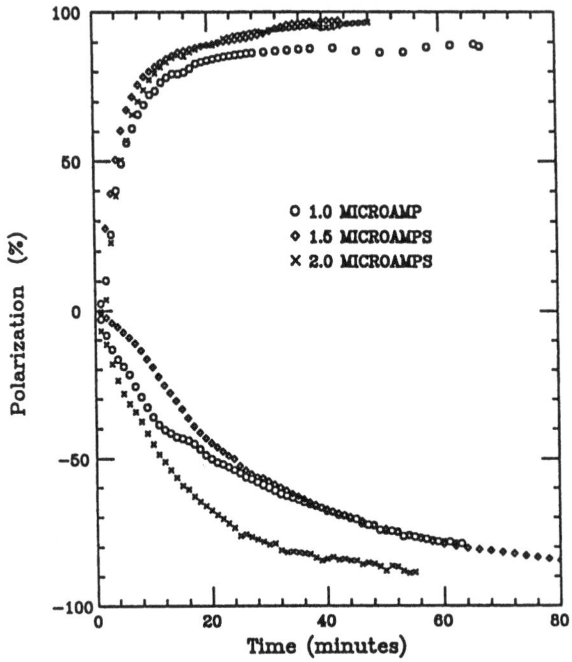

Fig. 12 Polarizing time as a function of initial irradiation current (5 T–1 K).

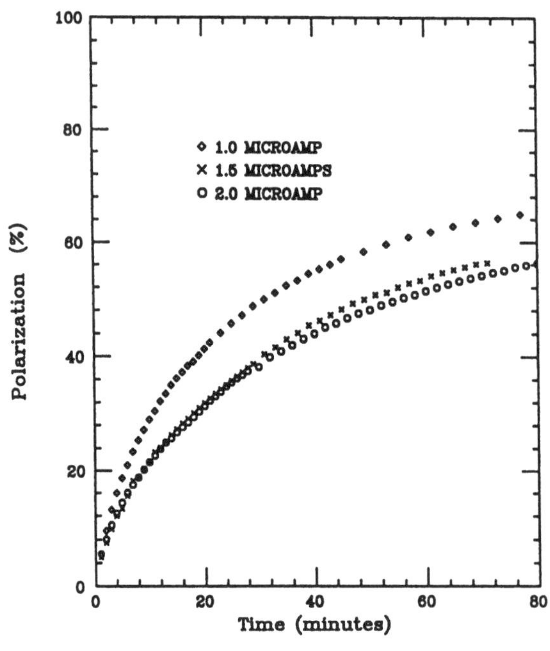

Fig. 13 Polarizing time as a function of initial irradiation current (2.5 T–0.5 K).

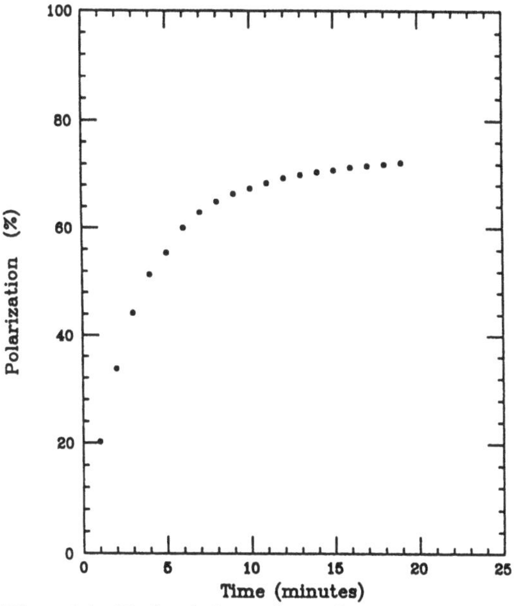

Fig. 14 Polarizing time for ammonia at 5 T–1.62 K.

2.5 T and 0.5 K. In Fig. 12 the best results are obtained from the 2μA material, with the 1μA clearly worse: the 1.5μA material is intermediate. From Fig. 13 the reverse is true; the 1μA material is best, the 2μA worst, and the 1.5μA again intermediate. If we assume the hypothesis that irradiating at higher beam currents produces less centers because of material heating and thus annealing, then these results suggest that less centers are required for optimum polarization in the 5 T/1 K case than in the 2.5 T/0.5 K situation.

Fig. 14 shows the polarization character of the 2μA sample at 1.62 K. The growth is rapid rising to ~72% in 20 minutes so even at this high temperature the NH_3 target is superior than when operated at 2.5 T/0.5 K.

Summary of Results

Finally we summarize the results of polarizing NH_3 at 5 T and 1 K.

a) NH_3 polarization very high with rapid growth; ~90% in 20 minutes to >95%.

b) Polarization relatively insensitive to temperature; >70% in 20 minutes at 1.6 K.

c) Optimum center density probably less than at 2.5 T/0.5 K.

d) Much higher radiation resistance than other materials, with an initial exponential fall of characteristic value ~ 4 x 10^{15} protons/cm^2.

This is even better than with NH_3 at 2.5 T/0.5 K[10] and the positive and negative enhancements are about the same.

e) Repeated anneals (2 minutes at 80 K) with no obvious loss in polarization. However there may be a threshhold after $\sim 10^{17}$ protons have passed through the target.

Conclusion

Finally we have a superb target for performing scattering experiments in high intensity beams with a factor of 10 improvement in the figure of merit (P^2I) over our 2.5 T/0.5 K polarized target.

References

1. D. C. Peaslee et al., Phys. Rev. Lett. **51**, 2359 (1983).
2. P. R. Cameron et al., Proc. of the 4th International Workshop on Polarized Target Materials and Techniques, Bad Honnef, ed. W. Meyer, 1984, p. 143.
3. Ibid, p. 7 to 53.
4. M. Seely et al., NIM **201**, 303 (1982).
5. D. G. Crabb et al., Phys. Rev. Lett. **64**, 2627 (1990).
6. R. S. Raymond et al., Proc. of the International Conference on Polarized Sources and Targets, Montana 1986, Helvetica Phys. Acta **59**, 777 (1986).
7. C. D. Jeffries, private communication.
8. K. A. Althoff et al., ibid (2), p. 23.
9. M. Seely et al., Proc. of the International Conference on High Energy Spin Physics, Minneapolis 1982, ed. G. Bunce, AIP Conf. Proc. **95**, 526 (1982).
10. D. G. Crabb et al., ibid (2), p. 7.

Polarization Behaviour of NH_3 at 3.5 Tesla and in the Frozen Spin Mode

K.H. Althoff, H. Dutz, S. Goertz, D. Krämer, W. Meyer, L. Reichertz, G. Reicherz, W. Thiel, A. Thomas, and J. Weber

Physikalisches Institut der Universität Bonn,
Nußallee 12, W-5300 Bonn 1, Fed. Rep. of Germany

With the new Bonn Frozen Spin Target we investigated the polarization behaviour of $^{14}NH_3$ *in a magnetic field of 3.54 Tesla and at a temperature of 70 mK. We got a maximum polarization of* $+94 \pm 2.5$ % *at a frequency of 98.750 GHz and -100 %* [1] *at 99.075 GHz. The polarization build up time* τ *was about 90 minutes. Examining the relaxation time at a holding field of 0.35 Tesla a superradiance effect was observed in the case of high negative polarization. The resulting signals show the same lineshape as signals during the polarization reversal by microwave irradiation. In case of positive polarization a relaxation time greater than 30 days was measured. Finally the maximum polarization of* $^{15}NH_3$ *was measured in a slightly different apparatus.*

Introduction

At the PHOENICS[2]-Detector, situated at the Electron Stretcher Accelerator ELSA single and double polarization observables in the single pion photoproduction ($\gamma p \to \pi^+ n$) and deuteron photodisintegration ($\gamma d \to pn$) will be measured. Especially for this reason a new frozen spin target of the Saclay type was built in Bonn. With this target a spin orientation in any direction as well as the use of a polarizing field up to 7.0 Tesla is possible. Beside the test of the whole equipment the maximum polarization at 3.54 Tesla and the relaxation time in a holding field of 0.35 Tesla for $^{14}NH_3$ was measured. The main components used for this measurements are:

- a vertical dilution refrigerator reaching a temperature of 70 mK ($\sim$ 250 mK during polarization)
- a superconducting polarization magnet with a maximum field of 7 Tesla and a field homogenity of $\frac{\Delta B}{B} = 10^{-5}$
- an impatt diode as microwave source producing a frequency of 99 GHz with a nominal power of 300 mW
- a phase sensitive NMR system, based on the Liverpool equipment and a VME Computer system, working at a frequency of 150.134 $\pm$ 0.2 MHz, which corresponds to a magnetic field of 3.54 Tesla

More detailed description can be found in the contribution of H. Dutz to this proceedings and the references therein.

---

[1] greater than 98.5% with a confidence level of 90 %

[2] Photon Experiments on Nuclei in Counter Setups

Target Material

An important property of ammonia is its high amount of free polarizable protons (0.17) compared to the alcohol materials. Therefore first examinations of ammonia was carried out already 1970 [Sche] with a chemical doped material. But only with the method of irradiation doping reproducibility can be achieved. The first time this method was used with $^{14}NH_3$ at CERN in 1979 [Nii] and in 1980 with $^{14}ND_3$ in Bonn [Här]. A second advantage of this irradiated material is its high radiation resistance compared to chemical doped material [Schi, See]. Therefore it is possible to use ammonia in an intensive electron beam which allows new kind of experiments with polarized targets.

The preparation of the material consists of two separate steps: In the first step white coloured crystalline chips of an approximately size of 2 mm are produced by slowly freezing gaseous ammonia using a dry ice methanol mixture. In the second step the paramagnetic centers have to be produced by irradiation. This irradiation was carried out at the 20 MeV electron injection Linac of the Bonn synchrotron at a temperature of 90 K. We used an argon cryostat [Bro] with a Faraday cup and a current integrator as irradiation monitor. With an integrated current of $350 \cdot 10^{-4}$ Cb we got optimal polarization results [Schi]. In this case the sample had a dark violet colour.

For the measurements 3 different samples were used: Two $^{14}NH_3$ samples irradiated 1986 and 1989 with an integrated current of $350 \cdot 10^{-4}$ Cb and $363 \cdot 10^{-4}$ Cb, respectively; as the paramagnetic centers are not completely stable at liquid nitrogen temperature, the first sample (1986) had changed its colour to pale violet. One $^{15}NH_3$ sample has been irradiated together with the second $^{14}NH_3$ sample.

Polarization Measurements of $^{14}NH_3$

The polarization was determined by comparing the enhanced signal with the thermal equilibrium (TE) signal. The calibration factor was calculated from the mean value of 10 to 15 TE signals measured with a statistical error of 2.5%. During this calibration the temperature was stable to ±5 mK. This temperature was measured with a calibrated 470 Ω Speer resistor. In the first sample (pale violet) we reached a maximum positive polarization of +94 ± 2.5% at a microwave frequency of 98.750 GHz and a negative polarization greater than 98.5% with a confidence level of 90% at a microwave frequency of 99.075 GHz. Simply speaking this corresponds to a negative polarization of -100%. This is the first time that an ammonia sample could be totally polarized. A comparable result with the material 1,2-propandiol was reported 1973 [deB]. The positive and negative polarization signals are shown in Fig. 1. Comparing both signals it can immediately be seen that the linewidth of the negative signal is smaller than the positive one. The corresponding full half line width are 47 kHz and 60 kHz. A second fact is the asymmetric line shape which depends on the polarization value. Details of this lineshape can easily be seen from Fig. 2 where the first derivative of both signals is shown. The lineshape is in agreement with the signals published by Niinikoski [Nii]. The asymmetry and the structure of the lineshape are most likely caused by the dipole-dipole interactions of neighbouring spins. G.Heyes calculated the lineshape of the proton signal taking into account the 3 protons within an ammonia molecule and using an analytical expression for the energy levels of triangular spin $\frac{1}{2}$ configuration [Hey]. The calculated signals are in good agreement with the observed ones. Nevertheless the difference between positive and negative polarization we found in our measurements can not be explained by this method and is still an open question. Comparing

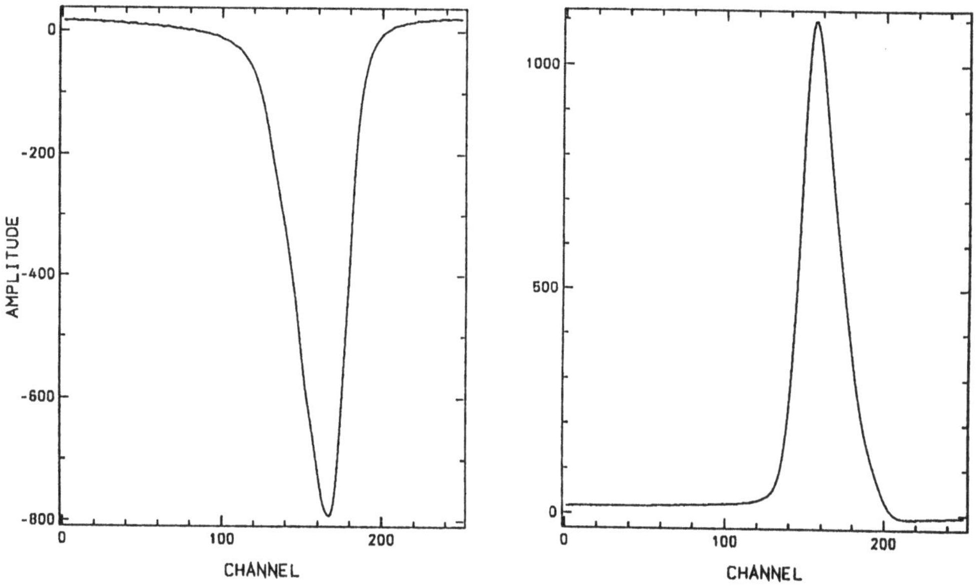

Figure 1: Maximum positive and negative Polarization in $^{14}NH_3$ at 3.5 Tesla

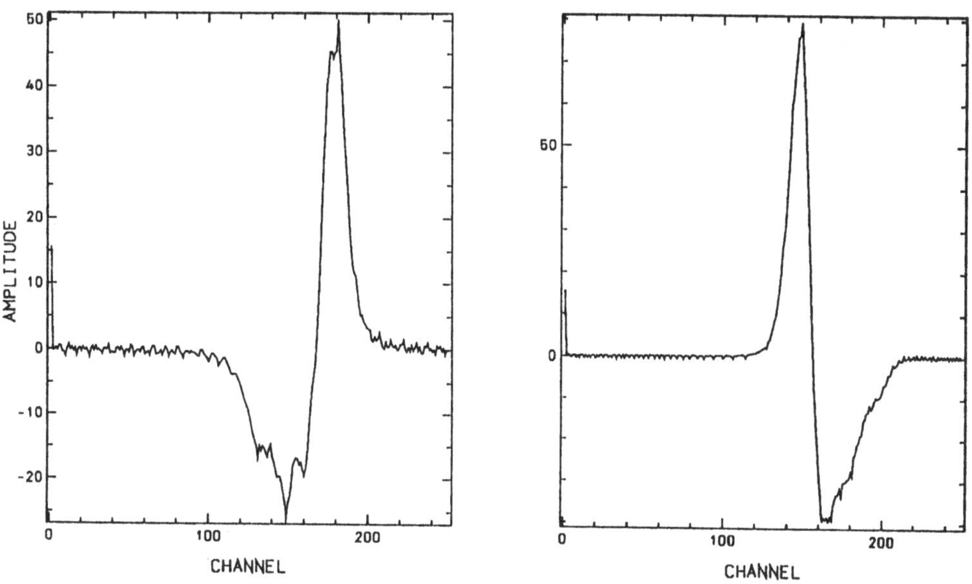

Figure 2: First derivative of the maximum polarization signals shown above

the polarization build up time[3] including older measurements with the same sample at 2.5 Tesla and a 10 W carcinotron as microwave source (see table 1) we found comparable build up times at 2.5 and 3.5 Tesla and a shorter build up time using higher microwave power. The influence of the higher amount of paramagnetic centers can be seen from measurement 4, giving a considerably faster polarization build up time.

[3]defined by the equation $P = P_0(1 - \exp(-\frac{t}{\tau}))$

Table 1: Polarization build up time τ under various conditions

no	B [T]	Sample	Microwaves	τ [min]
1	2.5	pale violet	10W-Carcinotron	104
2	3.5	pale violet	Impatt-Diode, reduced power	120
3	3.5	pale violet	Impatt-Diode, full power	90
4	3.5	dark violet	Impatt-Diode, full power	37

Relaxation Measurements

Combining the results at 1 K of our measurements at 2.5 and 3.5 Tesla (14 and 78 min) with the results at 5 Tesla (550 min) obtained by D. Crabb [Cra] gives an exponential behaviour according $\tau \sim e^B$. In the case of the PHOENICS experiment the target will be used in frozen spin mode with a holding field of 0.35 Tesla. The relaxation times at this field is therefore an important parameter determining the polarization reversal intervals. Starting with an positive polarization of +90% and measuring two successive values gave a relaxation time of $\tau \sim 30$ days. The relaxation time τ is here defined by the equation $P = P_0 \exp\left(-\frac{t}{\tau}\right)$. Starting with a negative polarization (-90%) we got the surprising result that the polarization changed immediately to a positive polarization of +7%. The corresponding signals are shown in Fig. 3. A further investigation of this effect showed a first decrease of polarization of 10% at a magnetic field of about 1.3 Tesla and a reversal below a field of 0.5 Tesla. Starting with a negative polarization of -18% nothing happened. During this measurements the NMR module was always connected to the cryostat. Replacing it by a 50 Ω resistor during lowering the field changed the situation in such a way that the effect disappeared. Although it was the first time that this effect was observed in Bonn it is well known under the name of supperradiance effect and was first

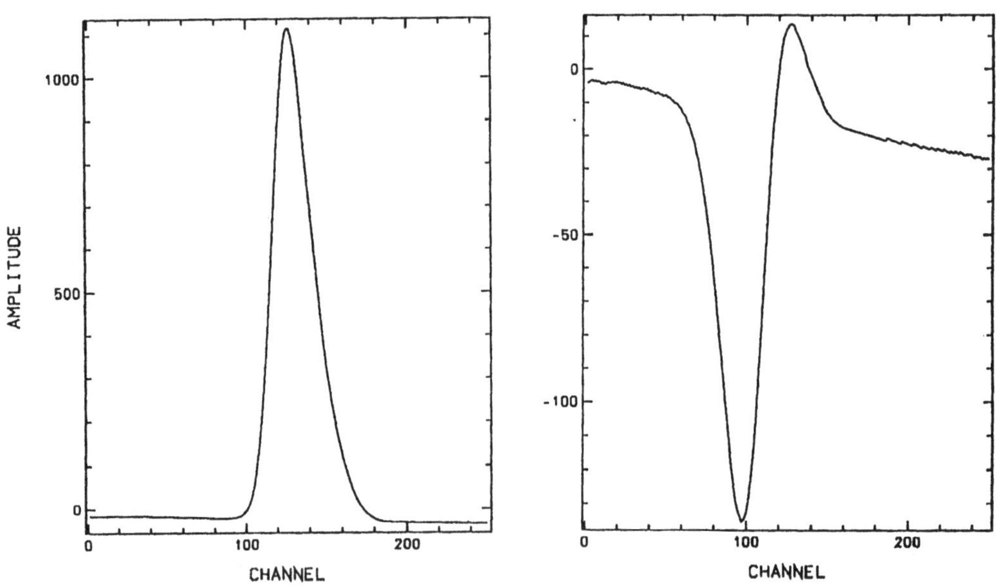

Figure 3: NMR signal before and after lowering the magnetic field to 0.35 Tesla

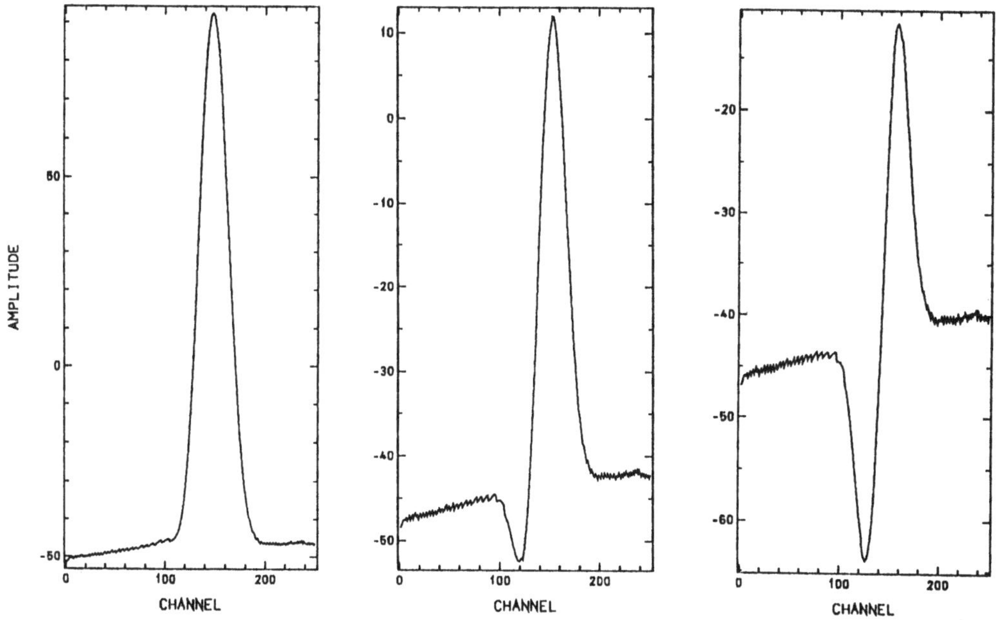

Figure 4: Polarization reversal by microwave irradiation from negative to positive polarization

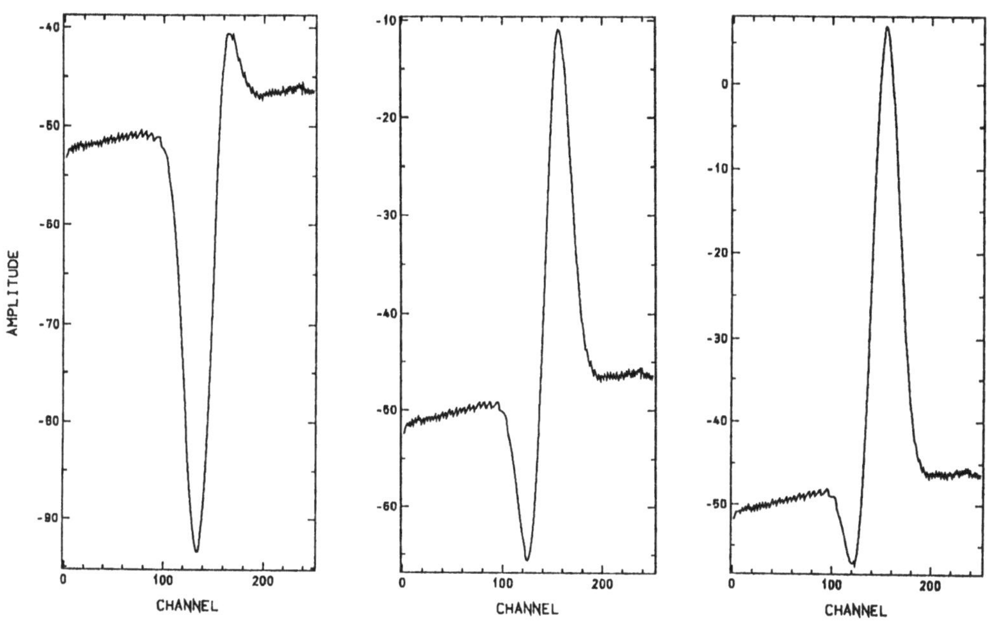

Figure 5: Polarization reversal by microwave irradiation from positive to negative polarization

described by Bösinger [Bös] and investigated in detail by Kovalev [Kov]. The results obtained by Kovalev could be confirmed with our equipment[Rei]. An interesting observation is the fact that the resulting NMR signal has a positive and a negative peak and does not change with time. A similar lineshape can be observed during polarization reversal by microwave irradiation, which is illustrated in Fig. 5 and 4 for negative and positive starting polarizations.

The $^{15}NH_3$ Measurements

The measurements with the $^{15}NH_3$ were carried out in a 4 Tesla Helmholtz magnet with a field homogenity of $\frac{\Delta B}{B} = 5 \cdot 10^{-5}$. Here we reached a maximum polarization of $+94 \pm 5$ % at 98.930 GHz and -90 ± 5 % at 99.260 GHz. The relative low negative polarization is due to instabilities of the microwave source at the required frequency. Again the lineshape shows the characteristic asymmetry and different line width for positive and negative signals. But in contrast to the $^{14}NH_3$ signals the $^{15}NH_3$ signals are narrower: 34 kHz for the positive and 24 kHz for the negative signal). This is probably due to the difference of the ^{14}N and ^{15}N spin.

Conclusions

Combining this results with the one presented by D. Crabb and S. Mango in this proceedings ammonia with its high polarization is an important and interesting target material offering still some unexpected behaviour. Further detailed investigations should be done.

References

[Bös] P. Bösiger, E. Brun and D. Meier, Phys. Rev. **A 18,2** (1978) 671

[Bro] S. Brown, Proc. 4th Int. Workshop Pol. Target Material and Techniques, Bad Honnef 1984, Ed. W. Meyer, page 66

[Cra] D.G. Crabb, this proceedings and
D. G. Crabb et al., Phys. Rev. Let. **64,22** (1990) 2627

[deB] W de Boer and T. O. Niinikoski, Nucl. Instr. and Meth. **114** (1974) 495

[Här] U. Härtel, 1980, Proc. of the Int. Symp. on High Energy Physics with Pol. Beams and Pol. Targets, Eds. C.Joseph and J.Soffer, Birkhäuser, Basel, page 451

[Hey] G. Heyes, Liverpool, Thesis 1985 (unpublished)

[Kov] A. Kovalev, personal information
N.A. Bazahanov et al., Sov. Phys. Solid State **31,2** (1989) 291

[Nii] T.O. Niinikoski and J.M. Rieubland Phys. Lett. **72A** (1979) 141

[Rei] L. Reichertz, Diplomarbeit, BONN-IR-91-04

[Sche] K. Scheffler, Nucl. Instr. and Meth. **82** (1970) 205

[Schi] E. Schilling, see Ref. [Bro] page 13

[See] M.L. Seely et al., Nucl. Instr. and Meth. **201** (1982) 303

DNP of $^{15}NH_3$

B. van den Brandt, J.A. Konter, S. Mango, and M. Weßler
Paul Scherrer Institute, CH-5232 Villigen PSI, Switzerland

New measurements cast doubts on the validity of the equal spin temperature (EST) theory in $^{15}NH_3$.

Introduction.

For the measurement of the analyzing power A_y in $\pi^{+15}\vec{N}$ elastic scattering at the PSI meson facility we have operated a $^{15}NH_3$ polarized target on the πM1 channel for an extended period of time. The experimental set up allowed to measure in parallel the analyzing power A_y in $\pi^+\vec{p}$ scattering, for which good experimental data were already available. During the experiment the ^{15}N polarization has been monitored continuously, and its value has been determined from the ^{15}N TEP signal at 1.02 K, with an estimated error Δ P/P $\simeq \pm$ 5%, using the PSI-NMR system, described elsewhere in these proceedings [1]. Initially, the proton polarization has been deduced from the measured ^{15}N polarization under the assumption that the two nuclear spin systems in the sample were in thermal equilibrium [2].
Disagreement with the published data and the PSA predictions for A_y in $\pi^+\vec{p}$ elastic scattering prompted us to perform a series of control measurements in the laboratory, of which we present a short report.

The experiment on the beam.

The material (99 at.% ^{15}N enriched NH_3 by MSD Isotopes, Canada) had been irradiated and prepared for us by W. Meyer and collaborators at Bonn, and it had the form of dark violet irregular chunks, 2-3 mm across. These chunks have been ground to a fine powder to improve the filling factor of our target container, a brass box of 18 x 18 x 5 mm across, with a copper double spiralled NMR coil on a glass fiber reinforced substrate in its median plane.
On the beam ^{15}N polarizations of $\sim$ 17% could be repeatedly achieved, albeit after days of microwave irradiation, at dilution refrigerator temperatures in a field of 2.5 Tesla with a homogeneity not better than 3.10^{-4} over the target volume. The optimum polarization frequencies were found to be $\simeq$ 350 MHz apart, and the ^{15}N spin-lattice relaxation time at 1.02 K, 2.5 T has been measured to be less than 10 minutes.

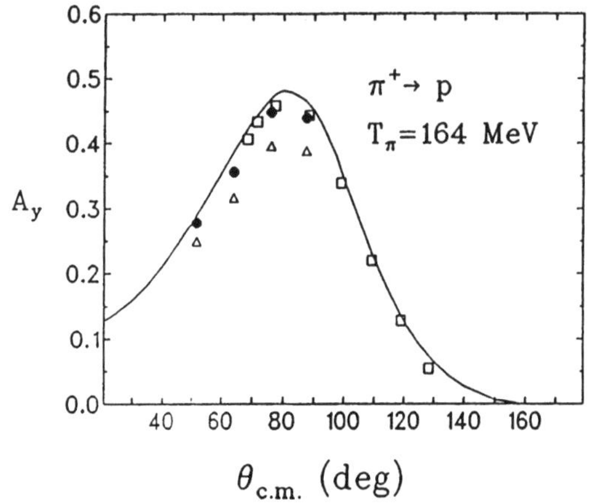

Figure 1. The asymmetry A_y for $\pi^+\vec{p}$ scattering at 164 MeV.
□ - M.E. Sevior, Phys. Lett.C (1989) 278.
△ - this experiment, for ^{1}H polarizations as predicted by the EST theory.
● - this experiment, for ^{1}H polarizations as experimentally correlated to the ^{15}N polarizations.

The off line analysis of the $\pi^+\vec{p}$ data, using proton polarization values derived from the measured ^{15}N polarization under the assumption of equal spin temperatures gave results in disagreement with the published experimental data and the PSA (s. fig. 1).
The disagreement could be wiped out assuming that the ^{15}N polarization had been systematically overestimated by more than 30%. This correction was definitely too large and in our opinion outside of all possible sources of errors.

The laboratory tests.

We decided to perform in the laboratory a control measurement of the same sample in the same set-up as used on the beam, but with an additional coil in the cavity to measure also the proton polarization. We took TEP signals of ^{15}N and ^{1}H at 1.03 K and 1.4 K for the calibration of our NMR system, and we measured in parallel the proton and the ^{15}N polarization. We estimate our relative measurement error to be for both nuclei not worse than ± 5% (s. fig. 2).
A plot of the proton polarization *vs.* the ^{15}N polarization showed significant deviations from the EST theory (s. fig. 3).
Use of the relationship between the proton polarization and the ^{15}N polarization measured in the laboratory to determine the proton polarization during the experiment on the beam, improved the agreement with the previous experimental data dramatically (s. fig. 2).

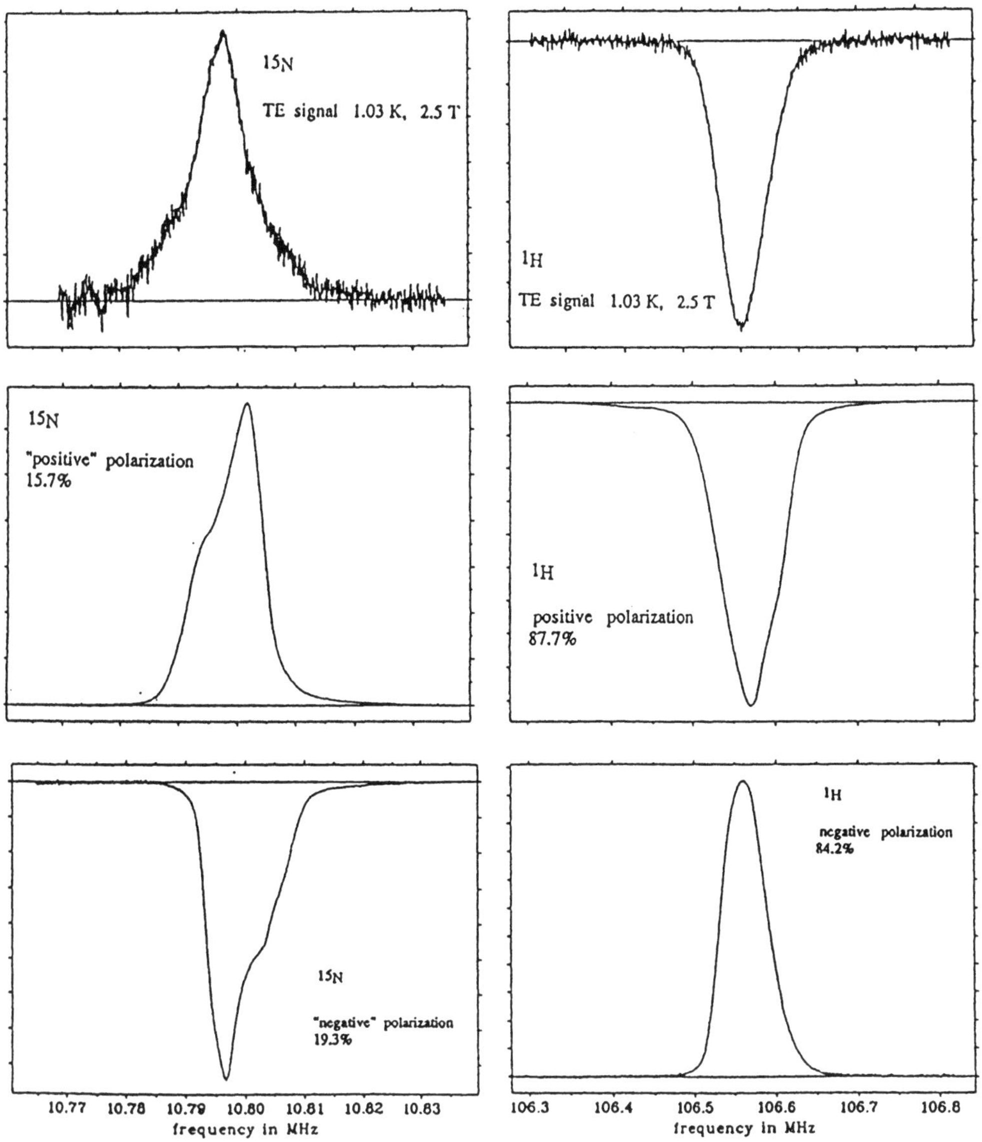

Figure 2. ^{15}N and ^{1}H polarization signals.

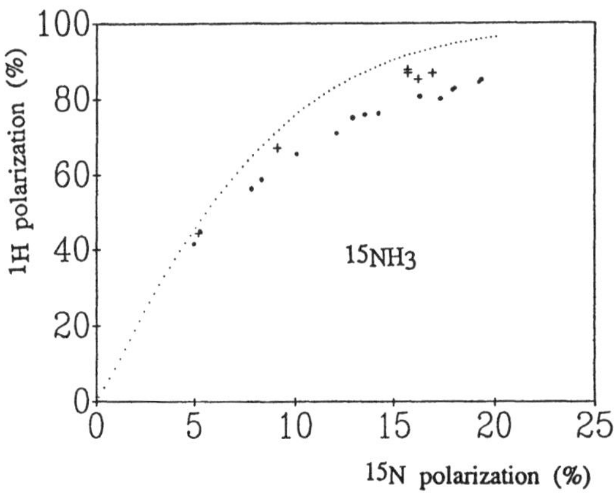

Figure 3. The proton polarization *vs.* the ^{15}N polarization.
+ - positive polarization, o - negative polarization.

Conclusions.

We conclude that either the two spin systems are not at the same temperature, or, being at the same temperature, one of them (or both) is (or are) not a free spin $\frac{1}{2}$ system. We think that more work should be devoted to investigate and understand NH_3 better, especially its ground state.

References.

1. M. Weßler *et al.*, these proceedings.
2. G.R. Court and W.G. Heyes, NIM **A243**(1986) 37-40.

Butanol in the Frozen Spin Mode

H. Dutz, S. Goertz, D. Krämer, W. Meyer, L. Reichertz, G. Reicherz, W. Thiel, A. Thomas, and J. Weber

Physikalisches Institut der Universität Bonn,
Nußallee 12, W-5300 Bonn 1, Fed. Rep. of Germany

Measurements of the polarization build-up in butanol at 3.5 T are presented. Measurements of the proton relaxation times in the frozen spin mode were performed at magnetic fields of 0.5 and 0.35 Tesla, leading to the conclusion that butanol is a suitable material for the Bonn Frozen Spin Target.

1. Introduction

The Bonn Frozen Spin Target (BOFROST) will be used in connection with the PHOENICS detector (ref.[1]) to measure single and double polarization observables in the single pion photoproduction and in the photodisintegration of the deuteron at ELSA. For a description of the target facility see ref. [2].
The dynamic polarization build-up at 3.5 Tesla has been performed using a 300 mW impatt diode with a phase locked loop frequency stabilization [3].
The proton relaxation times have been determined in the frozen spin mode at low magnetic fields.
Our basic consideration for this test is the fact that background substractions of asymmetry measurements should be much more modest and precise for butanol than for ammonia, whose excellent frozen spin properties have been shown [4].

2. Properties of Butanol and Ammonia

From the structure of ammonia (NH_3) and butanol (C_4H_9OH) we can deduce the so called k-factor, the ratio of free polarizable protons to the total number of nucleons, which is 0.176 for ammonia and 0.135 for butanol. The densities of both materials are rather similar (0.9 g/cm^3). For butanol the doping with free paramagnetic centres is made chemically according to the recipe of Mango et. al. with porphyrexide and water [5].
In the case of ammonia the centres were created by irradiating the material with 20 MeV electrons from the Bonn LINAC.
The different doping procedures lead to a radiation resistance of the polarization which is good for butanol but a factor of 30 better for ammonia. However, for a tagged photon beam of about $10^7\gamma/sec$, as it will be used at the PHOENICS experiment, the radiation damage is negligible.

3. The Target Preparation

The liquid butanol doped with the radicals drops down to a stream of liquid nitrogen, blown through a liq. N_2 cooled helix. The drops freeze to beads of about 1 mm size by passing through the helix and were collected in a vessel.
To analyse the influence of the spin density on the relaxation time we have used different samples with porphyrexide fractions of 1.5% and 0.5% by weight.

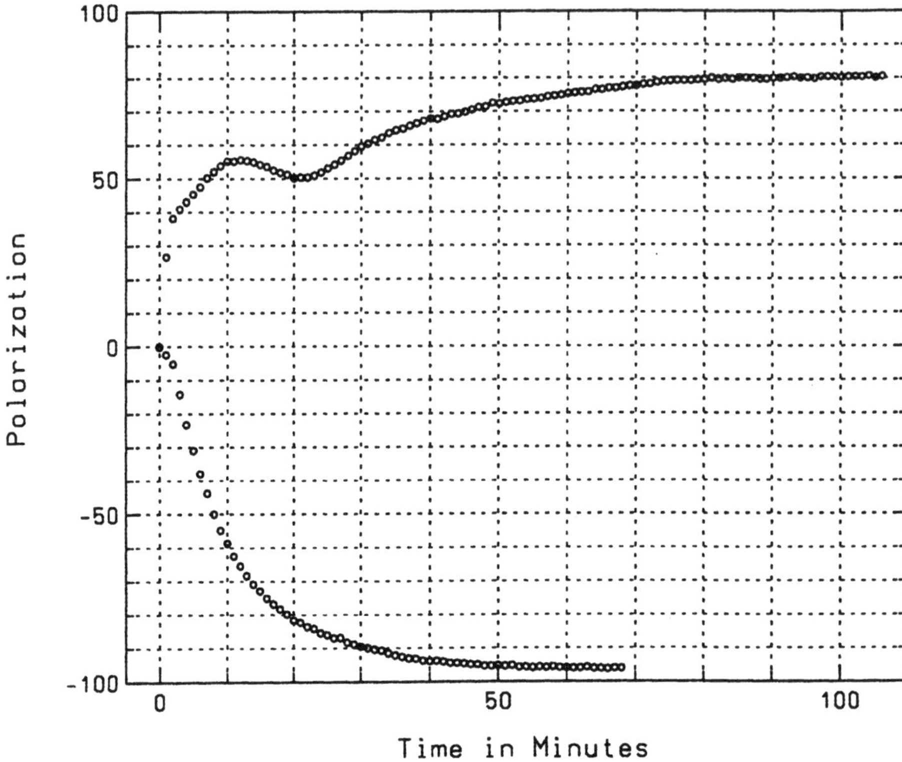

Figure 1 : Polarization build-up in butanol at 3.5 Tesla

4. Polarization Build-up at 3.5 Tesla

In fig. 1 the polarization build-up of butanol for positive and negative direction at 3.5 Tesla is shown. The dip in the upper curve is due to some problems with the microwave power.

The build-up process is nearly exponential, leading to characteristic build-up times of 12 min for the negative and 22 min for the positive direction. The microwave frequencies are 98.76 GHz and 99.20 GHz for positive and negative direction and the maximum polarization values are +90% and -96%, respectively.

The difference between the two directions are explainable by a mode jump of the impatt diode near the lower frequency (pos. polarization), leading to less stability and power output. However this can be suppressed by a slightly change of the magnetic field.

5. Relaxation Times in the Frozen Spin Mode

The relaxation measurements for butanol at holding fields of 0.5 and 0.35 Tesla at a temperature of 73 mK are presented in fig. 2.

For the higher doped sample the relaxation times are 88.0 h for positive and 72.5 h for negative polarization at 0.5 Tesla. At 0.35 Tesla the relaxation in positive direction was estimated to 20 hours. Using the sample with only 0.5% porphyrexide the relaxation time increases to a value of 116 h, although the magnetic field has been lowered to 0.33 T.

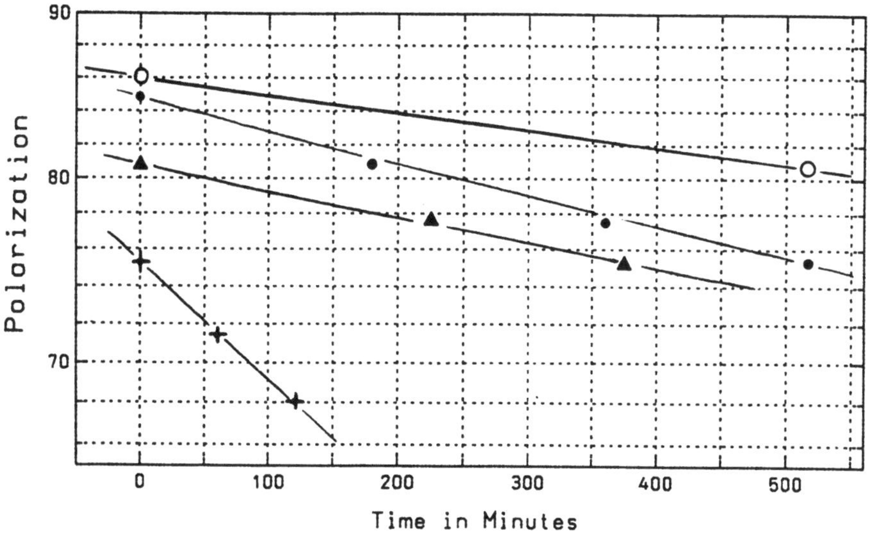

Figure 2 : Relaxation of butanol at 73 mK : (In brackets : doping of the sample)
• : 0.5 T, neg. pol. (1.5%); △ : 0.5 T, pos. pol. (1.5%); + : 0.35 T, pos. pol. (1.5%);
o : 0.33 T pos. pol. (0.5%)

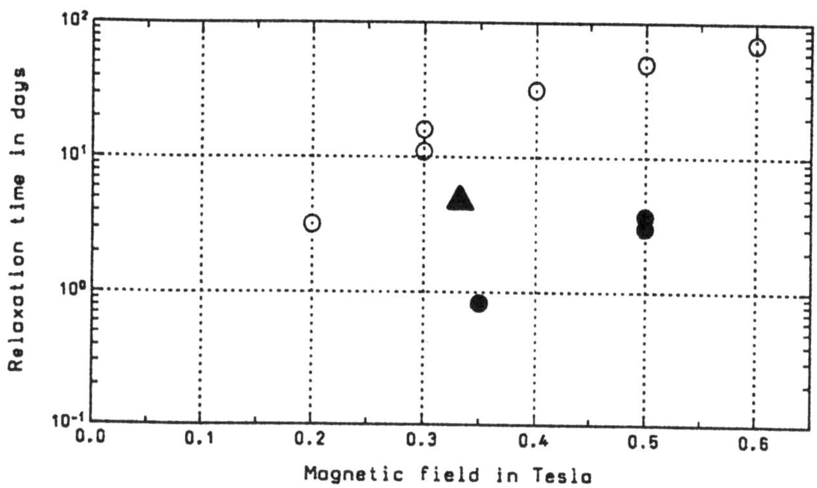

Figure 3 : Proton relaxation times in butanol in comparison with Saclay data :
o : Saclay at 55 mK; • : Bonn at 73 mK, 1.5% doped; ▲ : Bonn at 73 mK, 0.5% doped

Since our NMR system operates at 150 MHz, the field has to be raised up to 3.5 Tesla to measure each point at any given time. If we compare these results with those made in Saclay [6], one can clearly see some differences (fig. 3).

Out of the data follows that there are three different dependencies of the relaxation time: On the one hand it depends on the magnetic field, increasing by a factor of about 10 when doubling the holding field. On the other hand there is a strong dependency on the concentration of the paramagnetic inpurities, giving nearly a factor of 6 by using 0.5% porphyrexide instead of 1.5%. Third the relaxation time vary with the temperatur which the comparison to the Saclay data shows very well. We are confident that the estimated relaxation times will increase at lower temperatures in our cryostat by using a 3000 m^3/h pumping unit.

6. Conclusions

With the existing data we will have a polarization loss of 7% using a holding field of 0.33 Tesla after a time of 10 h which will be the usual time period between two relaxation reversals in an experiment. Therewith we would be able to run our cryostat with a butanol target in the frozen spin mode with only a slight loss of polarization. Further improvements to increase the relaxation time in future are at work.
Let us have at last a short view on the advantages of butanol :

One gain is the very fast polarization build-up in the order of a quarter of an hour together with a high maximum polarization. Another advantage we have to mention is that butanol is an easy to handle material. Loading of the cryostat and preparing of the target are not that critical than in the case of ammonia. Doping with a certain amount of free radicals is better to controll.
The most important profit are the spinless residual atoms carbon and oxygene in butanol so that there is no possible contribution to the measured asymmetry as it may be occure for the nitrogene spin in ammonia. According to the EST theory the nitrogen polarization is around 20% when the proton polarization reaches values above 90%. For high precision experiments this factor is not allowed to be neglected.

References :

1.) **G. Anton** : Progress in Particle and Nuclear Physics Vol. **24** (1990) 303

2.) **H. Dutz** : *The Bonn Polarized Target Facilities*, in these proceedings

3.) **J. Weber** : *Mikrowellensysteme zur dynamischen Nukleonenpolarisation bei 3,5 und 5 Tesla*, Bonn-IR-90-19 (1990)

4.) **W. Thiel** : *Polarization Behaviour of $^{14}NH_3$ at 3.5 T / Frozen Spin Mode*, in these proceedings

5.) **S. Mango, Ö. Runolfsson, M. Borghini** : *A Butanol Polarized Proton Target*, Nucl. Inst. Meth. 72 (1970) 45

6.) **J. Ball et al.** : *Improvements of the Polarized Target for Nucleon Nucleon Experiments at Saturne II*, in: Proc. of the 4th int. Workshop on pol. Tar. Material and Techn., ed. W. Meyer Bad Honnef (1984)

LAMPF Polarized ^{13}C Targets

S. Penttilä[1], J.J. Jarmer[1], N. Tanaka[1], M.L. Barlett[2,], K. Johnson[2], G.W. Hoffmann[2], W. Kielhorn[2], M. Purcell[2], G. Burleson[3], A. Klein[3,**], G. Kyle[3], B. Brinkmöller[4,†], D. Dehnhard[4], Yi-Fen Yen[4], J.R. Comfort[5], and J.J. Görgen[5]*

[1]Los Alamos National Laboratory, Los Alamos, NM 87545, USA
[2]University of Texas, Austin, TX 78712, USA
[3]New Mexico State University, Las Cruces, NM 88003, USA
[4]University of Minnesota, Minneapolis, MN 55455, USA
[5]Arizona State University, Tempe, AZ 85287, USA

Ethylene glycol, 1-butanol, and toluene highly enriched in ^{13}C have been used at LAMPF to produce dynamically polarized ^{13}C targets for scattering experiments with protons and pions. Preparation of the materials and characteristic properties of these targets are described.

1. Introduction

The interest in polarized nuclear ($A > 2$) targets at LAMPF began in 1986 with the *LAMPF Workshop on Physics with Polarized Nuclear Targets* [1]. This workshop was a catalyst for a series of experiments with polarized ^{13}C at LAMPF. It also stimulated polarized-nuclear-target proposals at other facilities as well as interest in performing experiments on other polarized nuclei like ^{6}Li, ^{7}Li, ^{14}N and ^{15}N. During the last years several polarization asymmetries have been measured in the reactions: $^{13}\vec{C}(\vec{p},p)^{13}C$ [2], $^{13}\vec{C}(\pi^{\pm},\pi^{\pm})^{13}C$ [3], $^{13}\vec{C}(\pi^{\pm},\pi^{\pm})^{13}C^{*}$, $^{13}\vec{C}(\pi^{+},\pi^{0})^{13}N$ [4], and $\vec{H}(\pi^{-},\pi^{0})n$ [5]. The motivation for the experiment with the proton beam was to test relativistic nuclear models. With the pion beam experiments the strength of spin effects in the spin-flip part of the pion-nucleon interaction was studied. For these experiments the carbon-13 nucleus was selected because it has spin-1/2, which minimizes the number of spin observables, and because a large amount of scattering data from unpolarized ^{13}C already exists in the LAMPF energy region. Also, theoretical calculations for polarization effects have been done [6] and ^{13}C has been demonstrated to be polarizable dynamically [7].

2. Target Materials and Material Preparation

The following ^{13}C enriched materials [8] doped with paramagnetic complexes have been used: EG=Ethylene Glycol; $^{13}C_2H_6O_2$ + EHBA-Cr(V), BU=Butanol; $^{13}C_4H_{10}O$ + 5% H_2O by wt. + EHBA-Cr(V), TO=Toluene; $^{13}C_6^{12}CH_8$ + DTBN (0.4% by wt.). The paramagnetic complexes were sodium bis(2-ethyl-2-hydroxybutyrato)oxochromate(V) monohydrate (EHBA) [9] and di-tert-butyl-nitroxide (DTBN) [10]. DTBN was selected because it was known to be soluble in toluene [11], which is not a polar solvent. The materials were deoxydized before and after adding the complex by pumping on

* Present address: *Applied Research Laboratories, University of TX, Austin, TX 78713, U.S.A.*

** Present address: *Department of Physics, Old Dominion University, Norfolk, VA 23529, U.S.A.*

† Present address: *Paul Scherrer Institute, CH-5234 Villigen, Switzerland.*

the mixture (EG) at room temperature or by the freeze-pump method [12] (BU and TO). The liquid was frozen into glassy beads about 1 mm in diameter by the charged-needle technique [13]. Polarizing tests were also made with two solid toluene slab targets. The toluene liquid with the dopant was poured into a copper microwave cavity and then slowly frozen, in a few minutes, or rapidly, in a few tens of seconds. The freezing speed was limited by the thermal conductivity of the toluene. In both cases the crystallization was not perfect; the frozen material was not transparent. Polarization results were essentially the same for both.

3. Purpose of the Development Work

The goal of the work has been to provide a polarized nuclear target for a medium-energy scattering experiment from which good statistical counting results with sufficient energy resolution can be obtained in a given amount of running time. At LAMPF, nuclear physics scattering experiments are generally performed at spectrometer facilities. Energy resolution, which is limited by the polarized target, is essential to separate the signals from each other as well as from background. Target thickness and nonuniformity of the target material (beads), the bending of scattered particles by the magnetic field as well as by cryostat wall material and liquid ^{3}He coolant degrades and limits the energy resolution to near 1 MeV [2]. This resolution was achieved in the proton experiment with EG target material where the real thickness of ^{13}C and ^{1}H was 280 mg/cm^2 and 60 mg/cm^2, respectively. These values can be compared to total background thickness (oxygen, cryostat walls and coolant) of 760 mg/cm^2 where almost half is from helium-3 liquid.

In a case where scattering occurs from a thin (a few millimeters) solid slab target, the resolution can be reduced to 0.5 MeV. We have built a 6-mm thick bead target for an inelastic $(\pi^{\pm},^{13}\vec{C})$ experiment. The energy resolution was improved by changing the target angle with the spectrometer angle, so the path of a scattered particle in the material stayed about the same. With this target an energy resolution of 0.7 MeV was achieved.

The content of the target material itself is important because in most cases it contains other nuclei, such as carbon-12 or oxygen, which are serious backgrounds for the ^{13}C signal. Extra running time is then needed for background measurements. Note that toluene does not contain oxygen. In addition to the points mentioned above the merit of a material and its effect on running time depends on the maximum polarization obtainable, polarization build-up time, and spin-lattice relaxation time (T_{1n}) at temperatures where the NMR calibration measurements are done.

4. Polarization Measurement

The ^{13}C and ^{1}H polarizations were continuously monitored with separate NMR circuits. The hydrogen signal was measured with a series-tuned Q-meter [14] and the ^{13}C signal with a bridge-type circuit [15]. With this arrangement the polarization behavior of the both spin species could be followed independently. The same coil was used to measure both NMR signals. For the different experiments the coil was either wrapped around the target or it was buried in with the target material. A computer-controlled RF switch, mounted at room temperature, enabled the data acquisition system to select either the ^{1}H or ^{13}C detector circuit.

The data acquisition system was a typical LAMPF type used in scattering experiments. It included a frequency synthesizer, CAMAC, MBD (microprogrammable branch driver), and Q-software in a μVAXII. In the polarizing mode, signals were saved on disk if there was a 0.5%-change in polarization. The uncertainty in the polarization measurement by NMR was 0.04 to 0.02.

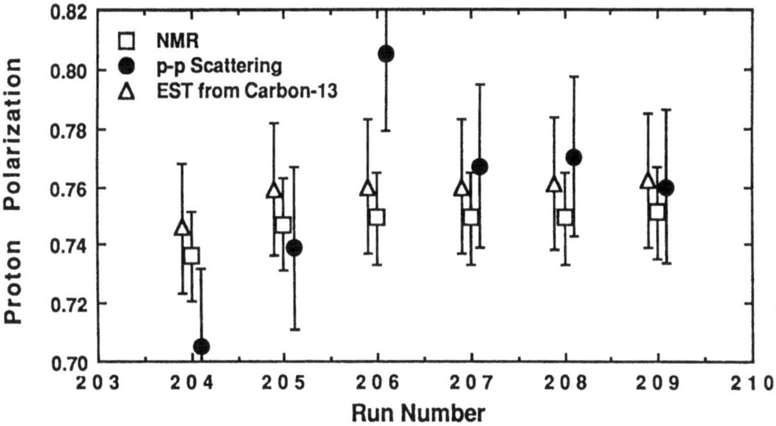

Fig.1. Comparison of ^1H polarization measured with NMR, p-p scattering, and calculated from measured ^{13}C polarization assuming EST.

The $^{13}\vec{C}(\vec{p},p)^{13}C$ experiment required high beam intensity to obtain reasonable counting statistics. Because the target material (EG) would be radiation damaged, the target hydrogen polarization was measured by detecting 500-MeV p-p elastic scattering at a center of mass angle of 46^0. This measurement provided the target hydrogen polarization in the beam interaction volume and checked the hydrogen NMR polarization measurements. Shown in Fig. 1 are the polarization results (over several averaged runs) measured with NMR and p-p scattering when the target material was not radiation damaged. The results show that the two methods agree within errors. Also shown are the hydrogen results calculated from the NMR measurements for carbon-13. It has been assumed that the two spin systems are in thermal equilibrium and the they have a common spin temperature [equal spin temperature hypothesis (EST)]. From the measured carbon-13 polarization a spin temperature was obtained and used to calculate the hydrogen polarization. The results show that the EST concept is valid for this EG material. NMR polarization data from radiation damaged material showed that EST was also valid under these conditions. This allowed us then to calculate, for radiation-damaged material, the polarization of carbon-13 from p-p scattering measurements [2].

Scattering data from pion experiments also allows one to compare measured hydrogen analyzing powers for elastic scattering with those calculated from the SAID [17] phase-shift analysis. For pion-proton elastic scattering at 132 MeV and 42.1° (lab.) we measured $A_y = 0.330 \pm 0.005 \pm 0.006$, where the first uncertainty is statistical and the second arises from the polarization measurement. The SAID analysis gave $A_y = 0.316 \pm 0.006$.

From our polarization results we conclude that target polarizations as measured by NMR are consistent and in agreement with values measured by particle scattering methods. Also, in steady state, the spin temperatures of ^1H- and ^{13}C-spin ensembles are the same within errors in these materials (EG, BU, TO).

Considerable effort was spent to understand and check the NMR polarization results. The bridge NMR circuit has been analyzed carefully [15]. Also, for each experiment, part of the running time was used to measure the analyzing powers for particle scattering from hydrogen to check the ^1H NMR measurements, as described below.

5. Target Parameters

In Table 1 polarization results and other target parameters obtained with the materials are summarized. All measurements were done at a magnetic field of 2.500 T. Both the ^{1}H and ^{13}C polarization was measured by NMR calibrated by measuring the thermal equilibrium signal near 1K. At a field of 2.500 T, the optimum microwave frequencies were for EHBA $f(+) = 69.048$ GHz and $f(-) = 69.472$ GHz and for DTBN $f(+) = 70.059$ GHz and $f(-) = 70.320$ GHz.

Table 1. $^{13}\vec{C}$ target material parameters.

Mat.	Dop.	Conc.	Cool.	Pol(^{13}C)	Pol.Time	T_{1n}
EG	EHBA	$6\text{-}7{\cdot}10^{19}$	He-3	.27 – .29[1]	20	8
BU	EHBA	$6{\cdot}10^{19}$	He-3	.27 – .28[1]	20	7
TO	DTBN	.4%b.w.	DR	.22[2]	60	20
			He-3	.14 – .16[3]	120	
			DR	.03[4]		

Keys: Mat.=Target material, Dop.=Paramagnetic complex, Conc.=Concentration (mole/cm^3), Cool.= Cooling refrigerator; dilution refrigerator=DR or He-3 evaporator=He-3, Pol(^{13}C)=Typical carbon-13 polarization, Pol.Time=Polarization build-up time in minutes from 10% to 90% of maximum, T_{1n}=Spin-lattice relaxation time in minutes at 1K.

1) Polarization averaged over the experiment, target size $\approx 5 - 10$ cm^3.

2) Target size ≈ 1 cm^3 and T=0.2 K, when T=0.47 K then Pol(^{13}C) ≈ 0.18.

3) Target size was 3.5 x 6.0 x 0.5 cm^3. Inhomogeneity of the magnetic field over the target was 12 gauss.

4) Size of the slab target was 3.5 x 6.0 x 0.5 cm^3.

References

[1] Proc. of *The LAMPF Workshop on Physics with Polarized Nuclear Targets*, edited by G. Burleson *et al.*, LA-10772-C, 1986.

[2] G. W. Hoffmann, *et al.*, *Phys. Rev. Lett.* (submitted).

[3] Yi-Fen Yen, *et al.*, Proc. of *the 7th Int. Conf. on Polarization Phenomena in Nuclear Physics*, Paris 1990, (les Éditions de Physique, in press).

[4] J. Görgen, *et al., ibid.* and J. J. Görgen, *et al.*, *Phys. Rev. Lett.* (submitted).

[5] J. J. Görgen, *et al., Phys. Rev.* D (accepted).

[6] Ref. 1 and L. Ray, *et al.*, *Phys. Rev.* C**37**, 1169 (1988).

[7] J. J. Jarmer, *et al.*, *Nucl. Instrum. and Methods* A**250**, 576 (1986).

[8] Merck and Co., Inc., Montreal, CANADA.

[9] M. Krumpolc and J. Roček, *J. Am. Chem. Soc.* **101**, 3206 (1979).

[10] Aldrich Chemical, P.O. Box 2060, Milwaukee, WI, 53201.

[11] R. C. Ferrow, *Nucl. Instrum. and Methods* **159**, 557 (1979).

[12] Rubin Battino and H. Lawrence Clever, *Chem. Rev.* **66**, 395 (1966).

[13] W. Ash, *Nucl. Instrum. and Methods* **134**, 9 (1976).

[14] G. Court, *et al.*, Proc. of *the 2nd Workshop on Polarized Target Materials*, ed. by G. R. Court *et al.*, (Rutherford, 1980), p. 76.

[15] W. F. Kielhorn, PhD Thesis, University of Texas, 1990.

[16] W. F. Kielhorn *et al.*, the *8th Int. Symposium on High-Energy Spin Physics*, edited by Kenneth J. Heller, AIP Conference Proceedings No. 187, (AIP, New York, 1989), p. 1254.

[17] R. A. Arndt and L. O. Roper, SAID Phase Shift Analysis Program.

Results from the PSI ^{6}LiD Target

B. van den Brandt, J.A. Konter, S. Mango, and M. Weßler
Paul Scherrer Institute, CH-5232 Villigen PSI, Switzerland

We report here on the first operational ^{6}LiD target. It has been used at the 600 MeV accelerator of the Paul Scherrer Institute for an experiment to measure the vector analyzing power iT_{11} in pion elastic and inelastic scattering from polarized ^{6}Li at different energies by means of the SUSI-spectrometer [1,2,3].

Introduction.

In 1978 Abragam and co-workers [4,5] achieved a ^{6}Li-polarization of 40% in samples of a few mm^3 at dilution refrigerator temperatures in a field of 2.5 T with a homogeneity better than 5.10^{-6}. The paramagnetic centres in these samples were obtained by irradiation with 3 MeV electrons at liquid argon temperatures.
At Saclay, DPhPE, the technique of the irradiation of larger quantities of ^{6}LiD and ^{7}LiH was investigated extensively by Chaumette *et al.* [6,7,8]. Their best polarization results for ^{6}LiD at 2.5 T in terms of growth time and polarization were ^{6}Li and D polarizations of 37% and 43%, respectively, in a sample irradiated in a special argon gas cryostat at 180 K with a total flux of about 2.10^{17} electrons/cm^2. These values could be reached after 8 hours in a dilution refrigerator. The field homogeneity over the sample was determined to be better than 2.10^{-5}. A part of this batch of ^{6}LiD (1 g) has been kindly provided to us by J. Deregel and G. Durand, and was used for the present experiment.

Experimental conditions.

We have polarized a sample of 1.6 cc., consisting of small chips sieved at Saclay to fit into our microwave cavity, a brass container 18mm high, 18mm wide and 5mm thick in beam direction, with a spiralled NMR coil positioned in the midplane as shown in fig. 1. This cavity was loaded with the sample holder into our dilution refrigerator, situated in a cryostat with a 2.5 Tesla superconducting split pair. The details of the cryostat and the dilution refrigerator are described elsewhere [9]. The field homogeneity is not better than 3.10^{-4} over the volume of the sample. A similar inhomogeneity should be expected for other magnets in the vicinity of a spectrometer with iron parts. The present polarization experiment is therefore also a feasibility test for the use of ^{6}LiD in experiments on a beam.

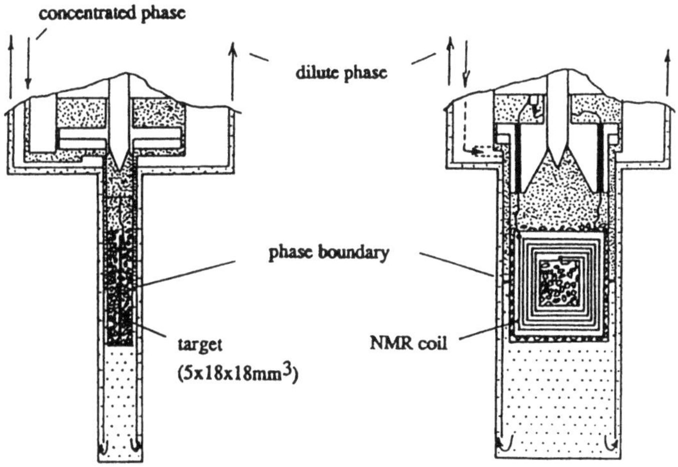

Figure 1. Brass microwave cavity with ^{6}LiD target material and NMR pick-up coil.

Results.

The polarizationsof the sample were measured at 15.74 MHz with the NMR system described elsewhere in these proceedings [10], using a strongly damped series resonance circuit. The TEP signal at 2.17 K, the positive and the negative polarization signals of ^{6}Li are shown in fig. 2a, 2b and 2c. At 2.17 K the relaxation time was 66 minutes. 2/3 of the full polarization were obtained in about 16 hours, the optimum polarizing microwave frequency being for positive polarization 70.410 GHz and for negative polarization 70.450 GHz. The difference of 40 MHz is close to the expected value for the solid effect, taking into account the magnetic field inhomogeneity over the sample volume.

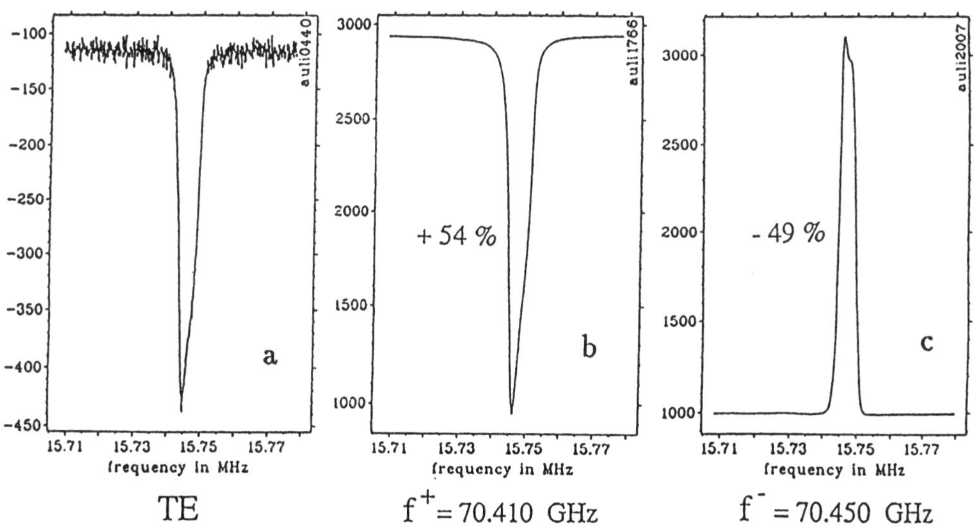

Figure 2. NMR signals of ^{6}Li in ^{6}LiD.

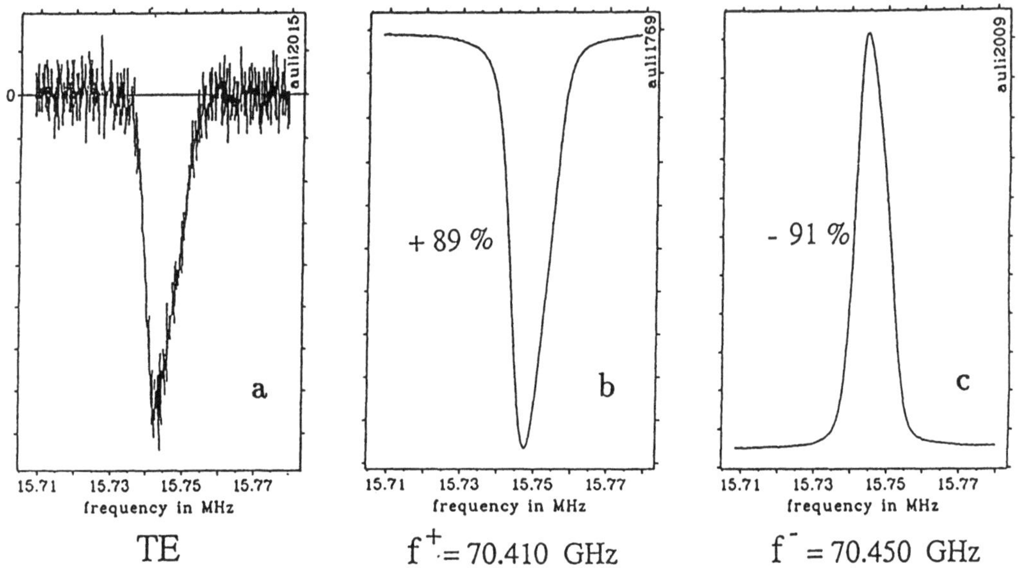

Figure 3. NMR signals of ^{7}Li impurities in ^{6}LiD.

Table 1: Polarization results

nuclei	*pos. pol.* %	*neg. pol.* %	*FWHM NMR sign.* *kHz*
^{6}Li	+54	-49	5.5
D	+53	-49	5.5
^{7}Li	+89	-91	9

The polarization of the deuteron and of the ^{7}Li impurity was also measured at 15.74 MHz. After putting the target in frozen spin state, the magnetic field was lowered until the D and ^{7}Li enhanced signals could be measured, while the corresponding TEP signals were taken at the field values corresponding to 15.74 MHz and at 2.17 K. See figures 3a, 3b, 3c and table 1.
The temperature determination for the TEP signals was done with two 100 Ω Matsushita carbon resistors, calibrated in zero field against a calibrated germanium thermometer, all positioned in the mixing chamber. The relative error in the value of the polarization is estimated to be $\pm 4\%$. At maximum polarization the spin temperatures of the three nuclear systems are equal within the error bars.
The polarization as a function of time during the scattering experiment in the πM1 area is given in fig. 4. It shows that it takes 2-3 days to achieve full polarization. Our estimate of the optimum microwave frequencies are based on tests on this timescale. The results of the scattering experiment are given by Ritt *et al.* [3].

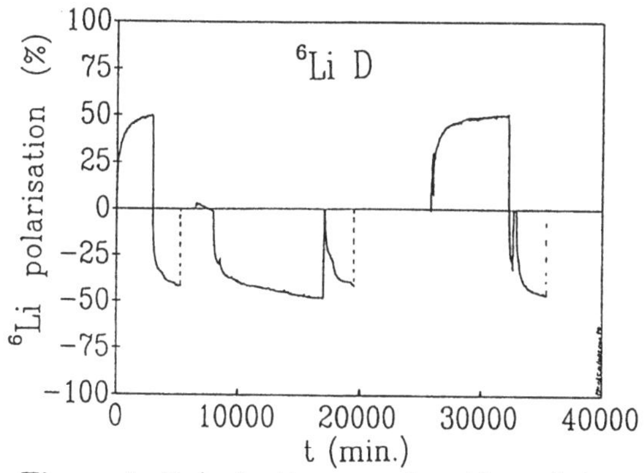

Figure 4. Polarization as a function of time, during a 3 weeks scattering experiment in the pion beam in πM1.

Conclusions.

The first operational ^{6}LiD target has been described briefly. ^{6}LiD has excellent properties as target material for polarization experiments. Polarization values of more than 50 % have been obtained for *both* ^{6}Li and D in a field of 2.5 Tesla with 3.10^{-4} homogeneity. The deuteron polarizations are higher than the values obtained before in operational targets of any other deuterated substance. However, the 2-3 days necessary to achieve full polarization are very long, and the irradiation procedure in an electron beam for the creation of paramagnetic centers needs further optimization. This work is planned to be done in the near future at Saclay.

Acknowledgements.

We wish to express our sincere thanks to Dr. J. Deregel and Mr. G. Durand for kindly providing us with irradiated target material.

References.

1. R. Tacik *et al.*, Phys. Rev. Lett. **63**(1989) 1784-1787
2. R. Tacik *et al.*, Proc. XII PANIC Conf., Boston, June 25-29, 1990
3. S. Ritt *et al.*, accepted for publication in Phys. Rev.
4. A. Abragam, Internat. Symp. on High Energy Physics with Polarized Beams and Polarized Targets, Argonne, Illinois, 1978, p.
5. V. Bouffard, Y. Roinel, P. Roubeau, A. Abragam, J. Physique **41** (1980) 1447-1451.
6. P. Chaumette *et al.*, Proc. 4th Internat. Workshop on Polarized Target Materials and Techniques, Ed. W. Meyer, Bonn (Germany), Sept. 3-6, 1984, p. 81-83.

7. P. Chaumette *et al.*. Proc. Internat. Workshop on Polarized Sources and Targets, Eds. S. Jaccard and S. Mango, Montana (Switzerland), Jan 13-17, 1986, p. 767-771.
8. G. Durand *et al.*, Proc. Internat. Symp. on High Energy Spin Physics, Ed. K.J. Heller, Minneapolis (USA), Sept 12-17,1988, (Am. Inst. Phys. Conf. Proc. N0. 187), p. 1275-1280
9. B. van den Brandt, J.A. Konter, S. Mango, NIM **A289** (1990) 526-531
10. M. Wessler *et al.*, these proceedings.

Comparison of Polarized Target Materials in Different Magnetic Fields

S. Trentalange[1], *G. Baum*[2], *E. Bunyatova*[3], *B. van den Brandt*[4], *P. Hautle*[5], *R. Ent*[6], *J.A. Konter*[4], *S. Mango*[4], *T. Niinikoski*[7], *A. Rijllart*[7], *S. Takala*[7], *and M. Weßler*[4]

[1]University of California at Los Angeles, Los Angeles, CA 90024, USA
[2]University of Bielefeld, W-4800 Bielefeld, Fed. Rep. of Germany
[3]Joint Institute for Nuclear Research, Dubna, Head P.O. Box 79, SU-101000 Moscow, USSR
[4]Paul Scherrer Institute, CH-5232 Villigen PSI, Switzerland
[5]ETH, CH-8093 Zurich, Switzerland
[6]University of Virginia, Charlottesville, VA 22901, USA
[7]CERN, CH-1211 Geneva 23, Switzerland

Tests were performed at the Paul Scherrer Institute (PSI) and CERN to determine the maximum polarization for two deuterated materials being considered for use in the SMC polarized target. The two materials studied were butanol and propanediol, both doped with Cr(V) complex. The maximum attainable polarization was found for magnetic field values of 2.5, 3.5 and 5.0 Tesla.

The Spin Muon Collaboration (SMC) is an experiment to measure the spin structure function of the neutron by use of a large polarized deuteron (and hydrogen) target. The volume of this target is approximately 2 liters and the desired deuteron polarization is in excess of 40%. Although this target is designed to operate at 2.5 Tesla, and has successfully produced *hydrogen* polarizations greater than 90% at this field, the operating parameters for deuterium were not as well-known, and it was decided to test the proposed deuterated materials at different magnetic field values, since significant improvements may be possible.

Studies of polarized target materials are often complicated by the wide variation in the facilities used by various groups. The ultimate polarization of "identical materials" is known to depend upon such (non-independent) factors as sample size, temperature, microwave irradiation power and magnetic field inhomogeneity. Unfortunately, data on the complicated variation with these parameters is often anecdotal in nature and a comparison of two materials which have been studied by different groups is somewhat risky. In this study, two materials were chosen for testing under conditions as identical as possible. The first material was fully-deuterated 1,2-propanediol (P-D8) doped with about Cr(V) complex to a concentration of about $6 \cdot 10^{19}$ spins/cc. EPR analysis of this complex at 298 K gave a width of 3 10^{-4} Tesla for the distance between points of maximum slope. This material has been used extensively in previous experiments with polarizations of about 40% reported[1]. The second material was a fully-deuterated butanol-D_2O (Bu-D10) mixture doped with fully deuterated EHBA-Cr(V) complex (EDBA) also to a concentration of $6 \cdot 10^{19}$ spins/cc. The dopant concentration and ratio of water to alcohol was roughly optimized in a separate series of experiments at CERN to produce a mixture with phase transitions well above the boiling point of liquid nitrogen in order to facilitate handling. The sample described below had the ratio 95%-butanol to 5%-water. Butanol was prefered as a target material over propanediol because of it's higher deuteron content, although this factor is largely offset by it's lower density.

The samples were placed in a ^3He-^4He dilution cryostat [2] with a base temperature of about 75 mK and a magnetic field whose homogeneity has been measured to be about $1 \cdot 10^{-4}$ over the $1 cm^3$ volume of the sample. The same sample holder geometry and NMR coil were used for the two samples to minimize differences due to magnetic field. The NMR coil consisted of conducting lines etched out of both sides of an annular piece of silver-clad G-10 with a square cross-section. The inductance was varied by shorting a different number of turns for each magnetic field value. Although the microwave sources for the dynamic nuclear polarization had to be changed with magnetic field, the input power was monitored by the same germanium diode and thus their relative values for both samples are roughly known. The input microwave power was limited by the cooling power of the dilution refrigerator.

The inherent noise in the system did not allow us to perform an absolute calibration of the deuteron polarization by measurement of the thermal equilibrium signal. However, the polarization as determined from the peak asymmetry is accurate and reproducible to about +/- 5%, and is therefore sufficient to identify significant variations between samples or magnetic field values [3]. The available microwave sources allowed us to try three different values of the magnetic field: 2.5, 3.5, and 5.0 Tesla. EIO sources were available for the two lower field tests while the test at 5.0 Tesla employed a carcinotron source.

Figure 1 summarizes the results for the maximum attainable polarization for each sample as a function of magnetic field. Although the precise regimen of applied microwave power and frequency was varied for each sample, the final values were found to be independent of this schedule to within the accuracy of the peak asymmetry measurements. The polarization of -53 +/- 5% obtained for the P-D8 sample at a field of 2.5 T is the highest deuteron polarization ever reached for this well-examined material. This might indicate the importance of a highly homogeneous magnetic field for high deuteron polarizations.

Further tests are planned, but we draw the following tentative conclusions from these data: Firstly, higher fields give lower polarizations. Since the polarization must decrease for very low field values, it would obviously have been of interest to determine the optimum field value. We were, however, prevented from testing at 1.5 Tesla by the small size of the cryostat waveguide. Secondly,the maximum attainable polarization is roughly 15% greater in the P-D8 sample than the Bu-D10 sample.

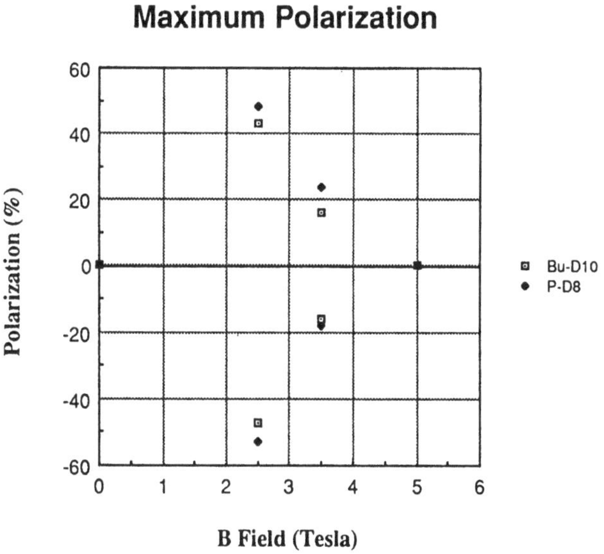

Figure 1 Maximum Polarization as a function of Magnetic Field.

We can define a first-order figure of merit for using the two deuterated materials for spin-physics scattering experiments as the product of the deuteron content squared, the maximum polarization squared and the density. Using the measured polarizations and densities measured at liquid nitrogen temperatures, the figure of merit for P-D10 would amount to 0.92 of the figure of merit for Bu-D10. The higher deuteron content of the Bu-D10 is largely offset by it's lower maximum polarization and density.

References:

1.) E.I. Bunyatova and N.N. Bubnov, Nucl. Instr. and Methods **A254** (1987) 252.
2.) B. van den Brandt, J.A. Konter, and S. Mango, Nucl. Instr. and Methods **A289** (1990) 526.
3.) O. Hamada et al., Nucl. Instr. and Methods **189** (1981) 561.

Dynamic Polarization of Solid ^{3}He

P.P.J. Delheij[1], *H. Postma*[2], *and K. Bindels*[2]

[1]TRIUMF, 4004 Wesbrook Mall, Vancouver, B.C., Canada V6T 2A3
[2]University of Technology, Delft, The Netherlands

The nuclear spins of solid ^{3}He are involved efficiently in the magnetic interaction accross phase boundaries. Furthermore, the spin diffusion coefficient in solid ^{3}He can be very large. These two features suggest the dynamic polarization of ^{3}He in an inhomogeneous mixture with a material, that provides the paramagnetic electron centers. The experimental equipment is described of a test of the dynamic polarization of ^{3}He at a temperature of 1 K with a sample cell, that withstands a pressure of 80 bar.

INTRODUCTION

Experiments on spin dependent observables at intermediate energies with polarized ^{3}He can provide unique information on the characteristics of a polarized neutron target. In elastic electron scattering the electric form factor of the neutron can be measured. With polarized protons in elastic scattering the Glauber model [1] or a model based on a momentum space potential can be tested [2]. In knock-out reactions spin momentum correlations can be determined which are essential for the interpretation of high-energy electron scattering [3]. The spin dependence in elastic scattering of pions, the absorption of pions and the production of pions on ^{3}He is experimentally an unexplored area. However, a number of theoretical calculations have been reported and should be tested.

It is possible to polarize ^{3}He by several techniques. Static polarization requires temperatures near 10 mK in a magnetic field of about 7 T. The small cooling power (tens of μW at most) limits this technique to neutron experiments. Another approach is the application of optical pumping. If the laser scheme avoids impurities by proceeding through the metastable state in ^{3}He, the density is limited to values near 10^{18} at/cm^3. With an alkali vapour acting as an intermediate stage in the polarizing scheme, the density can be increased to 2 10^{20} at/cm^3. These targets can absorb a power of many watts. But in the last set-up an appreciable contamination (about 10 % of the target mass) must be accepted from the nitrogen gas that quenches the re-emission of the laser light. An elegant solution would be the dynamic polarization of solid ^{3}He at a temperature of 1 K. A density of 3 10^{22} at/cm^3 is available and a power of several watts can be absorbed.

PHYSICS PRINCIPLE

In dynamically polarized targets typically a diluted ensemble of paramagnetic electron spins is irradiated with microwaves to cool their dipole - dipole interaction reservoir. A good thermal contact with the Zeeman reservoir of the nuclei of interest lowers the spin temperature of these nuclei to approximately +3 mK or -3 mK and therefore increseas their polarization.

With NMR techniques a good thermal contact has been observed between fluorine spins in small teflon spheres and the first layer of ^3He in contact with these spheres [4]. When the teflon spheres were irradiated with electrons to create paramagnetic microwave pumping centers, the ^3He followed the polarization enhancement of the fluorine spins [5]. In dynamic polarization the cooling of the nuclear Zeeman reservoir of is not very sensitive to the characteristics of the nuclei. It applies equally well for protons and deuterium. Therefore, it seems feasible to replace the nuclear fluorine spins in the teflon with an electron dipole - dipole reservoir that can be cooled with microwaves. After the ^3He is polarized near the material that contains the microwave pumping centers, the polarization must be transported over macroscopic distance of about 1 μm (or 10.000 Å) which is the size that can be obtained by mechanical grinding of a powder. This distance should be compared with a separation of typically 25 Å in the light alcohols. The problem is overcome by the unique value, 10^{-7} cm^2/s, of the spin diffusion coeffient in solid ^3He which is five orders of magnitude larger than in other solid materials. So, the polarization should become distributed homogeneously through the bulk ^3He.

APPARATUS

The ^3He cell was placed in the 1 K bath of a pool cryostat. A schematic drawing of the pressure cell is shown in fig. 1. The sample is contained in a copper housing with a inner diameter of 11 mm and a height of 10 mm. The bottem is sealed with an In o-ring of 0.7 mm diameter. A stainless steel capillary with an outer diameter of 1.5 mm and a wall thickness of 0.25 mm is silver soldered in the top. A 0.25 mm thick mylar window, for microwave access to the cell, is glued to the body with epoxy and clamped by a brass flange. The NMR feed-through required a custom solution. Initially a commercial semi-rigid cryogenic coax cable was soft soldered in the top of the pressure cell. After solidifying the ^3He, gas would start to escape through the cable and the SMA connector near the top flange of the cryostat as the boil-off lowered the helium level in the 1 K bath. This problem was solved by constructing, near the top of the cell, a feedthrough that remained in the helium liquid while the cell was pressurized. Over a length of 25 mm the center conductor was glued into a phenolic sleeve, which in turn was glued with epoxy in the outer conductor of the cable. Without any problem The NMR coil in the cell and the a SMA connector could be soft soldered to this feed through. The NMR coil was 4 mm long and consisted of three turns, with a diameter of 4 mm, that were wound from copper wire, covered by a thin teflon sleeve, with a diameter of 0.5 mm.

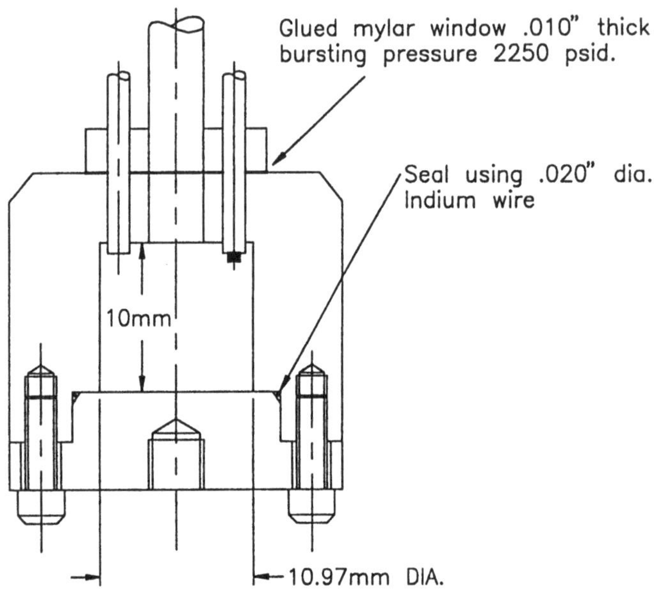

Figure 1: The ^{3}He pressure cell

The pressurizing system was based on a membrane compressor that could deliver an output pressure of 30 bar if an input pressure of 0.5 bar was maintained. A small forepump performed this task and it also returned the gas to the storage tank after an experimental run. The ^{3}He gas was always led through a conventional zeolithe cold trap. The compressor raised a stainless steel pot with a volume of 125 cm^3, in a liquid nitrogen bath to a pressure of 30 bar. The input of this pressure amplifier was closed and it was warmed up to room temperature. With this system a highest pressure of 78 bar could be applied to the cell.

A fixed frequency, pulse NMR system, that was tuned at 81 MHz, was available for detection of the free induction decay (FID) signal. After an 90 deg pulse, the decay of the magnetization perpendicular to the main DC field is observed on a storage oscilloscoop and photographed with a polaroid camera. Fig.2 contains a FID sweep for ^{3}He at a temperature of 1 K. For comparison also the signal from an empty cell is added. This FID signal is linearly proportional to the ^{3}He polarization. Due to a compression in the time domain the signal-noise ratio is excellent in comparison with CW systems. The disadvantage is that the measurements are destructive. By varying the repetition rate the longitudinal relaxation time can be determined.

Before insertion into the cryostat, the cell is evacuated through the fill capillary. While the cell is at a temperature of 4 K the ^{3}He is admitted and the pressure raised to about 75 bar. Then the 1 K bath is pumped off to a pressure of about 150 μbar.

The microwaves are produced by an Impatt source which delivers a power of 150 mW at 70 GHz. During a measuring run the frequency is varied over a range of about 1 GHz in steps of approximately 20 MHz. After each step the cell is ir-

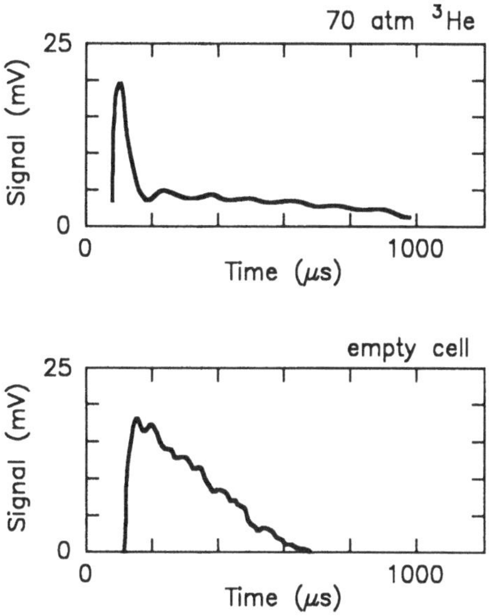

Figure 2: Free induction decay of ^{3}He

radiated for about 3 minutes. Then, a pulse NMR sweep is started and the next microwave frequency selected.

RESULTS

The light alcohols, that are used in the dynamically polarized targets, can be doped with EHBA [6], a compound that contains Cr^{5+} ions that act as paramagnetic centers. For the undiluted compound the susceptibility was measured down to a temperature of 1.0 K. Below 1.5 K a small deviation from the Curie behaviour was observed. Nevertheless, the cell was filled with EHBA. Also, two vanadyl compounds, vanadylsulphate and vanadylacetylacetonate, were measured. In the VO^{2-} vanadyl group the V^{4+} has the same electron structure as the Cr^{5+} in EHBA. However, with none of the compounds was a change of the polarization observed under the influence of the microwave irradiation. The sensitivity is approximately 5 % of the thermal equilibrium signal. This uncertainty level is determined by noise in the NMR system. With a small lanthanum magnesium nitrate crystall, that was doped with Nd, enhancements of the proton polarization by a factor 30 were observed which indicates that the pulse NMR system poses no problem. During slow decompression a jump of about 0.5 bar is observed in the pressure near 41 bar which is the melting pressure at a temperature of 1 K. This shows that the ^{3}He is solidified in the cell. It seems that all the equipment is performing satisfactorily and that only a material with a suitable paramagnetic electron structure must be found.

ACKNOWLEDGEMENT

The technical assistance of W. Schot, W. den Braver, W. Kellner, A. Kollen, A. de Vreede and D. Korbee is greatly appreciated. The discussions with Dr. J.H.N. Creyghton and Dr. H.C. Meier are greatfully acknowledged. The magnet was constructed by A. den Oude. The magnetization of the EHBA was skilfully measured at the Kamerlingh Onnes Lab. in Leiden by E. Lammers under the supervision of Dr. H. van Duyneveldt and Dr. J. Verstelle.

References

[1] D.K. Hasell et. al., Phys. Rev. C **34** (1986) 236.

[2] R.H. Landau, M. Sagen and G. He, Phys. Rev. C **40** (1990) 50.

[3] O. Hausser, 7-th Int. Conf.on Polarization Phenomena in Nuclear Physics, Paris, 1990.

[4] L.J. Friedman, P.J. Millet and R.C. Richardson, Phys. Rev. Lett. **47** (1981) 1078. P.C. Hammel, P.L. Kuhns, O. Gonen and J.S. Waugh, Phys. Rev. B bf 34 (1986) 6543.

[5] A. Schul, L. Sniadower, H. Aloul and M. Chapellier, Quantum Fluids and Solids-1983, Sanibel Island 1983, AIP Conf. Proc. vol.103, eds. E.D. Adams and G.G. Ihas, p. 191.

[6] M. Krumpolc and J. Roček, J. Am. Chem. Soc. **101** (1979) 3206.

Investigation of Organic Substances for Development of Targets with Polarized Hydrogen and Deuterium Nuclei

E.I. Bunyatova

Joint Institute for Nuclear Research, Dubna,
Head P.O. Box 79, SU-101000 Moscow, USSR

An important problem in development of the targets is the choice, investigation and synthesis of the target material where hydrogen (deuterium) nuclei can be polarized.

The basic substance of the target was chosen to be and organic compound. These compounds are rich in hydrogen which can be replaced by deuterium and have no heavy nuclei. The carriers of free electrons in those compounds were complex pentavalent chromium compounds Cr(V) and organic stable from radicals.

To investigate target substances, the method of electron paramagnetic resonance (EPR) was used. The method allows one to determine the most important characteristics of the target materials and the EPR parameters, e.g. the concentration of paramagnetic centres, HFI between electrons and nuclei, the width and position of the spectrum line, before polarization.

Target substance with polarized protons based on Cr(V) complexes in diols |I-3|

Complex Cr(V) compounds synthesized in diols (ethanediol) CH_2OH-CH_2OH and propanediol-I,2 $CH_2OHCHOHCH_3$) show a hyperfine structure from protons in the EPR spectra, so they are of great interest as target materials.

Complex Cr(V) compounds are an intermediate product of the oxydation-reduction reaction between Cr(VI) compounds (chromates) and diols. The kinetics of this process is best studied for ethanediol. In the liquid solution the EPR spectrum consists of a line form the ^{53}Cr isotope (its natural) content is 9.55%, 1=3/2) and a line from ^{52}Cr. At room temperature the line from the Cr isotopes in a diluted solution is split into 9 components of hyperfine structure (HFS). The HFS from ^{52}Cr is assumed to be caused by 8 equivalent protons, and the structure of the complex has the following form:

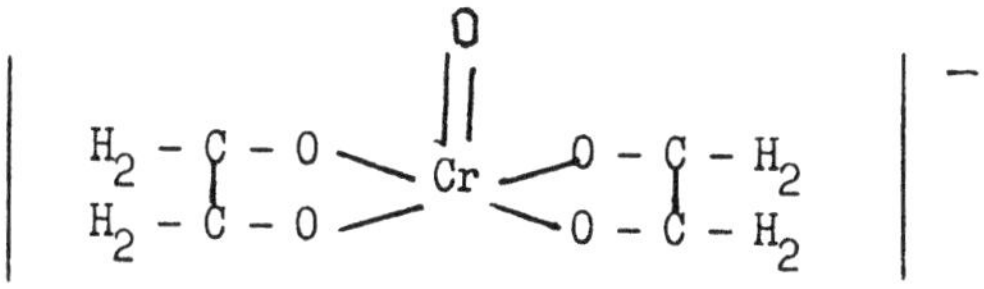

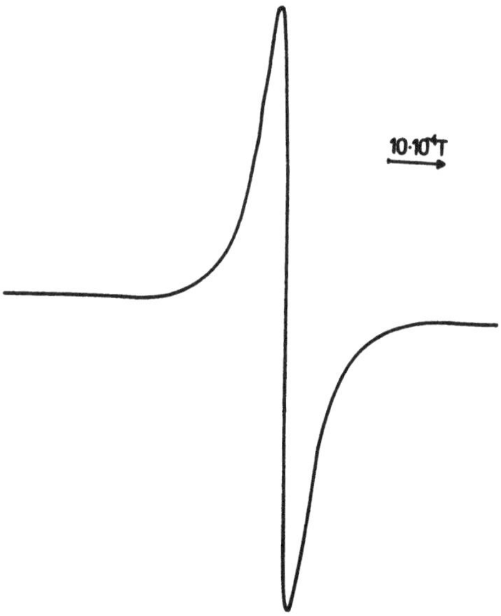

Figure I. EPR spectrum of the Cr(V) complex is propanediol-I,2 at T = 293 K.

The effective paramagnetic admixture for the target material is only a Cr(V) compound. So synthesis of the complexes must be carried out in such a way that the final product had a high Cr(V) content (from $2.5 \cdot 10^{19}$ to $1.8 \cdot 10^{20}$ cm^{-3}) with no or the smallest possible Cr(III) content. The Cr(III) compounds are not only ineffective in a polarization process but also are capable of significantly reducing the spin-lattice relaxation time of polarized protons.

We have managed to find suitable conditions for synthesis of Cr(V) complexes (absence of light, heating) and to get reproducible samples with the necessary characteristics. The best results were obtained with propanediol-I,2. The EPR spectrum of the complex is given in Fig. I.

A material is synthesized for targets with "frozen" polarization of protons, it is a complex Cr(V) compound in propanediol-I,2. Two polarized targets I5 cm and 60 cm in volume were made of this material and used in experiments at the accelerators of JINR (Dubna), IHEP (Serpukhov) and LINP (Leningrad). It is shown that the material allows one to obtain proton polarization close to I00% with relaxation time over I000 hours.

Synthesis and investigation of the Cr(V) complex in deuterated diols - ethanediol and propanediol - for a target with polarized deuterons |4-7|

The deuteron target material must have a high deuteron content and meet the same requirements as the proton target material. So it was reasonable to use deuterated diols for preparing the target material and synthesize Cr(V) complexes in them. There was an opinion, however, that it was impossible to obtain a Cr(V) complex in fully deuterated diols by reduction of Cr(VI) and that production of the complex required that hydrogen atoms should remain at least in hydroxyl groups. So Cr(V) complexes were usually synthesized in partly deuterated diols. When it was necessary to use fully deuterated diol, Cr(V) complexes with oxyacids were added as a paramagnetic admixture.

We have managed to synthesize Cr(V) complexes in fully deuterated diols - ethanediol CD_2ODCD_2OD (DED) and propanediol-I,2 $CD_2ODCDODCO_3$ (DPD) - by reducing chromate in them. However, the complex in DPD has a relatively larger number of deuterons per all nucleons as compared with the complex in DED. Besides, the EPR line width of the Cr(V) complex in DPD is I0% worse than that of the complex in DED. According to the DPN theory, the final temperature achieved by the nuclear spins during polarization is inversaly proportional to the width of paramagnetic admixtures responsible for DPN. So, it is preferable to use the Cr(V) complex in DPD as a target material. These complexes were not studied before. We were interested in the structure of the complexes, so we studied their EPR spectra. Samples with a low concentration of the complex were prepared. For comparison we recorded both the spectra of the deuterated complex and the spectra of Cr(V) in ordinary propanediol of low concentration (Fig. 2,3).

The Table lists the EPR parameters of the complexes. Now the Cr(V) complexes in diols are the most suitable material for a deuteron polarized target for beams of intensity up to $I0^{8}\ s^{-1}$ because this material has only one paramagnetic admixture. During polarization the target material is in a glass-like state retaining the structure of the liquid phase, which is characterized by a uniform distribution of paramagnetic centres with the practically isotropic g-factor. The Cr(V) complexes

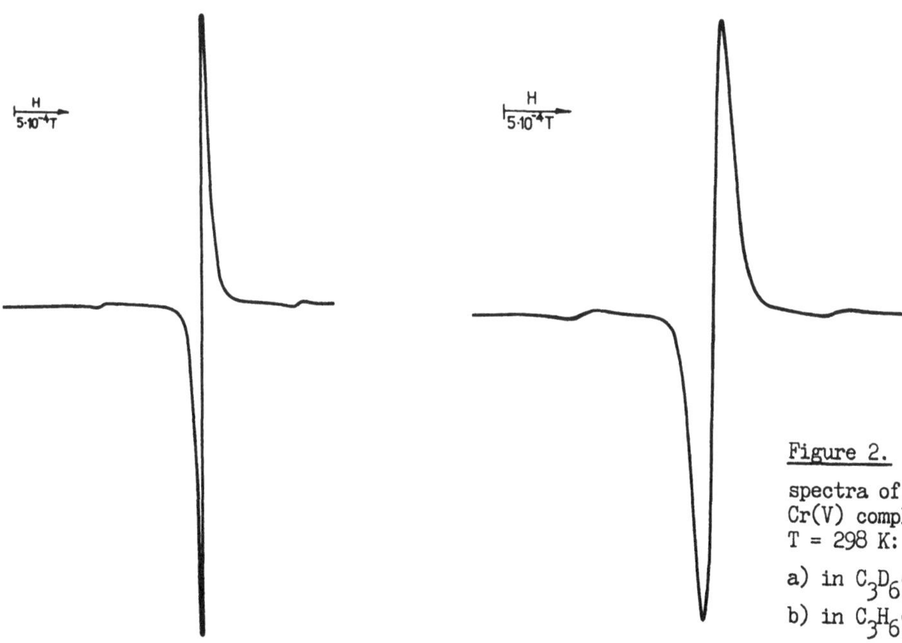

Figure 2. EPR spectra of the Cr(V) complex at T = 298 K:
a) in $C_3D_6(OD)_2$,
b) in $C_3H_6(OH)_2$.

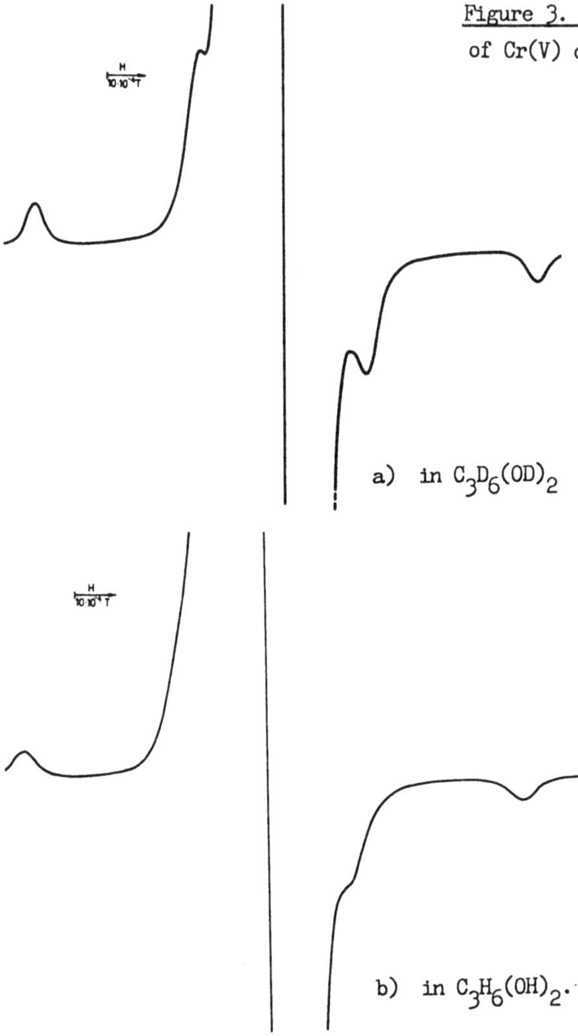

Figure 3. Central part of the EPR spectra of Cr(V) complexes at T=77 K:

a) in $C_3D_6(OD)_2$

b) in $C_3H_6(OH)_2$.

in diols have bidentant ligands based on organic alcohols. So, besides the deuterium nuclei and an insignificant number of hydrogen and Cr nuclei, the target material mainly contains ^{I2}C and ^{I6}O nuclei whose spin is zero. Thus, polarization of these nuclei does not complicate the situation, which is especially important in particle scattering experiments where it is necessary to discriminate between scattering by free protons (deuterons) and by bound nucleons.

The synthesized substances for targets with polarized protons and polarized deuterons on the basis of complex Cr(V) compounds turned out to be rather effective. Thus it was possible to make both ordinary targets with polarized nuclei and targets with "frozen" polarization. Non-equilibrium polarization can be "frozen" only if the spin-lattice relaxation considerably increases (to hundreds of hours) in the target material at very low temperatures. The Cr(V) complexes

Table. EPR parameters of Cr(V) complexes in $C_3D_6(OD)_2$ and $C_3H_6(OH)_2$

	T(K)	$g_{\perp}$	$g_{\parallel}$	g	$A_{\perp}$ ^{53}Cr (cm^{-1})	$A_{\parallel}$ ^{53}Cr (cm^{-1})	A_{iso} ^{53}Cr (cm^{-1})
$K_2Cr_2O_7$ + deuterated propanediol	298			I.98I*			
$C_3D_6(OD)_2$	77	I.977	I.984	I.98I**	$5.9 \cdot IO^{-4}$	$35.3 \cdot IO^{-4}$	$I5.7 \cdot IO^{-4}$
$K_2Cr_2O_7$ - propanediol	298			I.979*			
$C_3H_6(OH)_2$	77	I.976	I.984	I.978**	$5.7 \cdot IO^{-4}$	$35.6 \cdot IO^{-4}$	$I5.7 \cdot IO^{-4}$

* Experimental values

** Calculated values $g_{iso} = I/3(2g_{\perp} + g_{//})$

in diols are the best to meet this requirement as the relaxation time of their polarized nuclei is I000 hours. The obtained material - Cr(V) complex in fully deuterated propanediol-I,2 - ensured the maximal deuteron polarization of 40+2%. The deuteron polarization relaxation time was I000 hours at T=0.02 K and H=0.53 T. This material was used for a target with "frozen" polarization of deuterium nuclei developed in JINR in I986 and employed in the experiments at the IHEP accelerator (Serpukhov).

EPR-investigation of the stable Cr(V) complex with hydroxybutyric acid |8|

In I980s the stable Cr(V) complex with 2-hydroxy-2-methylbutyric acid (HMBA-Cr(V)) dissolved in alcohol was widely used as a target material in many laboratories of the world. The use of the complex made the process of target material preparation much simpler. It was considered that solutions of HMBA-Cr(V) are of homogeneous composition. We studied the behaviour of HMBA-Ce(V) in propanediol-I,2, in deuterated ethanediol (DED) and in ethanol. It was found that the complex is unstable in the solutions and reacts with the solvent producing several paramagnetic complexes. Three different paramagnetic complexes (I, II, III) were identified in the solutions of the complex in DED (Fig. 4).

Besides, during the cooling of the solutions formation of the glass-like phase can be accompanied with crystallization of the complex of the solvent, which results in aggregation of paramagnetic centres and their non-uniform distribution. Thus, the solutions of HMBA-Cr(V) in diols used as target materials provide lower and less reproducible polarization than the Cr(V) complexes synthesized in

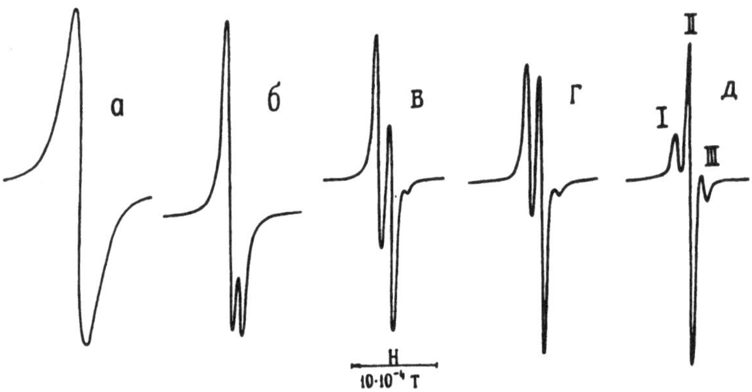

Figure 4. Central lines of the EPR spectra of HMBA-Cr(V) dissolved in DED with the concentrations: I) $0.7 \cdot I0^{20}$ cm^{-3}, 2) I:5, 3) I:I5, 4) I:50, 5) I:250 at T = 293 K.

diols. The presence of several paramagnetic admixtures in the target material complicates the relaxation processes.

Photochemical synthesis of Cr(V) compounds in pentanol for a target with higher resistance to damages due to radiation |9|

Pentanol $C_5H_{II}OH$ has a higher resistance to radiation than diols. Damages inflicted by radiation on the substance reduce the achieved polarization. This reduction is usually estimated by a characteristic target depolarization dose φ_0 , which is able to reduce target polarization by a factor of e . For a diol-based target $\varphi_0 = I\text{-}2.5 \cdot I0^{I4}$ particles/cm^2. For a pentanol-based target $\varphi_0 = 5 \cdot I0^{I4}$ particles/cm^2. Besides, diol-based targets contain I0% (by weight) of polarized protons, while pentanol-based targets contain I4% of protons. These characteristics of pentanol and properties of Cr(V) as a paramagnetic admixture taken into account, it

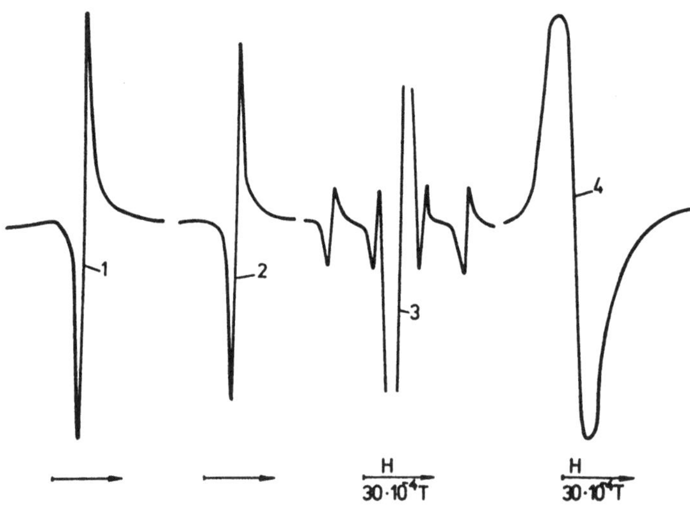

Figure 5. EPR spectra of Cr(V) compounds in pentanol: I - solution exposed to the laser beam (λ = 441 nm) for I min (concentration of ammonium bichromate (ABC) is $I0^{-2}$ mole/l); 2 - 0.5 min after irradiation stopped; 3 - solution exposed to laser beam for I min (ABC concentration is I mole/l, T = 293 K); 4 - solution exposed to laser beam for I min (ABC concentration is I mole/l, T = 77 K).

seemed reasonable to synthesize Cr(V) in pentanol. Yet, in contrast to diols, thermal reduction of Cr(VI) to Cr(V) failed in pentanol.

We have proposed a new method of synthesis - photochemical synthesis in laser radiation. Exposing the solution of ammonium bichromate in pentanol to helium-cadmium laser beam in the radiospectrometer resonator results in an EPR signal with g=I.973, which is typical of Cr(V), see Fig. 5. The intensity-width relationship of the HFS components in the EPR spectrum proves that they belong to the EPR spectrum of Cr(V) compounds of the same sort. The results obtained allow one to hope that the compounds synthesized can be used in polarized targets.

Solid target material - polyethylene with iminoxyl radical |IO|

In some cases simplification of target operation requires that a substance which is solid at room temperature should be used. With this purpose in view, we investigated polymers with stable radicals dissolved in them. We managed to obtain the proton polarization of 50% in low-density polyethylene with the stable radical 2,2,6,6-tetramethylpiperidine-I-oxyl introduced in the polymer by diffusion.

References

I. Borisov N.S., Bunyatova E.I., Kiselev Yu.F. et al. Preprint part I, I3-I0253, part 2, I3-I0257, JINR, Dubna, I976.

2. Borisov N.S., Bunyatova E.I., Volodin A.G. et al. Preprint I-80-98, JINR, Dubna, I980.

3. Borisov N.S., Bunyatova E.I., Kiselev Yu.F. et al. In: Proc.of the Int.Symp. on Polarization Phenomena in High Energy Physics, Dubna, I7-20 Nov.I98I, I99-2II.

4. Bunyatova E.I., Bubnov N.N. Nucl.Instr.and Meth. I984, 2I9, 297.

5. Kiselev Yu.F., Poljakov V.V., Kovalev A.I. et al. Nucl.Instr.and Meth., I984, 220, 399.

6. Bunyatova E.I., Bubnov N.N. Nucl.Instr.and Meth., I987, A254, 252.

7. Borisov N.S., Bunyatova E.I., Liburg M.Yu. et al. J.of Phys.E: Sci.Instr. I988, 2I, II79.

8. Bunyatova E.I., Galimov R.M., Luchkina S.A. In: Int.Workshop in Spin Phenomena in High Energy Physics, Protvino, I4-I7 Sept.I983, II2-II9.

9. Bunyatova E.I., Maloletov S.M., Sherstyuk N.P. Preprint PI2-87-796, JINR, Dubna, I987.

IO. Borisov N.S., Bunyatova E.I., Kiselev Yu.F. et al. Preprint P6-7408, JINR, Dubna, I973.

II. Bunyatova E.I., Sahling A., Sahling S. Preprint E8-90-I2I, JINR, Dubna, I990.

Progress in the Chemistry of Chromium(V) Doping Agents Used in Polarized Target Materials

M. Krumpolc[1], *D. Hill*[2], *and H.B. Stuhrmann*[3]

[1]University of Illinois at Chicago, Chicago, IL 60680, USA
[2]Argonne National Laboratory, Argonne, IL 60439, USA
[3]DESY-HASYLAB, W-2000 Hamburg 52, Fed. Rep. of Germany

We wish to report progress in two areas of the chromium(V)-based doping agents:

I. Two commonly used chromium(V) complexes, I and II, have been synthesized in perdeuterated form (i.e., all hydrogens replaced by deuterium). They are sodium bis(2-ethyl-2-deuteroxybutyrato)oxochromate(V) monodeuterate, IV, (acronym EDBA-Cr(V)), and sodium bis(2-deuteroxy-2-methylpropionato)oxochromate(V), III, (acronym DMPA- Cr(V)).

II. A synthetic route leading to the preparation of stable, chromium(III)-free solutions of chromium(V) in diols (1,2-ethanediol /ethylene glycol/ and 1,2-propanediol /propylene glycol/) has been outlined.

I. Synthesis of Perdeuterated Complexes of Chromium(V)

The role of chromium(V) complexes used as doping agents for the preparation of polarized target materials has met with success. The complexes have been used in a variety of experiments ranging from dynamic polarization of protons[1], deuterons[2], carbon-13[3] and fluorine-19[4] nuclei to biomolecules[5,6]. In recent years there has been a growing interest in polarized targets containing only the ^2H (deuterium) nucleus in the absence of any protons. For the purpose of these experiments we have decided to synthesize the most often used complex, sodium bis(2-ethyl-2-hydroxybutyrato)-oxochromate(V) monohydrate, II, (common acronym EHBA-Cr(V)) in perdeuterated form. In addition, we have also prepared sodium bis(2-hydroxy-2-methylpropionato)oxochromate(V), I, (acronym HMPA-Cr(V)) in perdeuterated form as its synthesis is less demanding and less costly. It was used in initial test experiments[7].

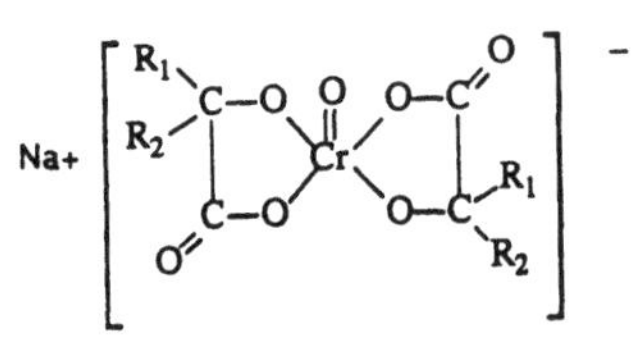

I. $R_1,R_2 = CH_3$
II. $R_1,R_2 = C_2H_5$
III. $R_1,R_2 = CD_3$
IV. $R_1,R_2 = C_2D_5$

Scheme I outlines individual steps leading to preparation of deuteroxy acids used as starting materials. As both acids are not commercially available they had to be synthesized from simple deuterated molecules. The first complex, EDBA-Cr(V), IV, was prepared from bromoethane-d_5 via a series of organometallic reactions (Grignard, organocadmium) to give 2-ethyl-2-deuteroxybutyric acid-d_{12}. The synthesis of the second complex, DMPA-Cr(V), III, was relatively fast. Acetone-d_6 was converted into a cyanohydrin which after hydrolysis yielded 2-deuteroxy-2-methylpropionic acid-d_8. Both deuteroxy acids were then oxidized to form chromium(V) complexes[8]. Experimental details of this procedure will be published elsewhere[9].

Scheme I

A. $$CD_3CD_2Br \xrightarrow[2.\ CdI_2]{1.\ Mg} (CD_3CD_2)_2Cd \xrightarrow[2.\ OH^-/D_2O]{1.\ ClCOCOOC_2H_5}$$

$$(CD_3CD_2)_2C(OD)COOD \xrightarrow{Na_2Cr_2O_7} \text{EDBA-Cr(V)}$$

B. $$CD_3COCD_3 \xrightarrow[2.\ DCl/D_2O]{1.\ KCN} (CD_3)_2C(OD)COOD \xrightarrow{Na_2Cr_2O_7} \text{DMPA-Cr(V)}$$

II. Complexes of Chromium(V) with Diols (Glycols)

Before the introduction of externally available chromium(V) complexes chromium(V) in diols was generated in situ.[10] Typically, a procedure described by Bontchev[11] was followed using potassium dichromate dissolved in excess ethylene glycol. The oxidation was allowed to proceed for some time in order to accumulate a chromium(V)-ethylene glycol complex, I, along with chromium(III), water and other reaction products. The complex was never isolated in pure form, the solution was used directly to produce polarized targets.

Encouraged by our successful synthesis of chromium(V) complexes based on hydroxy acids[8] we turned our attention to two diols, 1,2-ethanediol (ethylene glycol) and 1,2-propanediol (propylene glycol), in an attempt to prepare complexes I and II.

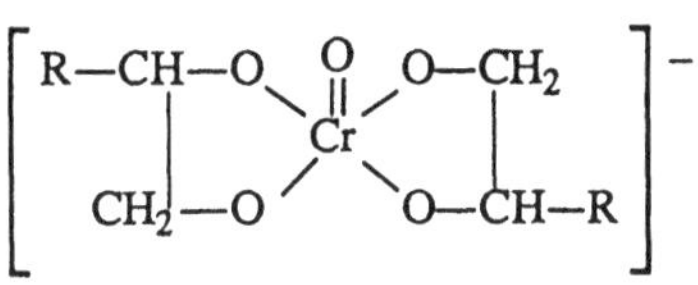

I. R=H
II. R=CH_3

First, we have noticed an unusually high reaction photosensitivity (Fig. 1), much higher than previously reported[11]. The maximum at 600 nm is indicative of the formation of chromium(III)[12,13]. Thus, all experiments were then performed in the absence of light. Having substantially changed reaction conditions — using a near stoichiometric ratio between the diols and sodium dichromate — we were able to considerably increase the amount of chromium(V) formed. The decomposition of chromium(V) is also accelerated by the presence of water formed in the course of oxidation[11]. Thus, we used a common dehydrating agent, anhydrous magnesium sulfate, to continuously remove water. These results are summarized in Fig. 2.

Curve 1 shows the formation of chromium(III) in the absence of a dehydrating agent (600 nm, maximum), curve 2 shows the spectrum of a stable chromium(V)-propylene glycol complex, II, free of chromium(III). These results were also confirmed by column chromatography using silica gel. Thus, under these conditions we were able to cease the second part of oxidation (eq. 1) which leads to decomposition of chromium(V) and formation of chromium(III). These results are also valid in case of ethylene glycol. A work is in progress to isolate both complexes in pure form.

$$\text{Cr(VI)} \xrightarrow{\text{diol}} \text{Cr(V)} \xrightarrow{\text{diol}} \text{Cr(III)} \qquad (1)$$

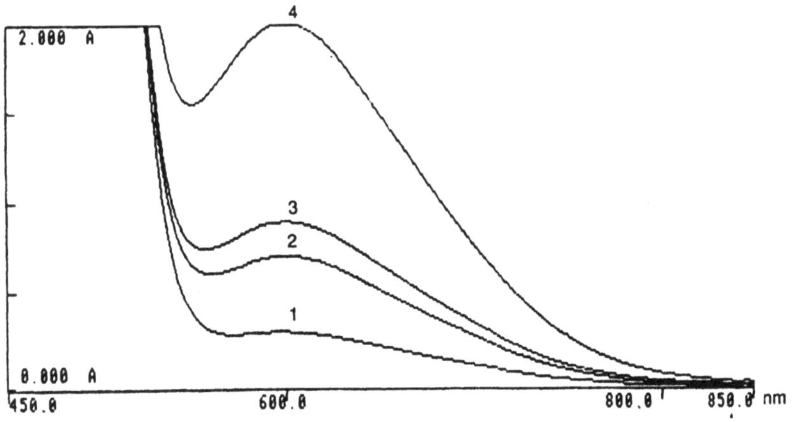

Figure 1. Visible spectrum scans of the initial stage of the oxidation of 1,2-propanediol by sodium dichromate (11:1, w/w, 25° C). Effect of light. Time: (1), 4 hr, dark; (2), 4 hr, incandescent lab light; (3), 10 min, sunlight; (4), 4hr, sunlight.

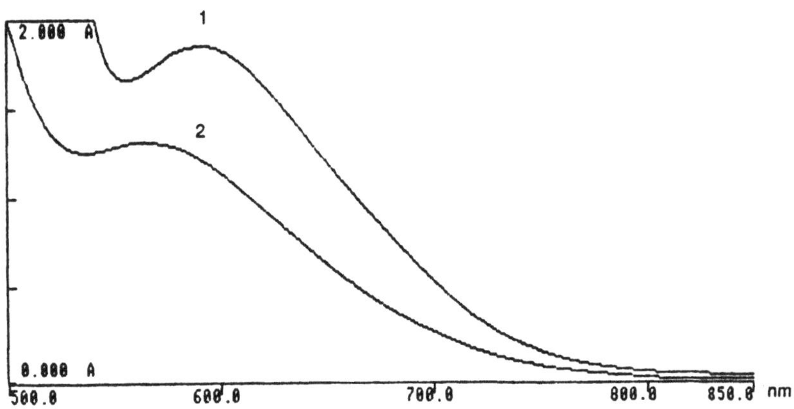

Figure 2. Visible spectrum scans of the initial stage of the oxidation of 1,2-propanediol by sodium dichromate (stoichiometry 1.2: 1 mol; 25° C). Effect of water. Time: (1), 6 hr; (2), 6 hr, water continuously removed by anhydrous magnesium sulfate.

Support of this work by the U.S. Department of Energy is gratefully acknowledged.

Literature References

1) D. Hill, R. C. Miller, M. Krumpolc and J. Roček, Nucl. Instr. and Meth., **150**, 331 (1978).
2) S. Hiramatsu, S. Isagawa, S. Ishimoto, A. Masaike, K. Morimoto, T. Adachi, M. Minowa and M. Krumpolc, Nucl. Instr. and Meth., **160**, 193 (1979).
3) J. J. Jarmer, S. Penttilä, D. Hill, T. Kasprzyk, M. Krumpolc, M. L. Barlett, G. W. Hoffmann, L. Ray, Nucl. Instr. and Meth., **A250**, 576 (1986).
4) D. Hill, T. Kasprzyk, J. J. Jarmer, S. Penttilä, M. Krumpolc, G. W. Hoffmann and M. Purcell, Nucl. Instr. and Meth., **A277**, 319 (1989).
5) H. B. Stuhrmann, O. Schärpf, M. Krumpolc, T. O. Niinikoski, M. Rieubland and A. Rijllart, Eur. Biophys. J., **14**, 1 (1986).
6) W. Knop, K. H. Nierhaus, V. Novotny, T. O. Niinikoski, M. Krumpolc, J. M. Rieubland, A. Rijllart, O. Schärpf, H.-J. Schink, H. B. Stuhrmann and R. Wagner, Helv. Phys. Acta, **59**, 741 (1986).

7) T. O. Niinikoski, private communication.
8) M. Krumpolc and J. Roček, J. Am. Chem. Soc., **101**, 3206 (1979).
9) M. Krumpolc and H. B. Stuhrmann, submitted for publication.
10) N. S. Garifyanov, B. M. Kozyrev and V. N. Fedotov, Dokl. Acad. Nauk SSSR, **178**, 808 (1968).
11) P. R. Bontchev, A. Malinovski, M. Mitewa and K. Kabassonov, Inorg. Chim. Acta, **6**, 499 (1972).
12) M. Krumpolc, B. G. DeBoer and J. Roček, J. Am. Chem. Soc., **100**, 145 (1978).
13) M. Krumpolc and J. Roček, J. Am. Chem. Soc., **98**, 872 (1976).

Some Metal Complexes as Free Radicals for Polarized Targets

N. Nakasuka

School of Science, Nagoya University, Furo-cho,
Chikusa, Nagoya 464-01, Japan

Some complexes of chromium(V) and titanium(III) were prepared, and their ESR spectra were measured together with TCNQ-Li, for the purpose of developing new polarized target materials.

1. ESR spectra of chromium(V) complexes with hydrogen peroxide, EHBA, HMPA and propanediol.

Kinetic studies on the reduction of chromate(VI) with hydrogen peroxide show that in general Cr(VI) is readily reduced to Cr(III) via a blue intermediate (Cr(VI)-H_2O_2) in acid solutions, without formation of a stable Cr(V) complex [1-3]. A peroxo chromium(V) complex, however, can be prepared from a strongly alkaline solution [4]. Although this complex shows a sharp ESR spectrum characteristic of Cr(V), it is sparingly soluble in common solvents.

Two Cr(V) complexes with tertiary hydroxy carboxylic acid (EHBA, HMPA) have been prepared according to Krumpolc's methods [5,6], and their ESR spectra were compared with those of other pentavalent chromium complexes.

ESR spectra of glycol complexes [7,8] have also been measured after cooling solutions of lithium chromate(VI) with propanediol (PD) heated at about 60 °C to ambient temperature.

All ESR spectra were recorded at 77 K on a JEOL JES-RE1X X-band spectrometer. Half-height widths of spectra are summarized in Table I, together with the solvents used.

2. Titanium(III) complexes with oxalate (ox) and urea.

Since tervalent titanium ion has a d^1 electron configuration and the literature indicates that Ti(III) ion shows ESR spectra, we tried to synthesize certain complexes with sufficient stability.

Caesium triaquabis(oxalato)titanium(III) dihydrate could be prepared by reacting titanium(III) chloride with oxalic acid in an atmosphere of nitrogen. The crystal is brown in color and stable at room temperature. A crystal structure determination by Drew and Eve shows that the central Ti ion is surrounded by seven oxygen atoms [9].

The ESR spectra have a fine structure, and their width is very wide (of the order of magnitude of 100 gauss). This compound is not soluble in common solvents, and may not be suitable for a polarized target material.

Some urea complexes of titanium(III) have been known as considerably stable and we prepared hexakis(urea)titanium(III) iodide [10]. This compound is blue in color, stable in air, and dissolves readily in water and some organic solvents. The crystal structure of this compound shows a typical configuration of six coordination [11]. The presence of iodide ion seems to be very important for the extraordinary stability of this complex. Its ESR spectrum is also wide, and has a fine structure.

Table I Comparison of half-height widths ($w_{\frac{1}{2}}$) of ESR spectra.

complex	$w_{\frac{1}{2}}$/G	solvents (concentration)
Cr(V)-H_2O_2	30	acetone (suspended)
Cr(V)-EHBA	20	PD, n-butanol (3 %)
Cr(V)-HMPA	20	PD, n-butanol (0.3 %)
Cr(V)-HMPA	19	n-butanol-d_{10} (0.3 %)
Cr(V)-PD	11	PD
Ti(III)-ox	120	aqueous acetone (3 %)
Ti(III)-urea	180	aqueous acetone (3 %)
TCNQ-Li	8	n-butanol

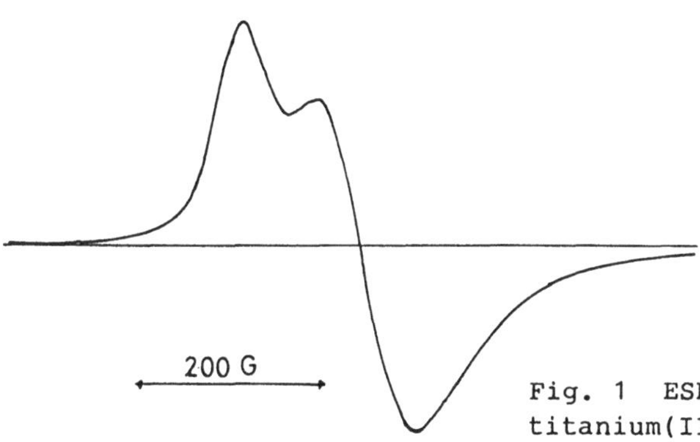

Fig. 1 ESR spectra of hexakis(urea) titanium(III) iodide.

3. Lithium(I) salt of TCNQ.

TCNQ (7,7,8,8-tetracyanoquinodimethane) is famous for an electron acceptor forming stable radical anions. Accordingly we measured ESR spectra of its lithium salt (Dojindo, Japan) in propanediol and n-butanol as solvents. The spectrum is very sharp and narrow, and thereby the compound is frequently used as a standard for determination of g values in ESR measurements (g = 2.00252). The solid powder is violet in color, and is very stable in air at room temperature.

The Li salt is readily soluble in common solvents, including water, alcohol and ethanediol. TCNQ-Li seems to be worth testing as a polarized target material, in view of the narrow ESR spectrum and other favorable properies.

This preliminary study has been carried out in collaboration with Dr. N. Horikawa, Dr. T.Iwata and Mr. J. Kishi in the School of Science, Nagoya University.

References

[1] F. A. Cotton and G. Wilkinson, Advanced Inorganic Chemistry, 4th edition, Interscience, New York (1980).

[2] S. B. Brown, P. Jones and A. Suggett, Prog. Inorg. Chem., **13**, 159-204 (1970).

[3] J. K. Beattie and G. P. Haight, Jr., Prog. Inorg. Chem., **17**, 93-145 (1972).

[4] E. H. Riesenfeld, H. E. Wohlers and W. A. Kutsch, Ber. Bunsenges. Phys. Chem., **38**, 1885 (1905).

[5] M. Krumpolc, B. G. DeBoer and J. Rocek, J. Amer. Chem. Soc., **100**, 145 (1978).

[6] M. Krumpolc and J. Rocek, J. Amer. Chem. Soc., **101**, 3206 (1979).

[7] N. S. Garifyanov, B. M. Kozyrev and V. N. Fedotov, Dokl. Akad. Nauk. SSSR, **178**, 808 (1968).

[8] E. I. Bunyatova and N. N. Bubnov, Nucl. Instr. Meth., **219**, 297 (1984).

[9] M. G. Drew and D. J. Eve, Acta Cryst., **B33**, 2919 (1977).

[10] H. Hartmann, H. L. Schlaefer and K. H. Hansen, Z. anorg. allg. Chem., **289**, 42 (1957).

[11] P. H. Davis and J. S. Wood, Inorg. Chem., **9**, 1111 (1970).

Measurements of Glass Properties and Density of Hydrocarbon Mixtures of Interest in Polarized Targets

S. Takala and T.O. Niinikoski

CERN, CH-1211 Geneve 23, Switzerland

The densities of binary and ternary mixtures of some alcohols, diols and water, measured at liquid nitrogen temperature, are reported. The glass transition and devitrification temperatures were determined by microcalorimetric techniques for a large subset of the materials. Basing on these, the optimization of target materials is discussed in view of the use of the pure stable Cr(V) complexes as dopants.

INTRODUCTION

Glassy hydrocarbon mixtures have been used with success in polarised targets. Early developments began with ethanol-based solutions[1)], but it was soon discovered that butanol-water gave better and more reproducible results[2-4)]. This work was also immediately extended to deuterated butanol-water[5-6)].

The reason for the success or failure with a given mixture of solvents was not well understood before it was realized in 1979 that the glass forming ability of the liquid was the key parameter contributing to the polarization results. After a series of tests at CERN ruling out gelification and polymerization, viscosity measurements[7)] gave first evidence that formation of a glass could prevent the phase separation of the paramagnetic molecules, required for the dynamic nuclear polarization (DNP), in the solid during rapid freezing. This finding was confirmed convincingly using microcalorimetric techniques[8)].

The advent of stable and soluble Cr(V) complexes[9)] and their deuterated counterparts have revived the interest in hydrocarbon glass polarized targets, in particular for applications requiring a high deuteron polarization. Fabrication of large amounts of uniform and reproducible material is thus made possible, with minimum losses of the expensive deuterated solvents.

The aim of this work was to study the glass properties of several hydrocarbon mixtures in a comparative way, for the purpose of possibly identifying new mixtures with improved stability of the glass phase and capability of dissolving more paramagnetic complexes while freezing the mixture of the solvent and solute. The glass properties were assessed using a microcalorimeter similar to that of Hill and Hill[8)]. We also report measurements of density of the solvents. The work is part of a development programme aiming at optimizing and constructing a large deuterated polarized target for the measurement of the neutron spin structure function in the NA47 experiment proposed by the Spin Muon Collaboration (SMC), using the polarized muon beam of the CERN SPS machine.

APPARATUS AND SAMPLE PREPARATION

A differential microcalorimeter was used to record the anomalies in the thermal behaviour of various butanol solutions. A series of solutions containing butanol isomers, water, pinacol and propanediol were prepared.

A drop of each solution (~3-5μl) was frozen in liquid nitrogen to a glassy bead which was placed on the sample loop, made of thermocouple wire, in the precooled microcalorimeter. The reference loop was left empty. The temperature was raised at a rate of ~5K/min and the temperature difference between the sample and the reference loops was recorded as a function of the sample temperature.

RESULTS

Non-deuterated solutions

The curve of pure 1-butanol (Fig.1/1) shows the glass transition (G), devitrification (D) and melting (M) anomalies.

The 2-butanol (Fig. 1/2) behaves differently when the temperature is raised. After glass transition there is no devitrification peak nor melting peak to be found. The 2-butanol bead softens gradually until it runs through the loop in the microcalorimeter. The third isomer tested was isobutanol (2-methyl propanol) (Fig.1/3). Its curve is similar to that of 2-butanol: there is no devitrification nor melting peak. The softening process is slightly displaced towards a lower temperature.

The solution containing 1- and 2-butanol still shows a devitrification and a melting peak up to the concentration of 10% 2-butanol in solution. When the 2-butanol concentration is increased, the curve

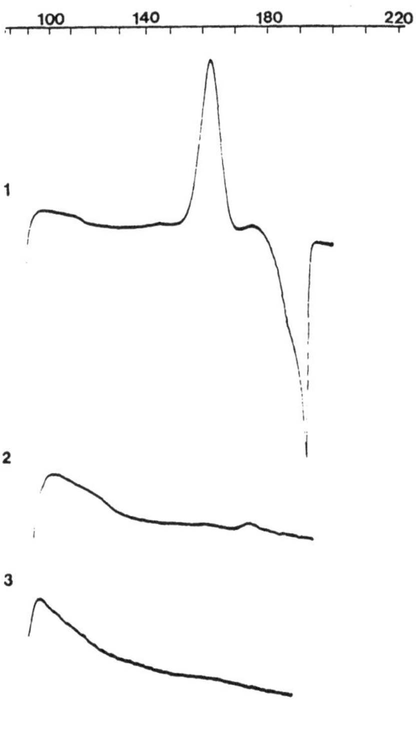

Fig.1. Butanol isomers
1. 1-Butanol
2. 2-Butanol
3. Isobutanol

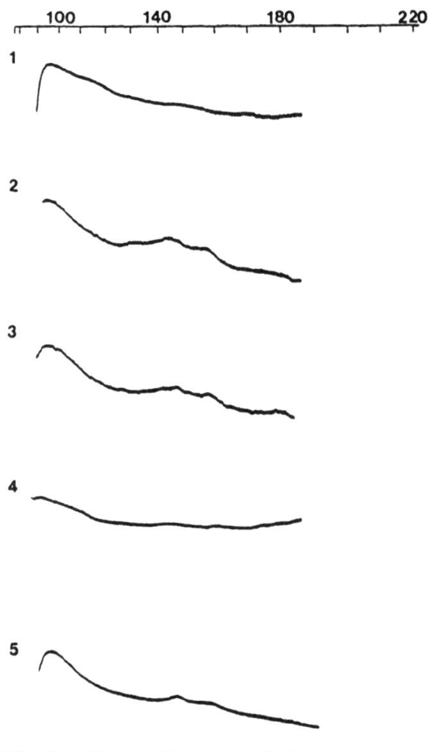

Fig.2. Butanol isomer solutions
1. 2-Butanol + Isobutanol 9.0%
2. 1-Butanol + 2-Butanol 8.9%
3. 1-Butanol + Isobutanol 10.3%
4. 1-Butanol + 2-Butanol 19.2%
5. 1-Butanol + Isobutanol 19.9%

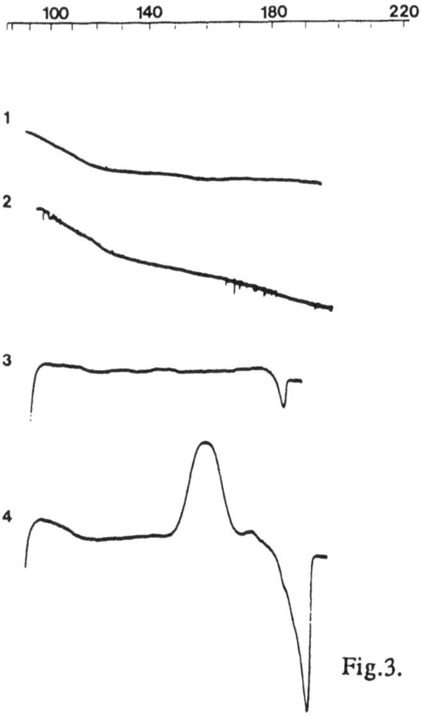

Fig.3.

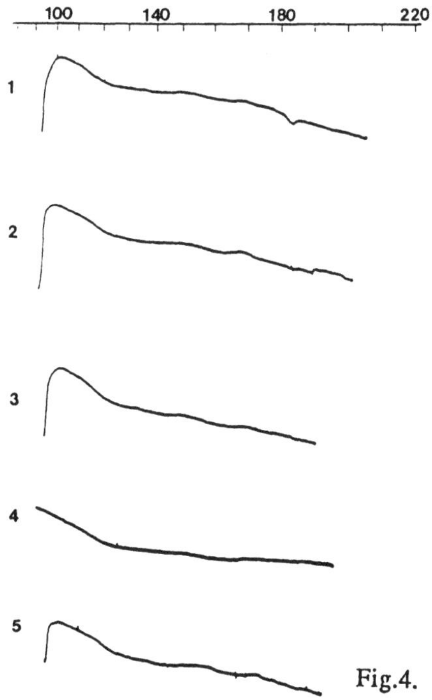

Fig.4.

Fig.3. The solution effect of additives

1. 1-Butanol + 2-Butanol 9.2% + Water 5.9% + EHBA-Cr(V) 4.4 10^{19} spins/g
2. 1-Butanol + 2-Butanol 12.7% + Water 5.5%
3. 1-Butanol + Water 5.9%
4. 1-Butanol

Fig.4. The 2-butanol concentration variation

1. 1-Butanol + Water 4.9% + EHBA-Cr(V) 3.7 10^{19} spins/g
2. 1-Butanol + 2-Butanol 5.0% + Water 4.7% + EHBA-Cr(V) 3.0 10^{19} spins/g
3. 1-Butanol + 2-Butanol 8.5% + Water 4.8% + EHBA-Cr(V) 2.8 10^{19} spins/g
4. 1-Butanol + 2-Butanol 9.2% + Water 5.9% + EHBA-Cr(V) 4.6 10^{19} spins/g
5. 1-Butanol + 2-Butanol 16.2% + Water 5.1% + EHBA-Cr(V) 3.1 10^{19} spins/g

resembles that of 2-butanol. Isobutanol acts in a similar way when added to 1-butanol to make a solution (Fig.2).

Solutions of 1-butanol and water in the range of 2.5%-6.5% water by weight do not devitrify. They show a reduced melting peak. The 1-butanol + 2-butanol + water solutions reach the saturation point at 8% concentration of water.

With EHBA-Cr(V) complex (sodium bis[2-ethyl-2-hydroxybutyrato(2-)]oxochromate monohydrate) in solutions containing butanol and 4.7%-5.9% water, no devtrification is detected (Fig.3). The viscosity of the solution is higher than that of solutions without the complex compound. None of the beads fall through the loop of the microcalorimeter (Fig.4).

The effect of possible phase separation was tested by cooling the melt back to liquid nitrogen temperature (Fig.5,6).

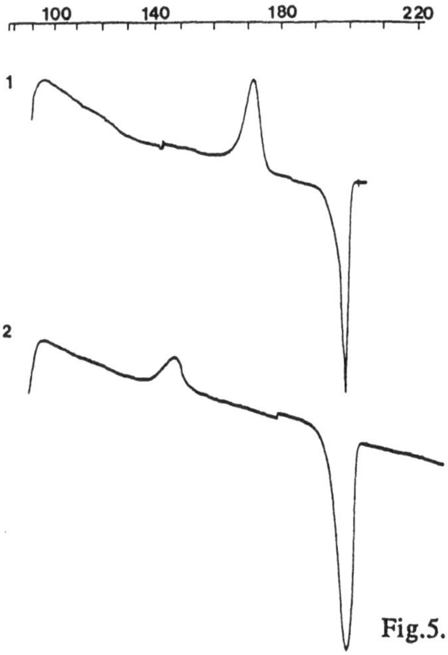

Fig.5.

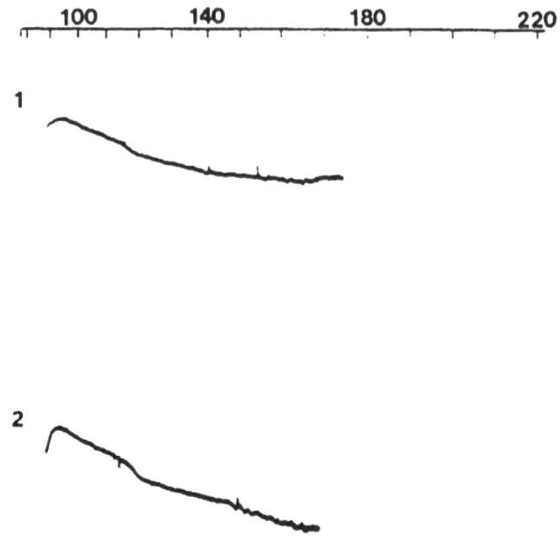

Fig.6.

Fig.5. Pentanol solution recooled
1. Pentanol + Water 5% + EHBA-Cr(V) 4 10^{19} spins/ml
2. The same pentanol solution bead after being recooled to 77K

Fig.6. Butanol solution recooled
1. 1-Butanol + 2-Butanol 9.2% + Water 5.9% + EHBA-Cr(V) 4.4 10^{19} spins/g
2. The same butanol solution bead after being recooled to 77K

Pinacol (2,3 - dimethyl - 2,3 - butanediol) is known to prevent devitrification in the same way as water does. A set of curves were produced from solutions of 1-and 2-butanol and pinacol. They form a glass with no phase transitions.

Propanediol has the same effect as water or pinacol when dissolved to 1-butanol.

Deuterated solutions

A solution of deuterated 1-butanol and deuterated water doped with deuterated EHBA-Cr(V) complex was prepared for a polarization test and calorimetric curves were recorded as well (Fig.7).

A glass transition is found at the same temperature as in the nondeuterated solution but no devitrification is visible before melting (185K), with complex concentration 6.4 10^{19} spin/g.

The corresponding curves were made for propanediol and deuterated propanediol - Cr(V) complex solutions (Fig.8).

Density measurements

The density is one of the parameters weighing in the figure of merit for polarized target materials. As two possible candidates for SMC deuterated target were estimated, the densities were needed.

The densities of various binary and ternary mixtures were measured at 77K temperature, by weighing samples submerged in liquid nitrogen. The results are displayed in Table 1.

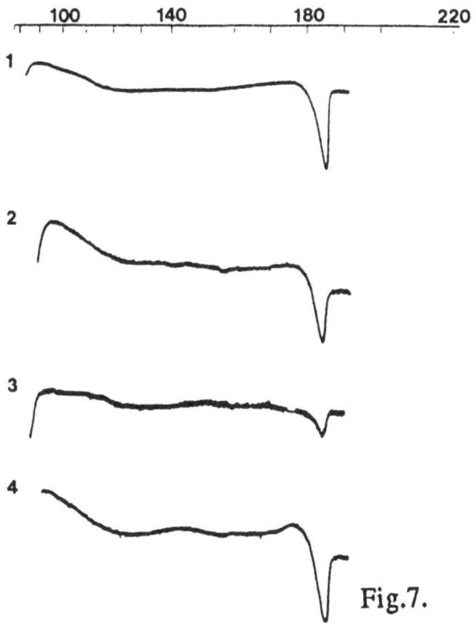

Fig.7.

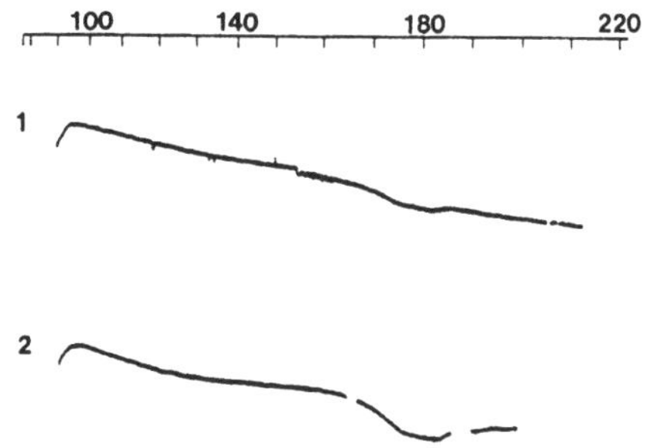

Fig.8.

Fig.7. Deuterated butanol solutions
1. 1-Butanol-D10 + Water (D_2O) 4.8%
2. 1-Butanol-D10 + Water (D_2O) 5.3%
3. 1-Butanol-D10 + Water (D_2O) 4.8% + EHBA-Cr(V)-D20 6.4 10^{19} spins/g
4. 1-Butanol + Water 5.9%

Fig.8. Propanediol and deuterated propanediol chrome(V) complex solution
1. Propanediol
2. Propanediol-D8 + Cr(V) 6 10^{19} spins/ml

Table 1. Densities of some hydrocarbon glasses used in polarized targets.

Material	Density (g/cm^3)
Propanediol-D8 + Cr(V) complex 6 10^{19} spins/ml	1.274 ±0.014
1-Butanol-D10 + 4.8% D_2O + EHBA-Cr(V)-D20 6.4 10^{19} spins/g	1.109 ±0.011
1-Butanol + Water 5.6% + EHBA-Cr(V) 7.5 10^{19} spins/g	0.966 ±0.003
1-Butanol + 2-Butanol 12.2% + Water 5.3% + EHBA-Cr(V) 6.6 10^{19} spin/g	0.985 ±0.001
1-Butanol + Pinacol 16.4% + EHBA-Cr(V) 6.1 10^{19} spins/g	0.973 ±0.012
1-Butanol + Propanediol 19.0% + EHBA-Cr(V) 6.5 10^{19} spins/g	0.994 ±0.009
Pentanol + Water 5% + EHBA-Cr(V) 4 10^{19} spins/ml (Ref. [10])	0.966 ±0.005

DISCUSSION

Polarized hydrocarbon target materials, with paramagnetic molecules dissolved in the glassy matrix, require a unique combination of material properties in the solid solvent. It is obvious that any crystallization, during solidification by cooling, may result in the separation and accumulation of paramagnetic molecules, either in microcrystals or on crystal boundaries. This can be observed visually by the transformation of the material from translucid, clear glassy phase into an opaque solid, and is always associated with a dramatic reduction of the DNP as well as in the increase of the nuclear spin-lattice relaxation rate.

The solubility of the stable paramagnetic Cr(V)-complex molecules at the room temperature and during solification is another important criterion for target materials, which need to have a relatively fast polarization rate. This is particularly important in frozen spin polarized targets operated at low magnetic fields, requiring a frequent repolarization.

The suitable glass is also limited by its ability to withstand thermal stresses upon rapid solification in liquid nitrogen. Glass with a large thermal dilatation and low stress resistance will fracture during preparation and handling.

The results of our tests suggest that the solutions of various butanol isomers and water could provide improved glasses for the use in polarized targets.

Polarization tests are underway in order to assess their figures of merit.

REFERENCES

(1) M. Borghini, O. Runolfsson and J. Vermeulen, Proc. Int. Conf. on Polarized Targets and Ion Sources, Saclay 1966, Ed. La Direction de la Physique, CEN Saclay (Gif-sur-Yvette 1966), p. 387.

(2) M. Borghini and K. Scheffler, Proc. 2nd Int. Conf. on Polarized Targets, Berkeley 1971, Ed. G. Shapiro, LBL-500, UC-34 Physics, TID-4500 (Lawrence Berkeley Laboratory, Berkeley 1971), p. 303.

(3) S. Mango, Ö. Runolfsson and M. Borghini, Nucl. Instr. Meth. **72** (1969) 45.

(4) P. Roubeau, J. Ezratty, H. Glättli, J. Vermeulen and M. Borghini, Nucl. Instr. Meth. **82** (1970) 323.

(5) M. Borghini and K. Scheffler, Nucl. Instr. Meth. **95** (1971) 93.

(6) M. Borghini, A. Masaike, K. Scheffler and F. Udo, Nucl. Instr. Meth. **97** (1971) 577.

(7) T.O. Niinikoski, Proc. 2nd Workshop on Polarized Target Materials, Abingdon 1979, Report RL-80-080 (SRC, Chilton, Didcot 1980), p. 69.

(8) D.A. Hill and J.J. Hill, ANL-HEP-PR-81-05, 1980.

(9) M. Krumpolc and J. Rocek, J. Am. Chem. Soc. **98** (1976) 872.

(10) T.O. Niinikoski, Nucl. Instr. Meth. **192** (1982) 151.

Low Temperature Thermal Properties of Pentanol-2 – A Perspective Polarized Target Material

S. Sahling

Technische Universität Dresden, Sektion Physik,
Mommsenstraße 13, O-8027 Dresden, Fed. Rep. of Germany

The low temperature heat capacity and the long-time energy relaxation of pentanol-1 and pentanol-2 were measured. The results demonstrate that pentanol-1 is crystalline and pentanol-2 is amorphous after cooling with a cooling rate of about 1K/min. From the power release data of pentanol-2 the distribution parameter of two-level systems $P = 4.1 \cdot 10^{44}/Jm^3$ was obtained.

1. INTRODUCTION

The high polarization of proton targets requires a uniform distribution of the paramagnetic centers. Therefore, the polarization of butanol (C_4H_9OH) or pentanol-1 ($C_5H_{11}OH$) in the glass state of the material is sufficiently larger than in the crystalline one[/1-3/]. The amorphous structure in these materials can be obtained only after dopening with 5% water and rapid cooling from room temperature to liquid nitrogen temperature[/1,2/]. In this case the polarization is probably dependent on the cooling rate or on the purity of the target material. Moreover, an accidental warming up of the target to a temperature near the glass transition temperature T_g (T_g of pentanol-1 is about 170K[/2/]), leads to a drastic reduction of the polarization. Thus the preparation, transport, storage and experiments with these materials are not very convenient.

It will be shown in this paper that there are not these problems with pentanol-2. Pure pentanol-2 (and probably pentanol-3) forms the glass structure independent of the cooling rate, e.g. even after a very weak cooling from 292K to 78K with a cooling rate of about 1K/min.

For the recognition of the glass structure the low temperature heat capacity and the long-time power release were measured. Since the long-time power release is caused by two-level systems (TLS) existing only in glasses, this recognition method is very sensitive. Moreover, according to the tunnelling theory[/4,5/], a direct determination of the distribution parameter P (and the corresponding density of states of TLS) is possible:

$$P = (24/\pi^2k^2)\ \dot{Q}\ t\ /\ V\ (T_1^2 - T_0^2)\ , \qquad (1)$$

where $\dot{Q}$ is the power released after a rapid cooling of the sample from the equilibrium temperature T_1 to T_0, t is the time after starting the cooling and V is volume of the specimen.

2. EXPERIMENTAL

A special calorimeter with a low heat leak ($\dot{Q}_B$ = 0.5 nW) to the sample was used in our experiments. A aluminium container (with or without pentanol) hung in a vacuum chamber on 8 kapron threads. The container was connected with a Ge-thermometer, a heater and a copper wire to provide the thermal contact with a mechanical heat switch.

For the heat release study the resistance drift $\dot{R}$ of the thermometer was measured as a function of time t after the rapid cooling of the specimen from the equilibrium temperature T_1 (where the specimen remained for 24 hr) to T_0 = 1.5K. The corresponding power release $\dot{Q}(T_1,T_0,t)$ was then determined from the measured power release $\dot{Q}_m$, the power $\dot{Q}_A$ due to the thermal contact between the specimen and the body of the calorimeter, and the heat leak $\dot{Q}_B$:

$$\dot{Q}(T_1,T_0,t) = \dot{Q}_m - \dot{Q}_A - \dot{Q}_B\ , \qquad (2)$$

where

$$\dot{Q}_m = \dot{R}C/(\delta R/\delta T)\ , \qquad (3)$$

$$\dot{Q}_A = A(T_K - T). \qquad (4)$$

C is the heat capacity of the container with or without the sample, $\delta R/\delta T$ is the sensitivity of the thermometer, and A is the total thermal resistance of the kapron threads and the electric wires to the heater and the thermometer. The sample temperature T was chosen close to the calorimeter temperature T_K, which was kept constant (to 10^{-4}K). When $T(t)-T_K=10^{-4}$K, the sample temperature was set to $T_0=T_K$.

3. RESULTS AND DISCUSSION

First the heat capacity and the power release of the hollow container (174.7g aluminium, 17.4g copper) were measured. As expected, no long-time heat release was observed. After the cooling of the container from T_1 = 10K to 1.5K in 8 min the heat leak $\dot{Q}_B$ = 0.5 nW was measured and the constant A = 0.1 nW/mK was determined. The accuracy of the power release measurement is given by the fluctuations of $\dot{Q}_B$, which are smaller than 0.1 nW (tested during one weak).

The container with pentanol-1 (V = 110 cm^3, m = 90.12 g, crystallization temperature: 190 K) or with pentanol-2 (V = 86.45 cm^3, m = 70.28 g) was cooled down from 292K to 78K with a cooling rate of 3K/min and 1K/min, respectively.

As in the case of the empty container no long-time relaxation was observed in pentanol-1, e.g. 20 min after cooling the container with pentanol-1 from 10K to 1.5K the heat leak $\dot{Q}_B$ = 0.5nW was measured again. Thus in agreement with the heat capacity data (s. fig. 1) pentanol 1 was crystalline. The cooling rate of 3K/min was not fast enough to form the glass structure.

The heat capacity of pentanol-2 is larger than that of pentanol-1 (s. fig. 1) as expected for the amorphous structure. However, determination of linear term typical of glasses requires heat capacity data at lower temperatures.

The power release measurements clearly demonstrate that pentanol-2 is indeed amorphous: the long-time power release typical of glasses was observed (s. fig,2). The power released after cooling from various T_1 to T_0 = 1.5K is proportional to t^{-a} with a = 0.69 ± 0.04 . This coefficient is much smaller than 1 (predicted by the tunneling theory and observed in usual amorphous solids/6/). A similar behavior

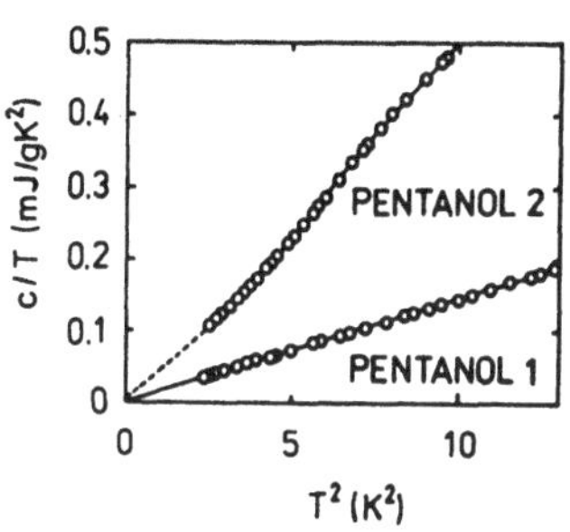

Fig. 1: The heat capacity of pentanol-1 and pentanol-2.

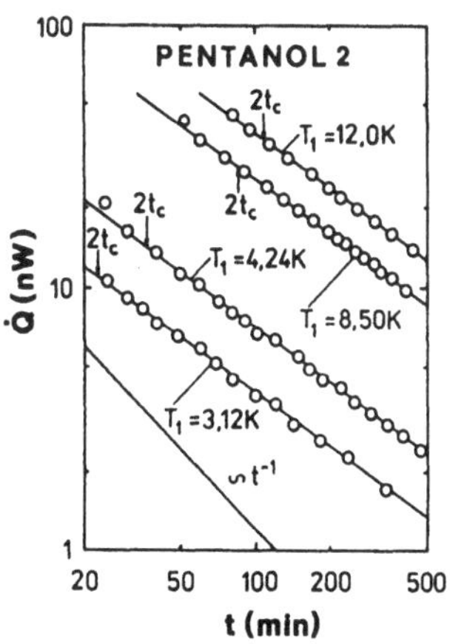

Fig. 2: The power Q released in pentanol-2 after rapid cooling from various equilibrium temperature T_1 to T_0 = 1.5K as a function of time t.

Solid lines: Q is proportional to $t^{-0.69}$.

t_C - time necessary for cooling from T_1 to 1.5K.

was observed in another organic material - epoxy resin - where a = 0.76[7]. Thus it is not excluded that the smaller a-value is a typical feature of noncrystalline organic materials. At least our experiment with pentanol-1 and pentanol-2 demonstrate that the replacement of the OH-group from the end of the chain to the side prevent the crystallization, e.g the amorphous structure is caused by the side group.

At $T_1 < 4K$ the power release is proportional to $T_1^2 - T_0^2$ and the distribution parameter $P = 4.1 \cdot 10^{44}/Jm^3$ of pentanol-2 can be determined from eq. (1) with $V = 86.45\ cm^3$ and $t = 1$ hr. This result is in a good agreement with the P-values of other glasses (s. table 1).

Table 1

Characteristic parameters of the long-time power release in various amorphous solids: a is the coefficient of the time dependence t^{-a}, P is the distribution parameter of TLS, E_f and T_b are constants used for calcuation of the temperature dependence (s. text).

material	a	P , $10^{44}/Jm^3$	E_f/k , K	kT_b/E_f	Ref.
vitr. silica	1.0	1.0 - 3.3	13	0	8,9
$Fe_{80}B_{14}Si_6$	1.0	1.3	20	0	10
LiCl $7H_2O$	1.0	10.1	48	0	11
epoxy resin	0.76	7.8	16.5	0.2	7
pentanol-2	0.69	4.1	23	0.2	t.w.

At $T_1 > 4K$ the power release data disagree with eq. (1). Similar deviations were observed for all investigated glasses. Using a distribution function with a cut-off energy E_f

$$n(E,t) = (P/2)\ln(4t/\tau_{min})/(1 + \exp((E - E_f/kT_b)) , \qquad (5)$$

where E_f and T_b are constants and τ_{min} is the shortest relaxation time of TLS, a good fit of the T_1-dependence can be obtained for all investigated amorphous solids[6] including pentanol-2. All obtained parameters a, P, E_f and T_b of pentanol-2 are not far from the corresponding values of other amorphous solids (see table 1), which demonstrates once more the universality of the low temperature anomalies of glasses.

REFERENCES

1. M. Borghini, S. Mango, O. Runolfsson and J. Vermeuten, Proc. of the Conf. on Polarized Targets and Ion Sources, Sacley, p. 387 (1966).
2. T. Niinikoski, Proc. of the 2. Workshop on Polarized Target Materials, Abingdon, p. 69 (1979).
3. D. Hill and M. Krumpolc, Proc. of the AIP Conf., New York, 479 (1983).
4. W. Anderson, B.J. Halperin and C.M. Varma, Phil. Mag. 25, 1, (1972).
5. A.W. Philips, J. Low Temp. Phys. 7, 351 (1972).
6. S. Sahling, A. Sahling and M. Koláč, Solid State Commun. 65, 1031 (1988).
7. M. Koláč, B.S. Neganov, A. Sahling and S. Sahling, J. Low Temp. Phys. 68, 285 (1987).
8. J. Zimmermann and G. Weber, Phys. Rev. Lett. 46, 661 (1981).
9. M. Schwark, F. Pobell, M. Kubota and R. Mueller, J. Low Temp, Phys. 59, 55 (1985).
10. S. Sahling, A. Sahling, B.S. Neganov and M. Koláč, J. Low Temp. Phys. 65, 289 (1986).
11. S. Sahling, Solid State Commun. 72, 497 (1989).

Determination of Polarization Parameters of a Polarized Deuteron Target by Experimental Data

M. Liburg

Joint Institute for Nuclear Research, Dubna,
Head P.O. Box 79, SU-101000 Moscow, USSR

1 Introduction

There are two ways of determining polarization of deuteron targets by the deuteron magnetic resonance (DMR) method. One of them (more traditional) is based on directly proportional relation of polarization to the integrated intensity of the DMR absorption spectrum [1]. But in this way calibration of the apparatus against the intensity of the thermal equilibrium spectrum is needed, although the procedure is carried out under low signal-to-noise ratio conditions. A more acceptable way is based on specific features of the DMR spectrum, which can be represented as a superposition of two lines corresponding to the transitions between neighbouring magnetic energy levels.

In structures, with the symmetry different from the cubic one, the observed spectrum consists of two peaks, widely separated as a consequence of quadrupole splitting. If the substance is not a single crystal, two broadened lines partially overlap each other. A procedure of calculation of these constituents of the whole spectrum in order to determine polarization parameters of a polarized deuteron target will be described below.

Let n_I, n_O, n_{-I} be the relative populations of the magnetic energy levels. According to usual definitions the expressions for polarization (P) and alignment (A) are as follows

$$P = n_I - n_{-I}\ ;\ A = 1 - 3n_O\ . \tag{1}$$

Let us suppose that the stationary state of a spin system can be described by a single spin temperature. Then the populations are given by

$$n_m \sim exp(E_m/kTs) \tag{2}$$

where k is Boltzman's constant, T_s is the spin temperature and E_m is the energy of the level whose magnetic spin quantum number is m. In the first order of perturbation theory the expression for the $m-th$ energy level can be written as [2]

$$E_m = E_O - E_Z \cdot m + E_q \cdot m^2 \tag{3}$$

The term proportional to m describes Zeeman splitting:

$$E_Z = \nu_D \cdot h$$

where ν_D is the central frequency of the DMR spectrum and h is Planck's constant.

The second term proportional to m^2 is a value known to be the quadrupole interaction energy. Usually, E_q is three or four orders less than E_Z. Neglecting E_q with respect to E_Z one can obtain formulas for polarization and alignment from (1):

$$P = 4\tau/(3+\tau^2); A = \tau \cdot P \tag{4}$$

with

$$\tau \equiv \tan h(E_Z/kT_s)$$

In other works the "asymmetry ratio" was used:

$$R = I_1/I_2 \tag{5}$$

where I_1, I_2 are the integrated intensity of two constituents of the DMR spectrum above mentioned. According to our notation:

$$\tau = (R-1)/(R+1)\,. \tag{6}$$

2 The polarization and DMR spectrum parameters

Let us assume (the detailed analysis of the experimental data agrees with this assumption [3] that the constituents are described by functions, each of them changing over to another when the mirror symmetry operation in the vertical axis passing through $\nu = \nu_D$ is applied. Then the experimental spectrum can be represented as

$$\Phi(x) = \varphi(-x) + R \cdot \varphi(x) \tag{7}$$

where $x \equiv (\nu - \nu_D)/\nu_Q$ (see Fig. I).

Dividing both parts of Eq. (7) by the integral $\int \Phi(x)dx$ we come to the equation, which consists of only normalized functions

$$F(x) = f(-x)/(R+1) + R/(R+1) \cdot f(x) \tag{8a}$$

or

$$F(x) = \frac{1}{2}[(1-\tau)f(-x) + (1+\tau)f(x)] \tag{8b}$$

It is easy to obtain the following expressions:

$$F^S(x) = f^S(x) \tag{9}$$

$$F^a(x) = \tau \cdot f^a(x) \tag{10}$$

where the symmetric and antisymmetric parts of the corresponding functions are denote by indices S and a respectively, i. e.

$$\begin{aligned} F^S &= [F(x) + F(-x)]/2 \\ F^a &= [F(x) - F(-x)]/2 \text{ etc.} \end{aligned}$$

Hence the experimental DMR spectrum is a linear combination

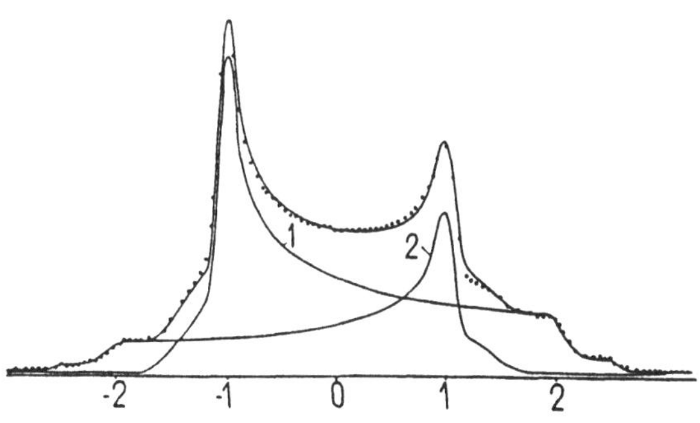

Figure I

DMR spectrum (the closed circles) approximated by the calculated function F(x), which in its turn is the sum of two components 1 and 2. The horizontal axis is given by $x = (\nu - \nu_D)/\nu_Q$ where ν is the current frequency, ν_D is the central one and $\nu_Q = (3e^2qQ/4h)$ is the quadrupole frequency.

$$F(x) = F^S(x) + \tau \cdot f^a(x) \tag{11}$$

The function $f(x)$ does not depend on the polarization of the target, being universal for the same target material.

Equation (9) shows that the sum of peak heights of the spectrum is proportional to the integral intensity of the whole spectrem, the fact being established in other works [3,5]. Substitutions $x = 1$ in (9) and (10) give

$$B_F = \tau \cdot B_f \tag{12}$$

where $B_F = F^a(1)/F^S(1)$ and $B_f = f^a(1)/f^S(1)$.

Another equation connects the first moments

$$M_F = \tau \cdot M_f \tag{13}$$

of the $F(x)$- and the $f(x)$- functions respectively. The ratio

$$B_F/M_F = B_f/M_f \equiv D \tag{14}$$

follows from (12) and (13) and D is the constant for the target material.

3 Calculation procedure

A calculation program for analyzing DMR spectra is described below. Completely deuterated ethanediol ($(CD_2OD)_2$) and 1,2-propanediol ($C_3D_6(OD)_2$) were used as basic substances for a frozen polarized target.

The essential part of the calculation procedure is the determination of the $f(x)$-function.

Every run begins with establishing a coordinate net. Zero point is placed half-way between the spectrum peaks. Then, adding a linear function $y = \beta x + \sigma$ (where β and σ are small quantities), to the spectrum one fixes the main parameters B_F, M_F, D so that to satisfy

$$M'_F = M_F$$

where

$$M'_F = \int x F(x) dx$$

and M_F is determined from (14) for a certain D. The run must be repeated for different D to obtain the best solution. In the $-2 \leq x < 1$ range $f(x)$ is approximated by

$$f(x) = K\left[C_1 \cdot (1 - x/\alpha_1)^{-1/2} + C_2(1 - x/\alpha_2)^{-1/2}\right]$$

The parameters of the last formula are not independent. Obviously, the following relations take place:

$$\begin{aligned} \alpha_1 \cdot C_1/\alpha_2 \cdot C_2 &= N_{CD}/N_{OD} \\ \alpha_1/\alpha_2 &= \nu_Q^{CD}/\nu_Q^{OD} \\ C_1 + C_2 &= 1 \end{aligned}$$

where N_{CD} and N_{OD} are the numbers of $C-D$ bonds and $O-D$ bonds in the molecule respectively, ν_Q^{CD} and ν_Q^{OD} are the quadrupole frequencies corresponding to these bonds.

A method of successive approximations is applied to obtain the values of K, C, α. For the first step they are determined by the equation

$$2\left[F^S(x) - F^S(0)\right] = f(x) + f(-x) - 2f(0)\,.$$

The current values of the parameters are used for calculation of to compare it with the experimental $F^S()$.

In the $1 < x \leq 2$ range $f(x)$ is defined by substracting

$$f(x) = 2F^S(-x) - f(x)$$

For other values of the argument $f(x)$ and $F^S(x)$ are assumed to be coincident.

Function $f(x)$, calculated for the least square adjusted K, C, α, is used to define τ from (12) and (13). The best value for τ is determined by fitting the calculated spectrum. The result of the fitting is quite satisfactory, as seen in Fig. 1.

4 Results

It is found that the ratio

$$\alpha_1/\alpha_2 = 0.83 \pm 0.03$$

and the constants of quadrupole splitting calculated for those α are in good agreement with the results of the detailed analysis [3].

As the calculations show, the parameter D varies in a rather wide range from one spectrum to another. Nevertheless, B can be found more accurately and the mean values obtained as a result of averaging over all treated spectra are

$$B_f = 0.57 \pm 0.01 \text{ for } (CD_2OD)_2 \tag{15a}$$

$$B_f = 0.63 \pm 0.01 \text{ for } C_3D_6(OD)_2 \tag{15b}$$

A convenient method of measurement of the polarization is applied in transient time of polarization build-up on reversal: after accumulation of the DMR spectrum the difference-to-sum ratio of peak heights is calculated to obtain B_F, τ, P and A in sequence by using (12), (15) and (4). This is the further development of the so-called "peak height method" [3].

The deuteron polarization of about 40 % is measured with an accuracy about ±2 % by means of DMR spectrum fitting. The difference-to-sum peak height ratio method gives results that differ from fitting by no more than 2.5 %.

Two essential remarks must be made.

1. The parameter of asymmetry of the electric field gradient in a molecule is not taken into account in the described calculations. As shown in [3], it does not influence the result of the polarization measurement because of being small enough for the substances considered. For other cases it can be easily introduced in calculation if necessary.

2. The assumption that the energy of the quadrupole interaction is negligible with respect to the Zeeman one, has been accepted above. According to [2,4] a general expression for τ can be given by

$$\tau_m = \tau - \frac{1-\tau^2}{2\tau} \cdot \frac{E_q}{kT_S} \tag{16}$$

where τ_m is determined by the described methods.

The second term of the right-hand part of the last equation is estimated to be 10^{-4} for diols, which is much less than the accuracy mentioned above. But in the case of the non-negligible contribution of quadrupole interaction one can calculate T_S from (16) and then the polarization parameters by the general formulas (1).

References

1. A. Abragam, M. Goldman. *Nuclear Magnetism: order and disorder*, Clarendon Press, Oxford (1982), Chapter V.

2. L. I. Lapidus. *Communication PI-84-267*, JINR, Dubna (1984)

3. O. Hamada, S. Hiramatsu, S. Isagawa, S. Ushimoto, A. Masaike, K. Morimoto. *Nuc. Instr. and Methods* 189, 561 (1981).

4. M. Yu. Liburg. *Communication PI3-88-308* JINR, Dubna (1988).

5. N. S. Borisov, E. I. Bunyatova, M. Yu. Liburg, V. N. Matafonov, A. B. Neganov, Yu. A. Usov. *J. Phys. E: Sci. Instrum.* 21, 1179, Printed in UK (1988)

DNP with Inhomogeneously Broadened ESR Lines

*T.O. Niinikoski and R. de Koker**

CERN, CH-1211 Geneva 23, Switzerland

This paper deals with the low-temperature theory of Dynamic Nuclear Polarization (DNP) with inhomogeneously broadened impurity electronic spin resonance (ESR), in the most favourable case of fast cross relaxation and complete saturation. The numeric solution of the steady-state equation at 2.5 T field was obtained for an axially symmetric g-tensor with principal components $g_{\|} = 1.9787$ and $g_{\perp} = 1.9877$, similar to that of propanediol-Cr(V). The calculation was performed without introducing nuclear polarization leakage in a phenomenological way, and was performed using the technique outlined in Ref. [1]. Comparison with earlier calculations [2,3] show that the previous numeric results were erroneous at the lowest lattice temperatures, due to an approximation made in the steady-state equation.

Our results indicate that the spin temperature in the rotating frame asymptotically approaches values around +0.50 mK and –0.25 mK when the lattice temperature is lowered to below 400 mK; these correspond to ultimate deuteron polarizations of +76.5% and –94.7%. The results also show quantitatively, that the optimum frequencies at which these polarizations can be achieved are increasingly narrowly peaked as the lattice temperature is lowered.

Based on the detailed results briefly reviewed above, conclusions on the desirable field homogeneity can be drawn, bearing in mind that the dipolar interactions were ignored in the numeric calculations, and that the internal and external fields may also have intrinsic variations. These variations may arise from the non-uniform distributions of nuclear and electronic magnetizations, and from the demagnetization factors of the non-spherical grains of the target material, in addition to the external magnetic dipole fields of the grains themselves. Such considerations lead us to suggest that deuteron polarizations in excess of -60% and +55% may be obtainable in practice in a well-optimized material, assuming a homogeneity of $<10^{-5}$ for the external field.

References

1. T. O. Niinikoski, High Energy Physics with Polarized Beams and Targets, AIP Conference Proceedings, Ed. M. L. Marshak (American Institute of Physics, New York 1976) p. 458.
2. W. de Boer, J. Low Temp. Phys., **22** (1976) 185.
3. W. de Boer, Dynamic Orientation of Nuclei at Low Temperatures, CERN Yellow Report CERN 74-11 (1974).

* Now at Stanford University

Polarization Reversal by Adiabatic Fast Passage in Deuterated Alcohols

P. Hautle[1], *W. Grüebler*[1], *J.A. Konter*[2], *A.I. Kovalev*[3], *S. Mango*[2], *B. Van den Brandt*[2], *and M. Weßler*[2]

[1]Institute for Intermediate Energy Physics, ETH, CH-8093 Zürich, Switzerland
[2]Paul Scherrer Institut, CH-5232 Villigen PSI, Switzerland
[3]Leningrad Nuclear Physics Institute, SU-188350 Gatchina, USSR

Introduction

In the study of spin-spin effects in nuclear interactions, the polarized target mainly sets the limits on the statistic and more importantly on the systematic errors of the experiment. To increase the measurement accuracy, it is important to reverse the sign of the target polarization frequently. This is practically impossible in a frozen spin target, as the reversal would take several hours by DNP methods.

One possibility to rapidly reverse the spin is in principle given by the method of adiabatic fast passage (AFP). It has been extensively studied for Cr^V doped 1,2-propanediol [1], where the results were in good agreement with theoretical calculations [2] but not very promising for application in a large polarized deuteron target.

It is aim to learn about the AFP mechanism in deuterated alcohols and to assess the feasability of an AFP polarization reversal in a large polarized target.

Principles of the AFP method

Polarization reversal by AFP consists essentially of irradiating a nuclear spin system situated in a strong magnetic holding field H_0 with a rf-field H_1 linearly polarized in the plane perpendicular to H_0. To perform a passage through the resonance line, either the rf-field frequency ω or the strength of the holding field is varied. In the frame of the spin-temperature concept one can visualize the passage as a transfer of order from the Zeeman system to the spin-spin system (quadrupole system in deuterons) and back to the Zeeman system with reversed orientation. When a suitably large rf-field is applied, and the sweeptime τ is slow enough so the coupling between the two subsystems established via the field H_1 is good enough, the passage results in a reversal of the magnetization orientation with some loss in magnitude. The condition for adiabaticity, i.e. that the subsystems are in equilibrium with each other at any time so the entropy is approximately constant, can be expressed by [3]

$$\frac{dH}{dt} \ll \gamma H_1^2, \tag{1}$$

where $\frac{dH}{dt}$ is the passage rate. However, in practice, the slowness of the sweep is limited by the condition that the entire passage has to take place in a time τ much shorter than the spin-lattice relaxation time T_{1ss} in the the field H_{eff} in the rotating coordinate frame (RCF)

$$\tau \ll T_{1ss}. \tag{2}$$

T_{1ss} is essentially the spin lattice relaxation time of the spin-spin system (dipole and quadrupole system resp.).

Disregarding the Zeeman spin lattice relaxation time T_{1z}, the saturation occurs mainly during the passage of the line. It is larger the slower the passage, whereas the increase of entropy due to nonadiabaticity of the passage decreases as the passage rate is decreased. Thus there is an optimum passage rate for given values of H_1 and T_{1ss} for which the entropy increase, and therefore the magnetization loss is minimum. (See also figure 2). More details on the AFP principles can be found in refs. [2] and [4].

Apparatus

The experiments were performed in a toploading ^3He evaporation cryostat [5]. A superconducting split pair magnet gave a field H_0 of 2.5 Tesla with a homogeneity of $\frac{\Delta B}{B}$=1-2$\cdot 10^{-4}$ over the sample volume. The sample consisting of small spheres was placed in a teflon cylinder of 12 mm inner and 14 mm outer diameter and a total depth of 30 mm. The cylinder was filled to a height of about 15 mm, which amounts to a sample volume of roughly 1.5 cm^3. Two different coils of the solenoid type were used for measuring the polarization and for rf irradiation respectively. The 13 turn NMR coil was placed on the inner wall of the cylinder around the sample and the rf coil with 29 turns was wound around the full length of the cylinder. To keep this cylinder in position, it was fitted into another cylinder which filled the space of a 30 mm diameter brass cup serving as a liquid ^3He collector and its bottom as a microwave cavity.

The polarization was measured by a Q-meter with a parallel resonance tuned by a variable capacitor diode and a GaAs FET preamplifier. The signals were digitized in a PC controlled ADC system which also operates the frequenzy synthesizer. An independent rf unit produced the field H_1. This unit consisted of a voltage controlled crystal oscillator (VCXO) an amplifier with a maximal output power of 10 watt and a programmable wave generator for gating the rf and triggering a ramping voltage to sweep the frequency of the oscillator. Figure 1 shows a block diagram of the whole apparatus. Only the rf or the NMR circuit could be in operation at one time while the other had to be disconnected.

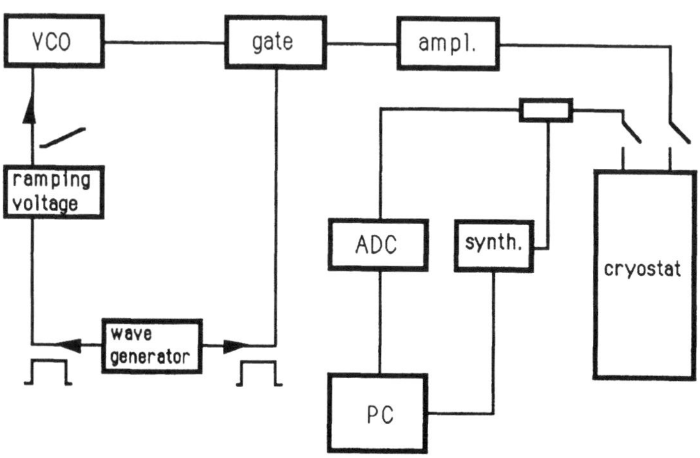

Figure 1: Experimental setup.

Measurements and results

All measurements were carried out at a temperature of 0.5 K in a 2.5 Tesla field. The procedure was the following. First the polarization was pumped up by DNP, then the microwaves were turned off and the polarization signal taken. After the linear sweep of the rf frequency had been performed with the NMR circuit disconnected , the NMR circuit was connected back and the signal was recorded again. The effect of the passage was estimated from the ratio of the final and initial intensities of the signal.

a.) Protons

To learn about the experimental problems of AFP, the first tests were carried out with a proton sample (95% n-butanol, 5% H_2O doped wtth 1% porphyrexide by weight) of 1.6 cm^3 volume . In this case the frequency was swept linearly by 360 kHz through the resonance. The signal had a half width of 20 kHz which gives a local field in the rotating frame of $H_L' = 2.3$ Gauss. The reversible fraction of the polarization

$$\delta P = \frac{P_{final}}{P_{initial}} \tag{3}$$

was investigated as a function of the passage rate $\frac{dH}{dt}$ for a fixed value of $H_1 = 0.7$ Gauss. The results are shown in Figure 2 as a function of the sweep rate $(H_1^{-2})dH/dt$. They confirm the studies carried out by Parfenov [1].

b.) Deuterons

The deuterated material used was 95% n-butanol d-10 (98% deuterated), 5% D_2O doped with 1% porphyrexide by weight. The spin lattice relaxation times at 2.5 Tesla and 0.5 K were measured to be $T_{1z}^{+} = 57$ min for the positive and $T_{1z}^{-} = 48$ min for the negative polarization. The sweep width was 480 kHz and typical passage times τ were in the order of 1-2 seconds which gave a passage rate of roughly 400-800 G/s. The power fed to the sample was about 5 mW, which was less than what was used in the proton system for an optimum reversal at comparable sweep times.

It was found that one full sweep of 480 kHz through the resonance is not enough to reverse the polarization. An additional half sweep in the same direction to the centre frequency (cf) had to be performed. Figure 3 shows such a sequence.

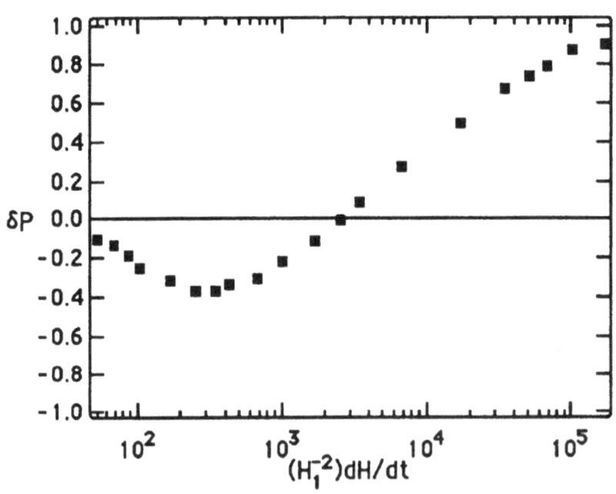

Figure 2: Dependence of δP on the sweep rate $(H_1^{-2})dH/dt$. The points show results averaged over both signs of initial polarization and both directions of passage. The best value of δP reached was about -36%

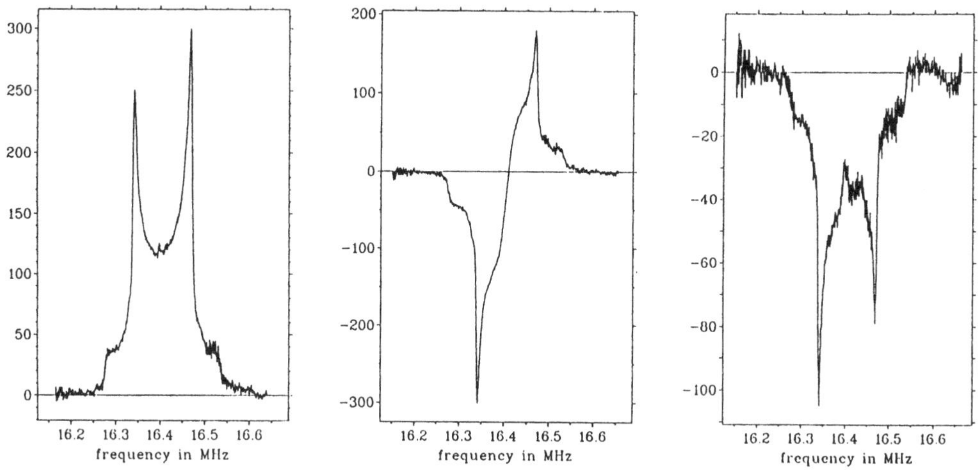

Figure 3: left: initial signal (positive polarization), middle: after a full sweep from the high to the low frequency, right: after an additional half sweep from the high to the centre frequency (negative polarization)

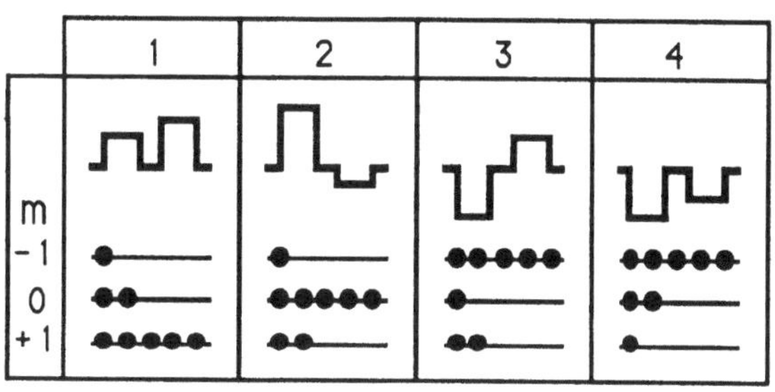

Figure 4: Sequence of 3 half sweeps with corresponding level populations. 1: Initial state , positive polarization. 2: After a sweep from the high frequency (hf) to the centre frequency (cf). 3: After a sweep from cf to the low frequency (lf). 4: Final state after a sweep from hf to cf, negative polarization

In the deuteron system, the quadrupole interaction P lifts the degeneracy of the two identical resonances of the pure magnetic Hamiltonian which leads to two resonances separated by $2P$ [3]. If we perform now an AFP with an rf-field $\gamma\hbar H_1 \ll P$, we can no longer consider it as an ensemble rotation through 180^0 but rather as a sequence of two successive population inversions between adjacent m-substates. Figure 4 shows a sequence of half sweeps through the individual resonances with the corresponding population inversions and explains the observed behavior of the spin system. Furthermore this simple picture was able to predict the signal shapes in a whole variety of sweeps performed.

Even though the apparatus was not optimal concerning the elapsed time between the recording of the initial signal and the taking of the final signal, a maximal δP = -57% was reached (see figure 5) as compared with -36% for protons. This fact may have the following reasons. The spin lattice relaxation time of the quadrupole system is longer than the one of the dipole system in the proton case, and the mixing rate between the subsystems might be higher in the deuteron system, which could explain the considerably smaller rf power needed. But so far no theoretical calculations of the entropy increase have beeen carried out. A similar dependence of δP on the sweep rate, as shown in figure 2 for

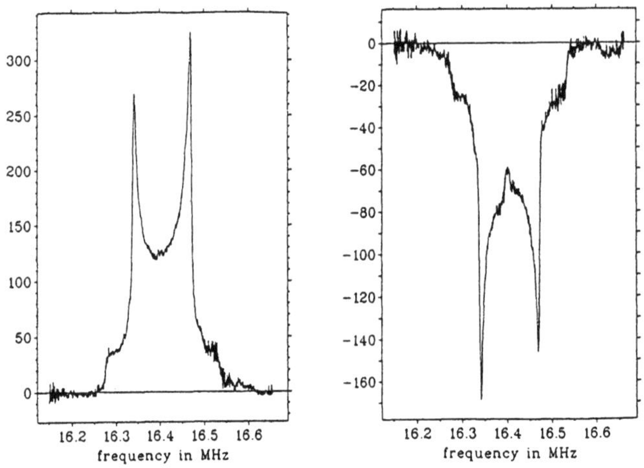

Figure 5: Initial signal and final signal after a full and a half sweep both from hf to lf. δP is about -57%.

the proton system, was also observed by varying the irradiation power or the sweeptime τ. No significant dependence of δP on the sign or magnitude of the initial polarization nor on the direction of the passage was observed. Slight differences in the results obtained could have been caused by the not yet optimized irradiation procedure. To overcome this drawback, tests with a computer controlled gate and frequency synthesizer were carried out. Unfortunately the outcome of a sweep depends heavily on the step size with which the frequency of the synthesizer is ramped. The maximal obtainable resolution of 50 Hz did not give acceptable results and 500 Hz was even too coarse to reverse the polarization.

The relaxation time of a signal reversed by 50% was only $T_{1z}^{rev} = 10$ min. In practice, DNP would be necessary after a reversal, to ensure that the populations of the magnetic sublevels correspond to one spin temperature. Tests showed no problems to pump up a reversed polarization.

Conclusions

The unexpected results, i.e. the higher $|\delta P|$ for deuterons at lower rf power compared to AFP in a proton system, could bring this method into the region of feasability in a polarized deuteron target, especially considering the gains of an improved irradiation system (computer controlled) and lower temperatures (dilution). On the other hand, a clear theoretical understanding of the processes involved has yet to be worked out.

References:

[1] L. B. Parfenov and A. F. Prudkoglyad, Sov. Phys. JETP **60**, 123 (1984)
[2] M. Goldman, M. Chapellier, and Vu Hoang Chau, Phys. Rev. **168**, 301 (1968)
[3] A. Abragam, The Principles of Nuclear Magnetism, Oxford 1961
[4] B. N. Provotorov, Sov. Phys. JETP **14**, 1126 (1962)
[5] B. van den Brandt et. al., 8th Int. Conf. on High Energy Spin Physics, 1988 Minneapolis

Computer Controlled Microwave System for CERN-SMC Polarized Target

S. Ishimoto[2], *R. Ent*[3], *J.F.M. van Gelder*[4], *T. Hasegawa*[5], *N. Hayashi*[6], *N. Horikawa*[6], *T. Iwata*[6], *T.O. Niinikoski*[1], *S. Okumi*[6], *H. Postma*[4], and *A. Rijllart*[1]

[1]CERN, CH-1211 Geneva 23, Switzerland
[2]KEK, 1-1 Oho, Tsukuba-shi, Ibaraki-ken, Japan
[3]University of Virginia, Charlottesville, VA 22901, USA
[4]Delft University of Technology, Lorentz Weg 1, NL-2628 CJ Delft, The Netherlands
[5]Miyazaki University, Miyazaki, 889-21, Japan
[6]Nagoya University, Furo-cho, Chikusa, Nagoya, 464, Japan

The design of computer controlled microwave system for SMC polarized target is described. The microwave system consists of two Extended Interaction Oscillator(EIO) tubes, power attenuators, frequency counters and power meters. A personal computer(PC9801) is interfaced directly to the microwave components for control and feedback, using ADC's, DAC's, digital read-outs and GP-IB. The operator controls the elements and monitors them from micro-VAX which is connected to PC9801 through RS232C interface.

1 Introduction

The SMC(Spin Muon Collaboration) polarized target consists of two, 1 liter target halves in opposite directions [1]. Two high power microwave systems delivering about 1 W at the mixing chamber are required. Frequent and rapid reversal of the target polarizations is essential to reduce systematic errors for long term measurement, and requires computer control. For obtaining higher polarization, optimization of microwave power and frequency is important. Such optimization would be available with computer controlled system. Besides this, as SMC is a long run experiment and a large collaboration, the polarized target sometimes should be operated by the persons who are not experts in a polarized target. In such case, the computer controlled system would be useful for us.

2 System Overview

SMC polarized target has two target halves. Thus, we need two independent microwave circuits for operating each target halve. Fig.1 shows the block diagram of the microwave system. As a microwave sauce, two Extended Interaction Oscillator's(EIO) having frequency of 70 GHz and power of 20 W will be used. The frequency can be changed in following two ways. One way is to change the cavity size of EIO tube mechanically by driving a motor. In this manner, one can change the frequency over wide range. The motor is controlled by a personal computer through digital-out. This method will be used when one needs to reverse direction of spin. Another way is for fine tunning of

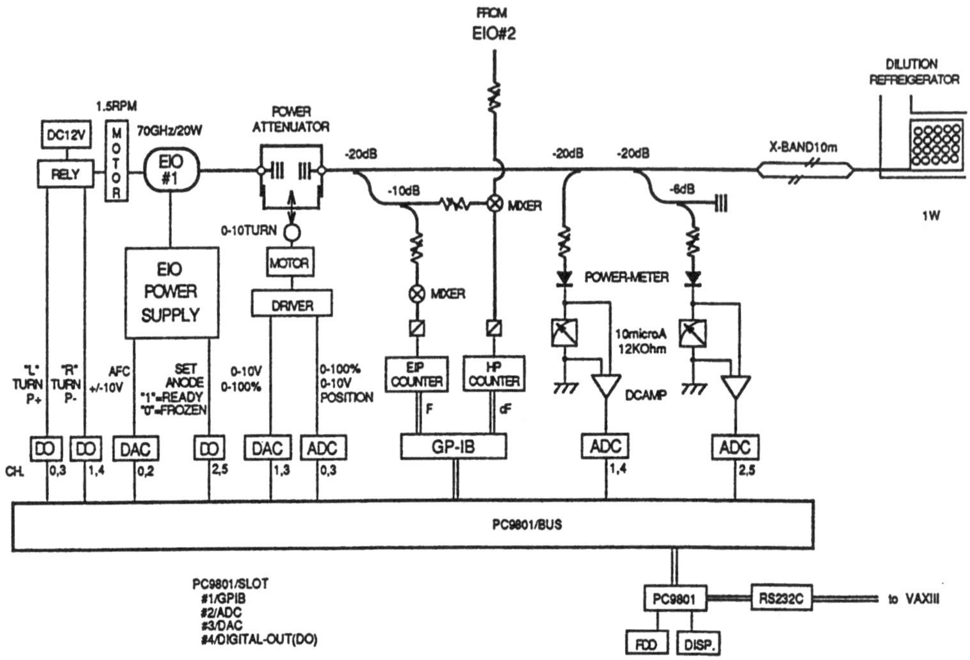

Fig.1 Block diagram of the microwave system

frequency. One can change the cathode voltage by appling DC level to AFC input of the EIO power supply to make a little variation of the frequency (see Fig.2). $\pm$ 10 V level makes frequency variation of $\pm$ 30 MHz. The DC level is fed by DAC which is also controlled by the personal computer. When cutting off the microwave power is required, we do not cut off the cathode voltage or do not cut off the heater current but only change the anode voltage out of the resonance condition to the direction of higher voltage. This procedure is performed by digital-out from the personal computer. In this method, as beam current in the tube is held without cutting the heater off, the life time of the tube would be long and quick restart would be possible.

A high power attenuator, which must receive microwave power as much as 20 W, is located following the EIO tube. Fig.3 shows a design of a high power attenuator of trombone type which has been developed in Delft University. Incomming power is split into two arms in a hybrid Tee. One of the arms has a variable length. Then, the two arms are recombined in another Tee. The outgoing power will depend on the relative phase of the two arms, which can be changed by varying the length of the arm. The personal computer controls the length through DAC and monitor it through ADC.

The system is equiped with two frequency counters: one is EIP counter and another is HP counter. The EIP counter, which can measure the frequency up to 110 GHz, directly measures the frequency of EIO#1. Since HP counter is, however, a low frequency type, combining the two waves from EIO#1 and EIO#2 in a mixer, only the frequency difference between two EIO tubes is measured. The frequency of EIO#2 can be derived as F $\pm$ dF from the frequency value of EIO#1 F, and the frequency difference dF. The sign of dF could be obtained by detecting the responce of the frequency difference when a little variation of EIO frequency is made. These counters are controlled by the personal computer through GP-IB interface. Power meters consist of diode and DC-amp. The output of the DC-amp is fed to ADC and is monitored by the personal computer.

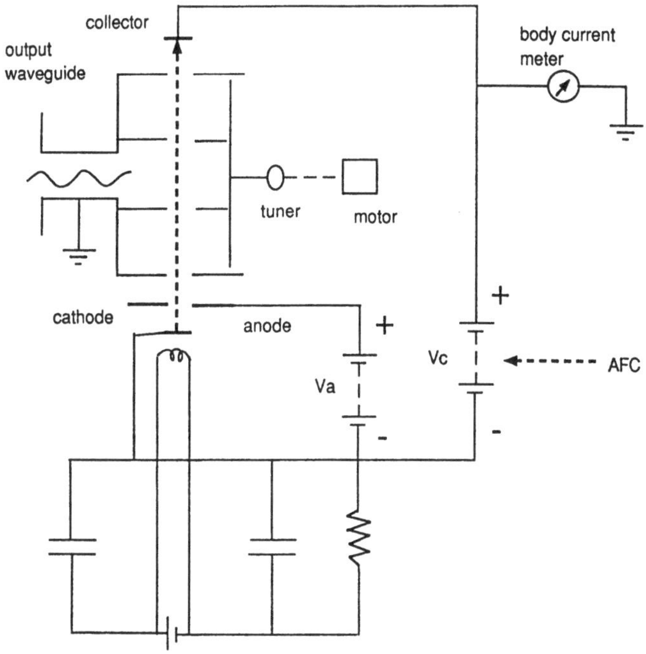

Fig.2 The structure of the EIO tube

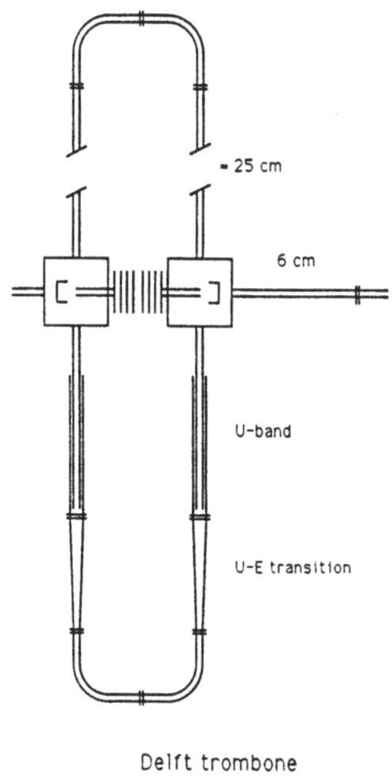

Fig.3 The design of a high power attenuator

3 Computer Connection

The personal computer, PC9801(NEC), performs control and monitoring of the system with software written in BASIC. It is connected to micro-VAX III system through RS232C interface. The VAX sends parameters of microwave system to the personal computer in direct-mode. Receiving these parameters, the personal computer begin to control the system to set the parameters at the required values. In another mode, automatic-mode, frequency and attenuator parameters alone are sent to the personal computer and it independently controls the system to set the parameters at the required values. On the other hand, the personal computer transfers all of the data to the VAX when it is required.

The VAX is a host computer for polarized target system [2]. It controls NMR data taking system, cryogenic system, magnet control system and microwave system. Thus, obtaining the status of the polarized target from NMR data taking system, one can control the microwave system according to the status. Final optimization of microwave power and frequency would be performed by a certain kind of hill-climbing software on the VAX.

References

[1] "CERN Experiment NA47 (SMC)", presented by G.Baum in this conferenc.

[2] "NMR system concept of the SMC experiment", presented by A.Rijllart in this workshop.

NMR Data Taking System by Personal Computer PC9801 (NEC)

S. Ishimoto[2], *R. Ent*[3], *J.F.M. van Gelder*[4], *T. Hasegawa*[5], *N. Hayashi*[6], *N. Horikawa*[6], *T. Iwata*[6], *T.O. Niinikoski*[1], *S. Okumi*[6], *H. Postma*[4], *and A. Rijllart*[1]

[1]CERN, CH-1211 Geneva 23, Switzerland
[2]KEK, 1-1 Oho, Tsukuba-shi, Ibaraki-ken, Japan
[3]University of Virginia, Charlottesville, VA 22901, USA
[4]Delft University of Technology, Lorentz Weg 1, NL-2628 CJ Delft, Netherlands
[5]Miyazaki University, Miyazaki 889-21, Japan
[6]Nagoya University, Furo-cho, Chikusa, Nagoya 464, Japan

A NMR data taking system, incorporating PC9801 personal computer, is described. A dedicated ADC board is used for automatic ADC sampling synchronizing with an internal clock. Integration of the area of NMR resonance curve is obtained, with appropriate background subtraction. NMR signals obtained with this system using Nagoya polarized target are presented.

1 Design Concept

This system was designed to determine target polarization. Required characteristics to the system is following :
(1) High speed NMR data taking is required for determination of polarization with a high accuracy.
(2) The system involves personal computer with a low price, for example PC9801, which is commertially available and widely spreads in Japan.
(3) The previously developed Q-meter circuit is also used in this system.

2 System Overview

The previous NMR circuit, which we developed and used in several experiments, ws used also in this system [1]. Fig.1 shows the block diagram of the system. The system in-

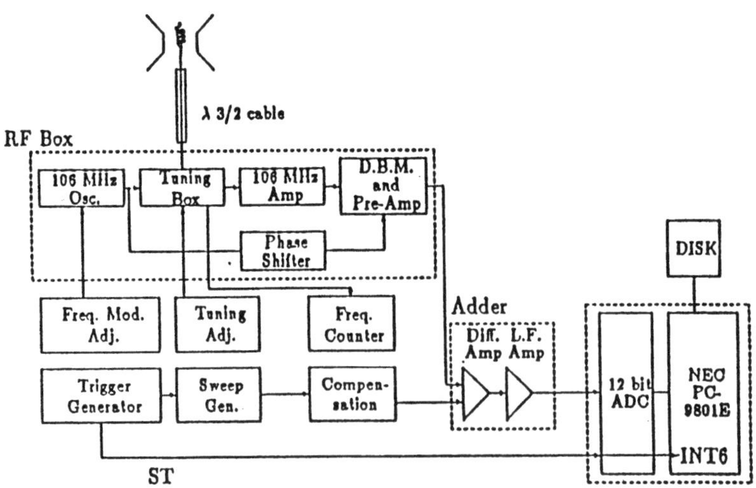

Fig.1 block diagram of the NMR system

cludes VCO(Voltage Controled Oscillator) as a NMR oscillator, comensation device and a phase sensitive detector. The phase sensitive detector consists of a double baranced mixer and a phase shifter [2]. A 16 bit personal computer of NEC, is used for data taking.

Its cpu is INTEL 8086 and the clock frequency is 8 MHz. The personal computer is commertially available at a low price and widely spreads in Japan. A dedicated ADC board is installed in one of the extention slots on the back plane of the computer. 30 KHz automatic ADC sampling is available synchronizing with an internal clock on the board. The ST signal generated by Trigger Generator is a trigger for NMR frequency modulation and ADC sampling.

Once it is triggered, ADC sampling automatically continues synchronizing with the internal clock. ADC sampling is performed 150 times according to NMR frequency modulation for one ST signal in a period of about 5 m sec as shown in Fig.2. The sequence is repeated 128 times for each two ST signals. After about 1.5 sec, signal averaging is achieved and averaged NMR signal is displayed on a monitor screen. The signal integration is calculated followed by appropriate background subtraction.

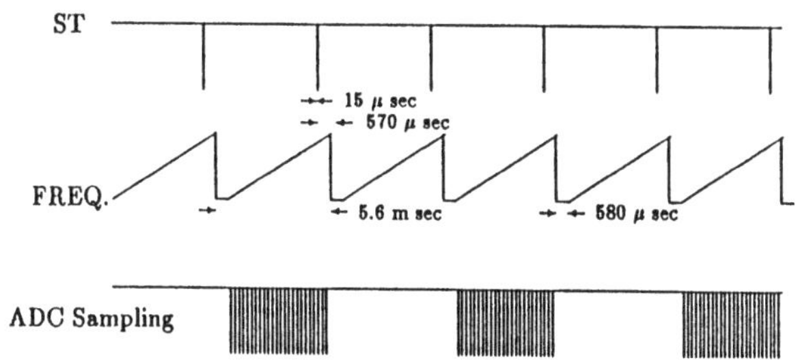

Fig.2 Time chart of the NMR data taking

3 ADC Board

The ADC board, PCN-2098A (Neologu Electronics), which is dedicated to PC9801 personal computer series, is used. This ADC board converts an analogue signal of the range from $-5V$ to $+5V$ into 12 bit digital data. The frequency value of the internal clock was set to be 30 KHz (33 μ sec for one sampling). Its conversion time is 15μ sec. Converted data is transfered to the 16bit data register(AX register) of the personal computer.

Table1.
Neolog Electronics PCN-2098A specification

A/D converter	ADC84-12 resolution: 12bit (precision : $< \pm 2$ LSB) linerity: $\pm 0.012\%$ drift: $< \pm 3$ppm/°c
analogue input	$0 \sim +10V$ $-5V \sim +5V$ $-10V \sim +10V$
conversion time	15μsec
internal clock	SEIKO 8650A 60KHz $\sim$ 0.0005Hz presision: 0.01 % drift: 0.03ppm/°c
interrupt	3 level(Int6,Int5,Int4) external interrupt(Int6) timer interrupt(Int4) EOC interrupt(Int5)

When the ADC data are transfered, a conversion end flag EOC, is also transfered to the MSB bit of the register. The EOC holds 1 over a period of conversion and it changes to 0 when conversion ends. Other general chracteristics of this board are listed in Table 1.

4 NMR data taking program

The NMR data taking program consists of 4 parts, MONITOR, MAIN ROUTINE, ROUTINE 1 and ROUTINE 2. Almost part of the MONITOR is written in BASIC language, however, partially in assembler where high speed transaction is needed. It performs signal display, background subtraction, signal integral calculation, data save on disk and so on. ROUTINE 1 and ROUTINE 2 are transaction routines for external interrupt. They are written in assembler for high speed transaction. The ROUTINE 1 only clears interrupt signal which entered the interrupt processor in the period of the last sampling sequence. The ROUTINE 2 sends a instruction of sampling start to the ADC board and afterwords reads out the sampling data containing the EOC flag. The EOC flag is checked to decide whether one takes the data or not. If it is still under conversion(EOC=1), then the program flow jumps to a waiting loop. When conversion end flag appears(EOC=0), the sampling data is taken and is stored on memory. This sequence is repeated 150 times according to the NMR frequency modulation and at last total of 150 word data are stored on memory after one scan.

Since correspondence between ADC data and NMR frequencies is significant, one should suppress time fractuation between the external interrupt and sampling start timing. Thus, MAIN ROUTINE, which calls these transaction routines by external interrupt, is not written in BASIC but in assembler to avoid software interrupts due to BASIC interpreter. The time fractuation was reduced to be less than 5 μ sec with these assembler routines. After 128 calls of ROUTINE 2, the program flow returns to MONITOR. The schematical flow of the program is shown in fig.3.

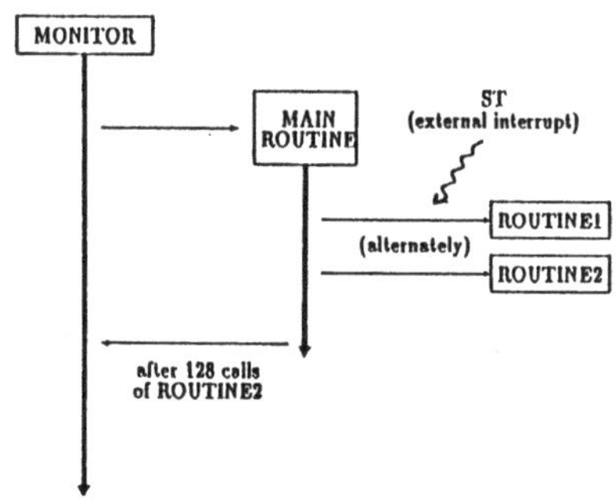

Fig.3 The schematical flow of the data taking program

Memory space of 256 word was allocated for one scan data (one NMR frequency modulation) and thus totally 64 KB for 128 sacns as shown in Fig.4. The average values are stored on another place in the memory with almost same format as the data.

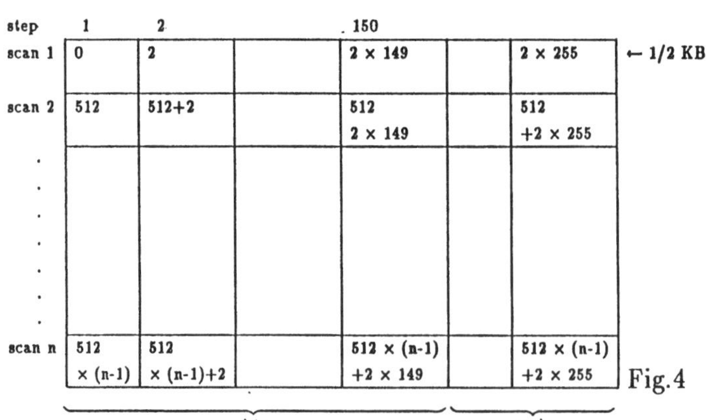

Fig.4 memory map of the data

5 NMR signal obtained with this system

The averaged NMR signals of NH_3 target obtained by using this system are shown in Fig.5(a)(b). (a) is a TE signal at 0.7K. Measurement of the TE signal was repeated with same condition and consequently the signal integration, S_0 , was calculated. The variance of S_0 was obtained to be less than 3% when the scan of 128 sweeps was repeated 20 times. Fig.5 (b) shows a typical enhanced NMR signal.

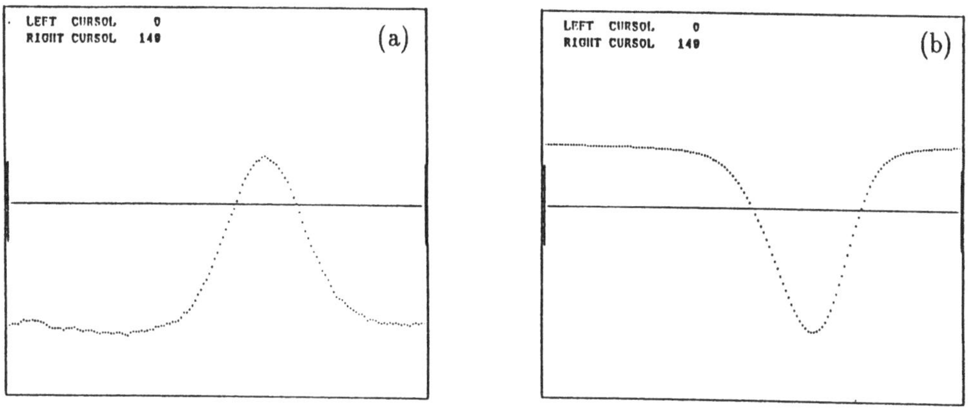

Fig.5 NMR signals for NH_3 target (a)TE signal (b) enhanced signal

6 Conclusion

We have developed a NMR data taking system incorporating a personal computer which is commertially available with a low price and widely spreads. A dedicated ADC board is used for automatical ADC sampling by an internal clock on the board. ADC sampling is performed 150 times according to NMR frequency modulation in the period of 5 m sec. Within about 1.5 sec, 128 NMR signals are obtained then signal integration is calculated followed by appropriate background subtraction.

References

[1] Hayashii et. al. Genshikaku-Kenkyu, vol. 25, no.1
Fukushima et. al. Genshikaku-Kenkyu, vol.25, no.5

[2] T.Matsuda et. al. DPNU-89-61, 1989

A Remark on Nuclear Spin-Lattice Relaxation

M. Liburg and V. Matafonov

Joint Institute for Nuclear Research, Dubna, Head P.O. Box 79, SU-101000 Moscow, USSR

We shall consider the experimental results on nuclear spin-lattice relaxation at low temperature from the point of view of its application to the problem of a polarised frozen-spin target.

Fig. I,2 show generally well known results obtained at CERN, for proton relaxation time in I,2-propanediol doped with Cr(V)-complex |I|. The smooth curves obey the equation

$$T_{In}^{-I} = | A \cdot T_{Ie} H^2 \cosh^2 (\frac{h \nu_e}{2\ kT}) |^{-I} + |0.I9 \frac{H^{I.32}}{T^{I.64}} |^{-I} \qquad (I)$$

where T_{In} is the proton relaxation time (expressed in hours), T_{Ie} is the electron one, H is the magnetic field (expressed in kG), T is the temperature, ν_e is the electron spin resonance frequency, h is the Plank constant, k is the Boltzman constant and A is a constant.

The first term in (I) describes the case when nuclear relaxation occurs via forbidden transitions. This term causes a steep increase of the relaxation time in Fig. I. However, for $T_{In} > 50$ h, this steep temperature dependence is no longer observed and the second purely empirical term must be added |I|.

In Fig. I,2 we have also plotted a few experimental points that were obtained in deuterated diols |2,3,4,5|. One can see the deuteron and proton relaxation times to differ from each other by more than an order of magnitude in the temperature range $0.03K < T < 0.08K$. This difference decreases with temperature, so that proton and deuteron relaxation times became about: $T_{In} \approx I0^3$h, at $T < 0.03K$.

The results permit the conclusion that the holding temperature is the most important parameter of any polarised frozen-spin target.

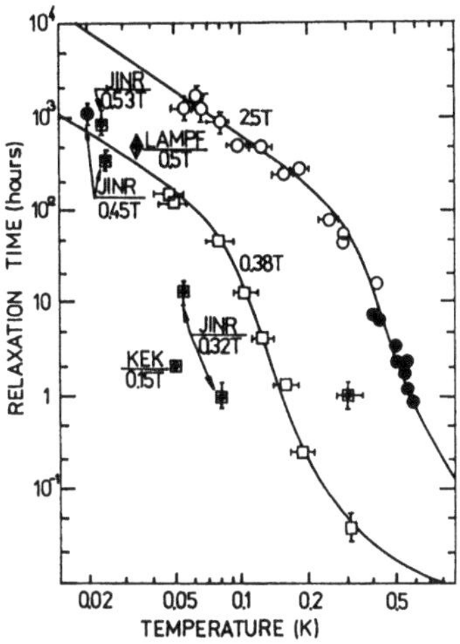

Figure I. The temperature dependence of the proton spin-lattice relaxation time in I,2-propanediol for magnetic fields of 2.5 T (O and ●) and 0.38 T (□) (Ref.I). The experimental points of the proton (⧫ and ◈) and deuteron (▣) spin-lattice relaxation times in diols are added. The captions are used to indicate the values of magnetic field and the source of the information: KEK |2|, LAMPF |5|, JINR |3,4|.

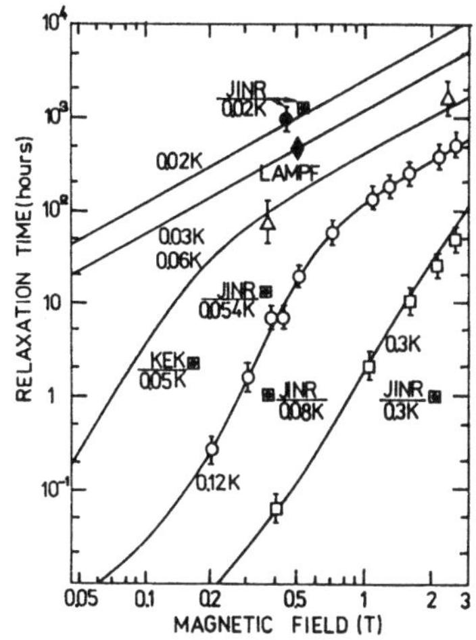

Figure 2. The magnetic field dependence of the proton spin-lattice relaxation time in I,2-propanediol at the temperatures of 0.3 K (□), 0.I2 K (O) and 0.06 K (Δ) |3|. The experimental points of the proton (⧫ and ◈) and deuteron (▣) spin-lattice relaxation times in diols are added. The captions are used to indicate the values of magnetic field and the source of the information: KEK |2|, LAMPF |5|, JINR |3,4|.

References

I. De Boer W. CERN 74-II, Geneva, I974.

2. Hiramatsu S., Isagawa S., Ishimoto S., Masaike A., Morimoto K. In: High Energy Spin Physics, I982 (AIP Conf.Proc. 95) N.Y. I983, p. 5I2.

3. Borisov N.S., Bunyatova E.I., Liburg M.Yu., Matafonov V.N., Neganov A.B., Usov Yu.A. Sov.Phys. JETP 60 (6) I984.

4. Borisov N.S., Bunyatova E.I., Liburg M.Yu., Matafonov V.N., Neganov A.B., Usov Yu.A. J.Phys.E: Sci.Instrum. 2I (I988) II79-II82. Printed in the UK.

5. Jarmer J.J., Penttilä S. In: VII Int. Symposium on High Energy Spin Physics, Serpukhov I987, v. 2, p. 228.

Heat Sinking of Cryogenic Coaxial Cables in a Dilution Refrigerator

L. Klostermann[1], *S. Trentalange*[2], *B. Ritzi*[2], *H.C. Wen*[2], *and T.O. Niinikoski*[3]

[1]NIKHEF, NL-Amsterdam, The Netherlands
[2]University of California at Los Angeles, Los Angeles, CA 90024, USA
[3]CERN, CH-1211 Geneve 23, Switzerland

INTRODUCTION

High-quality coaxial lines are used for transmitting signals between room temperature and the mixing chamber of a dilution refrigerator in many applications; among these are various NMR experiments, for example the measurement of the nuclear polarization in polarised targets. The thermal conduction along these cables often becomes problematic, particularly in frozen-spin targets, which require a low base temperature and therefore a low heat leak.

Low-loss coaxial cables are made of materials with relatively good electrical and thermal conductivity, and it is clear that the heat sinking of such lines is important for reaching the lowest base temperatures in the mixing chamber. The design of heat sinks is very critical if the lines pass through an evacuated channel, such that only localized thermalization points can be provided.

The European Muon Collaboration (EMC) constructed a large polarized target[1] requiring the use of 8 NMR coils for the mapping of the target polarization in the two-litre volume of the target. The target refrigerator[2,3] features direct horizontal cold access into the mixing chamber for the purpose of loading the target material at liquid nitrogen temperature. The target material, held in perforated plastic cartridges, is supported in the end of a target holder consisting of a vacuum chamber and the target support structures. The target holder and the access channel are shown in Fig. 1 schematically indicating the positions of the 5 heat sink rings, designed to thermalize the 9 coaxial lines (one of them serving for non-NMR instrumentation).

The coaxial lines traverse the vacuum chamber from room temperature into the mixing chamber; the detail of one heat heat sink point is shown in Fig. 2. The thermalizer rings are made of titanium alloy having a thermal dilatation lower than that of the stainless steel wall of the vacuum chamber. Thermal conduction from the coaxial cable to the helium gas flow may be restricted by the transverse thermal resistance from the inner to the outer conductor of the cable itself, by the thermal resistance along the titanium alloy ring, by the thermal resistance of the stainless steel of the target holder wall, by the thermal resistance at possible contact spots between the titanium and the stainless steel structures (helped by applying contact grease), and by the thermal resistance of the ^{3}He exchange gas introduced in the gap between the access tube and the target holder vacuum chamber.

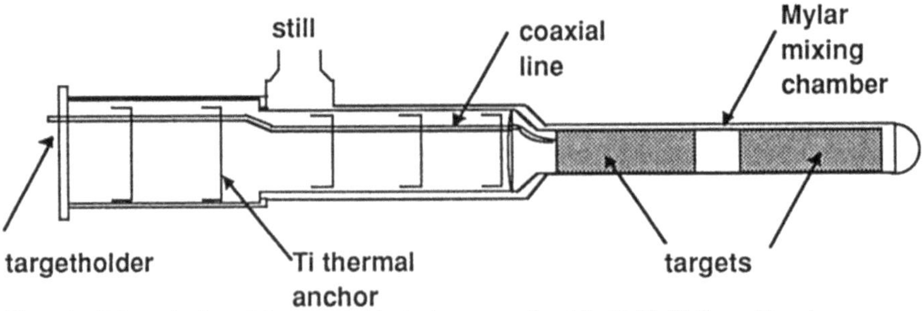

Figure 1: Schematic view of the target holder in the access tube of the EMC-dilution refrigerator.

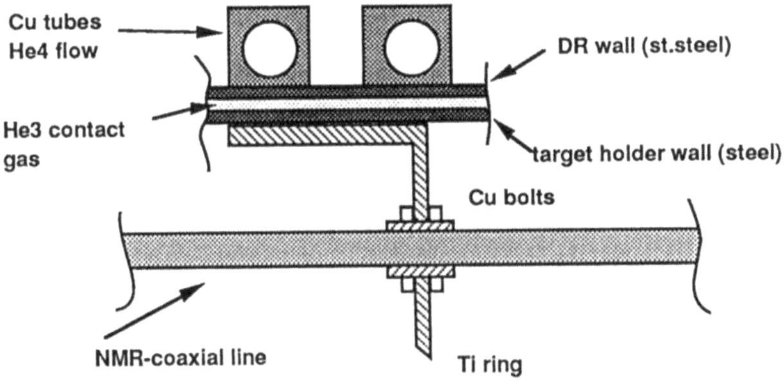

Figure 2: Details of the heat sinks of the semirigid coaxial lines.

The experimental thermal performance[2] of the refrigerator enables one to determine [4] the heat load to the mixing chamber. The result was about 10 mW at the base temperature of 80 mK. The heat load to the still was determined to be 0.5 W. These numbers were surprisingly large, and remained unexplained untill the thermal conduction characteristics of the coaxial lines were determined recently.

The aim of this work is to calculate the heat load to the mixing chamber of the refrigerator of Ref. [1] under various assumptions, basing on the measured thermal properties of the cables. Understanding the mechanism of the heat leak could enable to remove the limitation and therefore substantially improve the performance of the refrigerator, planned for use in the frozen spin mode in the forthcoming CERN NA47 experiment of the Spin Muon Collaboration.

THERMAL CONDUCTION

The basic equation for uniaxial thermal conduction in isotropic materials is

$$\frac{\dot{Q}}{A} = \kappa(T)\,\frac{dT}{dx}\,, \tag{1}$$

where A is the cross-sectional area, perpendicular to the axis of conduction and $\kappa(T)$ is the thermal conductivity of the material. Because we are concerned with a large temperature variation we cannot take κ to be a constant. In case of a constant cross-section, the conductivity integral I(T) is a practical variable:

$$I(T) = \int_0^T \kappa(T')dT'. \tag{2}$$

Using this expression the integrated equation (1) yields:

$$\dot{Q}\,\frac{L}{A} = \int_{T_1}^{T_2} \kappa(T)dT = I(T_2) - I(T_1), \tag{3}$$

where L is the length of the heat conductor and T_1 and T_2 the temperatures of its ends.

When the conductor consists of several different materials in parallel, the quantity used is the *conduction* integral D(T). This integral is not a material property like the conductivity integral, but is a property of a specific cable, for example. The expression for the conduction integral of a cable made of different materials with conductivity $\kappa_i(T)$ is:

$$\dot{Q}\cdot L = \sum_i A_i \int_{T_1}^{T_2} \kappa_i(T)dT = D(T_2) - D(T_1) \tag{4}$$

In this expression it is assumed that the components of different materials i, with constant cross-section A_i, are assembled in parallel. Furthermore it is assumed that the transverse heat contact between the materials is very weak.

For cylindrical symmetry the conductivity integral can also be used for conduction in the *transverse* direction. However, the geometrical factor L/A in expression (3) must be replaced by another geometrical factor G/L:

$$\dot{Q}\,\frac{G}{L} = \dot{Q}\,\frac{\ln(r_o/r_i)}{2\pi L} = I(T_2) - I(T_1). \qquad (5)$$

MEASUREMENTS OF THERMAL CONDUCTANCES

Because a possible cause for the large heat input could be a bad transverse heat conduction of the coaxial lines, both the transverse and the longitudinal conduction integrals were measured. Figure 3 gives the dimensions of the cable. The inner and outer conductors of the cable are made out of beryllium copper, and the isolation material is Teflon. Our hypothesis was that because of the difference in the thermal expansion coefficients, the Teflon could become detached from the outer conductor at low temperatures, which could impair the transverse heat conduction.

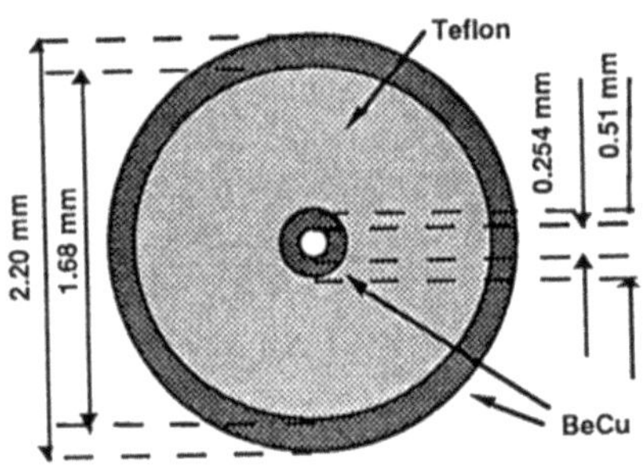

Figure 3: Dimensions of the coaxial line UT-T-85-B-B.

Longitudinal conduction integral

The inner and outer conductors of the sample cable were soldered together at each end. Heat was supplied at one end of the cable, the other end was connected to a heat sink in contact with a bath of liquid helium. To avoid heat leaks through convection, and to be able to have a lower heat sink temperature, the samples were placed in vacuum. The temperature was measured at both ends of the sample. For higher heat flows the heat sink temperature was slightly above the 1.3 K bath temperature; a correction was made for this effect.

The assumption was made that the conduction integral D(T) behaves as $\alpha \cdot T^m$; the data were fitted to this function, yielding $\alpha = 0.0122$ mW·cm and $m = 2.25$:

$$D_L(T) = 0.0122 \cdot T^{2.25} \text{ mW} \cdot \text{cm} . \qquad (6)$$

Transverse conduction integral

The outer conductor of a coaxial line of one centimetre length (approximately the length of the heat sinked part of the cable) was soldered to a piece of copper that was is contact with the liquid helium bath. Heat was applied to the inner conductor of the cable at one end (upstream). Because there is only one heat sink, all the heat applied has to go through the Teflon. The temperature was measured at the downstream end of the inner conductor. The heat sink temperature was also measured and corrected for.

For reasons related with the realization of practical heat sinks to be used, the conduction integral is determined as a function of the measured *downstream* temperature. Assuming a behavior $\beta \cdot T^n$, the fit resulted in the values $\beta = 0.013$ mW/cm and $n = 2.28$:

$$D_T(T) = I(T)/G = 0.013 \cdot T^{2.28} \text{ mW/cm.} \tag{7}$$

MODELS TO CALCULATE HEAT FLOWS IN COAXIAL LINE

In order to find an explanation for the large heat input in the mixing chamber, various models were set up and examined. All our models are schematically described by Figure 4, which displays a network of heat conductances or resistances, showing also the five heat sink temperatures. The longitudinal heat flows are given by Eq. (6), and the transverse flows are determined by the assumptions of each model. In equilibrium the following equations

$$\begin{aligned} \dot{Q}_{L1} &= \dot{Q}_{L2} + \dot{Q}_{T1} \\ \dot{Q}_{L2} &= \dot{Q}_{L3} + \dot{Q}_{T2} \\ \dot{Q}_{L3} &= \dot{Q}_{L4} + \dot{Q}_{T3} \\ \dot{Q}_{L4} &= \dot{Q}_{L5} + \dot{Q}_{T4} \\ \dot{Q}_{L5} &= \dot{Q}_{L6} + \dot{Q}_{T5} \end{aligned} \tag{8}$$

enable us to determine all the heat flows, which only depend on the temperatures of the intermediate nodes. So, Eq. (8) is a set of five non-linear equations with five unknown parameters T_1 to T_5, the five downstream temperatures. Once these temperatures are determined, the quantity of interest $\dot{Q}_{L6}$, the heat flow to the mixing chamber, is also known.

The transverse heat flow from the heat sink posts on the cable to the helium coolant is limited by a number of heat resistances, which we approximate by a series network as shown in Fig. 5. The various models differ from each other by the conductances (or resistances) included in the expression for the transverse heat flow. By plotting the conductances $\kappa(T) \cdot A/L$ for titanium [5], helium gas [6], stainless steel [5] and Apiezon grease [7], and the Kapitza resistance between grease and metal [7], using the relevant geometric factors, one may already conclude that the most likely cause for the limited transverse heat flow is the low heat conductivity of the titanium alloy, associated with spot contacts between the heat sink ring and the stainless steel vacuum chamber tube.

In the first model it was assumed that the intrinsic transverse conduction in the coaxial line itself was the only limiting factor. Over the whole length of the cable the inner and outer conductor are in finite thermal contact through the Teflon. At the position of the heat sinks the coaxial line is heat contacted over a certain length. The temperature of the inner conductor and thus the heat flow from the inner to the outer conductor is changing continuously over this length.

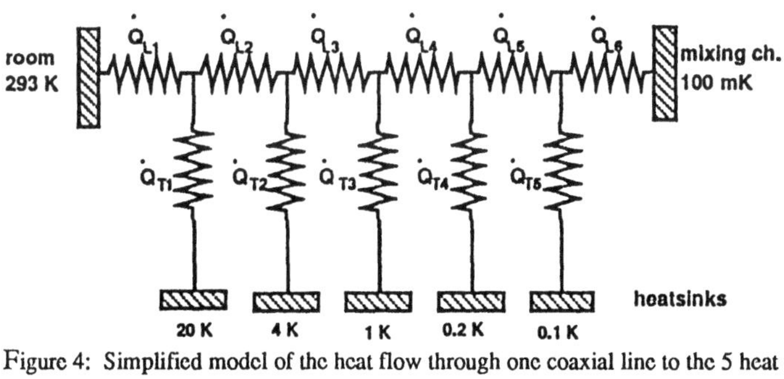

Figure 4: Simplified model of the heat flow through one coaxial line to the 5 heat sinks and to the mixing chamber.

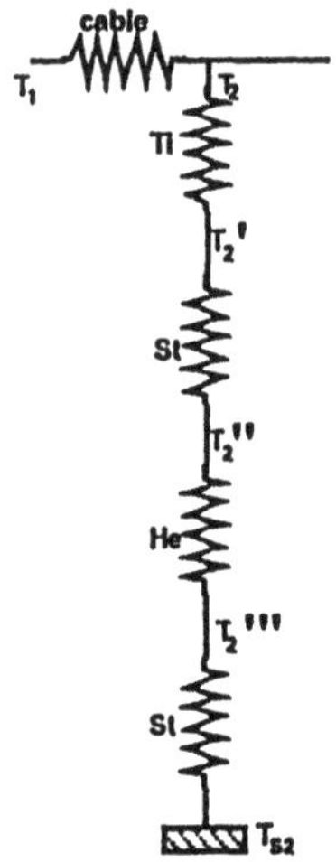

Figure 5: Schematic view of a the transverse heat flow at one sink. For model 2 the contact grease resistance should be added.

In order to be able to solve the Model 1, some simplifications were made. Firstly, the outer conductor was assumed to be perfectly well heat sinked at each post. Secondly, the assumption was made that contact is only made between the inner and outer conductor at the heat sink positions. Furthermore, it is assumed that the heat conducted axially through the cable is given by Eq. (6), and that it arrives to the heat sink area through the central conductor only; the temperature of the central conductor downstream the heat sink can now be obtained by applying the empirical transverse conduction given by Eq. (7). The axial heat flow downstream each contact is now determined by assuming that the temperature of the cable, throughout its cross section, is equal to the the downstream temperature of the inner conductor.

A computer program was written to solve the set of equations (8), and the results are shown in Table 1 together with those from the remaining Models 2 to 5. It turns out that in our Model 1 the calculated temperatures T_1 to T_5 are not significantly higher than the heat sink temperatures. We conclude that the transverse heat conduction of the cable itself is sufficient, and that for the remaining models the cable may be approximated as a conductance with axial heat resistance only, defined by the empirical Eq. (6).

Referring to the series of thermal resistances as shown in Fig. 5, the Table 1 lists the geometric factors used in the Models 2 to 5; each model assumes a plausible geometric factor, reduced in the effective values of cross sectional are A and length L for conduction. The Model 5 assumes that the transverse heat resistance is dominated by the spot contacts between the stainless steel wall of the target holder vacuum chamber and wall the of access tube.

The Model 2 is based on the hypothesis that there is a thermal contact between the titanium and the stainless steel through a layer Apiezon grease around the entire circumference of the titanium alloy

Table 1: Geometrical parameters and the results of the different models. The heat input $\dot{Q}_{L6}$ is the total heat input to the mixing chamber, and ΔT_1 and ΔT_5 are the differences of the temperature of the cable at heat sink positions 1 and 5 and the temperature of the thermal ground (helium coolant). For a description of the models see text.

Model	Geometrical factors								Results		
	Titanium		Steel		He gas		Grease		ΔT_1	ΔT_5	$\dot{Q}_{L6}$
	A(cm^2)	L(cm)	A(cm^2)	L(mm)	A(cm^2)	L(mm)	A(cm^2)	L(mm)	(K)	(K)	(μW)
1	Equations (6) and (7)								1.9	0.002	0.003
2	0.2	0.7	1.5	0.5	1.5	.25	1.5	0.2	24	0.017	0.04
3	0.3	7	1.5	0.5	1.5	.25	N/A	N/A	104	16.3	3500
4	0.3	21&7	1.5	0.5	1.5	.25	N/A	N/A	143	48.3	25000
5	Spot diameter 0.25 mm (3 spots/ring)								60	0.25	0.5

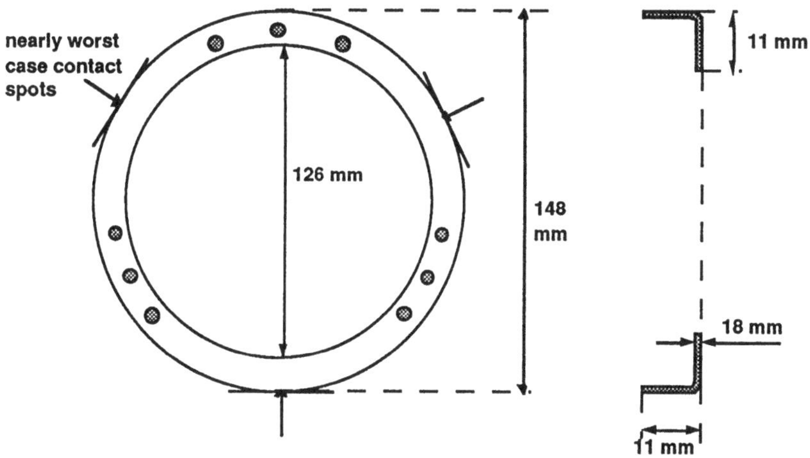

Figure 6: The titanium contact ring and the positions of the contact spots in Models 3 (all three) and 4 (only one).

ring. In Models 3 and 4 only spot contacts are made between the two metals at unfavorable positions sketched in Figure 6, which also gives the dimensions of the titanium ring. The spots themselves are assumed not to be the bottleneck. For the calculations the model has been simplified by including only the dominating transverse heat resistance.

The Model 5 is based on the hypothesis that the contact spots between the titanium and the steel are very small. The heat will first spread into the steel, a poor heat conductor, and then pass through the helium gas. If the contact spot is of the same order of magnitude or bigger than the thickness of the steel, the problem is two dimensional. Neglecting the heat resistance of the titanium alloy, calculations based on a spot of radius 0.25 mm, half the thickness of the steel, show that in this case the difference between T_5 and the heat sink temperature is not significant.

The Table 1 lists the temperatures T_1 and T_5, and the resulting heat load $\dot{Q}_{L6}$ to the mixing chamber, obtained from the Models 1 to 5 described above.

CONCLUSIONS

We conclude that the thermal properties of the UT-T-85-B-B coaxial lines themselves are so good that 5 perfect heat sinks are adequate for reducing the heat load to the mixing chamber well below 1 μW. Within the accuracy of the models employed, the plausible leading transverse heat resistance would seem to be related with the poor heat conductance of the titanium alloy heat sink rings, chosen for their low thermal dilatation upon cooling. The conductivity of the alloy, however, would be sufficient if thermal contact could have been provided on the whole circumference of the ring (Model 2), which turned out to be difficult to accomplish.

Substituting the titanium with copper in the contact rings would solve the problem. Such rings were used in a similar but smaller machine [2] with success. Furthermore, the Model 5 would indicate that thermal contact restricted to rather small spots, between a ring of high conductivity and the stainless steel wall with limited conductivity, would not seriously compromise the thermalization of our coaxial lines, although it would be advisable to aim at large contact spots well distributed over the circumference of the heat sink rings.

ACKNOWLEDGEMENTS

The authors would like to thank Prof. Hans Postma for his interest on this work and for his helpful advise on this manuscript. The technical support of Mr. J.-M. Rieubland is gratefully acknowledged.

REFERENCES

1. S. C. Brown, G. R. Court, R. Gamet, O. Hartmann, T. O. Niinikoski, J.-M. Rieubland and A. Rijllart, Proc. 4th Int. Workshop on Polarized Target Materials and Techniques, Bad Honnef 1984, Ed. W. Meyer (Physikalisches Institut, Universität Bonn, Bonn 1984) p. 102.
2. T. O. Niinikoski and J.-M. Rieubland, Proc. 9th Int. Cryogenic Engineering Conference, Kobe 1982, Ed. K. Yasukochi and H. Nagano (Butterworth, Guilford 1982) p. 580.
3. T. O. Niinikoski, Nucl. Instr. and Meth., **192** (1982) 151.
4. T. O. Niinikoski, Proc. 6th Int. Cryogenic Engineering Conf., Grenoble 1976, Ed. K. Mendelssohn (IPC Science and Technology Press, Guilford 1976) p. 102.
5. A Compendium of the Properties of Materials at Low Temperature (Phase I) Part II. Properties of Solids, WADD Technical Report 60-56, Ed. V. J. Johnson (National Technical Information Service, Springfield 1960) .
6. B. Colyer, Cryogenic Properties of Helium-3 and Helium-4, Rutherford Laboratory Report RHEL/R 138 (1966).
7. O. V. Lounasmaa, *Experimental Principles and Methods below 1 K* (Academic Press, London 1974).

Density Measurement of Solid Butanol by γ-ray Attenuation

S. Robinson[1], R. Henneck[1], I. Sick[1], B. van den Brandt[2], J.A. Konter[2], and S. Mango[2]

[1]Institut für Physik, Klingelbergstraße 72, CH-4056 Basel, Switzerland
[2]Paul Scherrer Institute, CH-5232 Villigen PSI, Switzerland

The γ-ray attenuation in a thick butanol sample has been measured at 293 K and 4 K. The hydrogen density in the butanol target at 4 K was determined with an uncertainty of $\pm 1\%$.

The measurement of the spin dependent cross section difference $\Delta\sigma_L$ (and $\Delta\sigma_T$) in nucleon-nucleon scattering with both beam and target polarized requires precise knowledge of the number of hydrogen nuclei in the target. An accuracy on the level $\leq 1\%$ is desirable in order to make this error contribution negligibly small compared to other error sources. To our knowledge there is no information on the density of frozen butanol below 184 K [1]. Previous determinations of the target thickness at the operating temperatures of polarized targets were based on the conventional density determination by volume and weight measurement. This method suffers from the fact that the sample, which is contained in a very thin-walled container, may change its form in an unpredictable manner during the process of freezing. In particular, the open surface necessary for the microwave irradiation does not remain flat and thus makes a volume measurement very difficult. Another problem is that the structure of the glass can change as a function of temperature.

The method we have used is based on the γ-ray attenuation in a thick butanol sample. The temperature dependent density of the sample will result in a temperature dependent attenuation, which can be measured accurately. The attenuated γ-ray intensity N is given by the incident flux N_0, the attenuation coefficient $A = \frac{\mu}{\rho(293K)}$ and the target thickness $\rho(T)x$

$$N = N_0 e^{-A\rho(T)x}$$

Although the γ-ray absorption occurs predominantly by C and O in the hydrocarbon, the determination of the hydrogen density is possible as the molecular composition is independent of temperature.

The experimental setup is given in Fig.1. The source was a 30 mCi ^{241}Am-source. The radiation was collimated to a 5 mm thin beam. For detection a 2 mm thick silicon surface barrier detector was used. It was cooled to -37°C

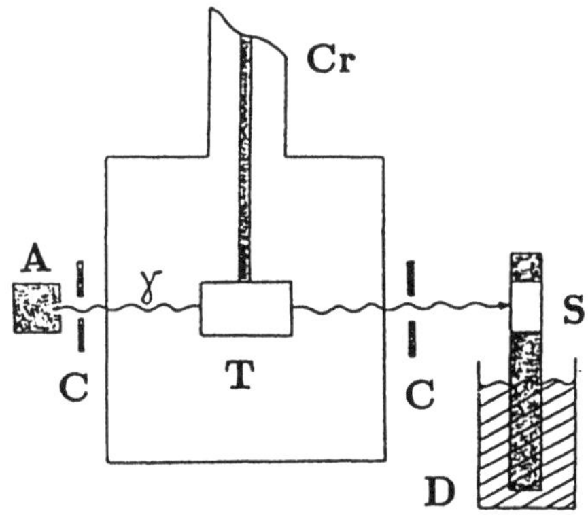

Fig.1: Experimental setup: Am-source (A), Collimators (C), Target(T), Cryostat(Cr), Dewar(D) with LN_2/alcohol-bath, Si-detector (S) in copper bar

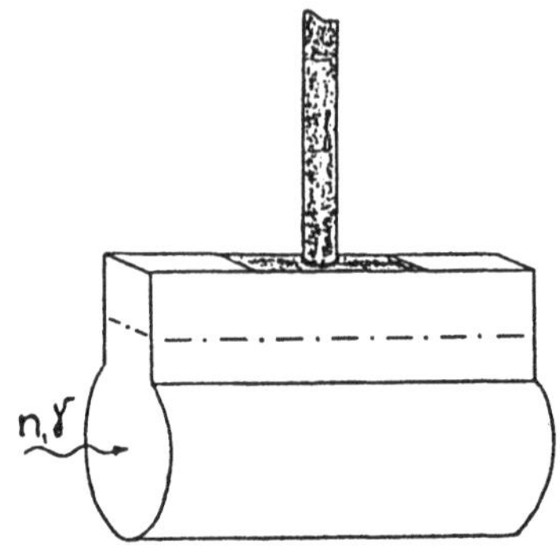

Fig.2: Target cell. The butanol level is indicated by the dash-dotted line.

by thermal contact with a LN_2-alcohol mixture. Its efficiency was 14% for 60 keV γ's, the resolution $\sim$ 5%.

The target cell used in the $\Delta\sigma_L$-experiment [2] was a copper cylinder, 24 ± 0.01mm long and 20mm in diameter (at room temperature), with an opening at the top over the whole length, about 15mm wide. It was fabricated by galvanic deposition on an aluminium mould, resulting in a fairly uniform wall thickness of $\sim 50\mu$m.

The γ-ray yield was accurately measured for the following conditions:

1. Frozen butanol at 4 K (CF)
2. Empty target at 4 K (CE)
3. Butanol at 293 K (WF)
4. Empty target at 293 K (WE)

From these measurements we can then deduce the density ratio

$$R_D = \frac{\rho(4K)}{\rho(293K)} = \frac{ln(\frac{N_{CF}}{N_{CB}})}{ln(\frac{N_{WF}}{N_{WB}})}$$

where e.g. N_{CF} is the time-normalized integral of the 60 keV γ-line for the CF-measurement.

The attenuation in condition 'empty' target was $\sim$ 30%, with the filled target $\sim$ 50%.

For a sample consisting of 95% butanol, 4.3% water and 0.7% porphyrexid we obtained

$$R_D = 1.1472 \pm 0.0024 \pm 0.01$$

The systematic error (±0.01) contains estimates of the uncertainties due to background under the peak ($\leq 0.07\%$ of the peak integral) and slight changes in detector resolution. By far the largest systematic error originated from the fact that the target had to be removed for filling and emptying. Therefore the location of the cell with respect to the source/detector arrangement may have differed slightly between 'full' and 'empty' measurement. A non uniform wall thickness thus may lead to an erroneous determination of R_D. Our error estimate is based on a 3 μm variation of the wall thickness.

Based on the calculated value ρ(95% butanol, 5% water) = 0.82 g/cm^3 at 293 K we thus evaluate the density at 4 K as 0.94 ± 0.01 g/cm^3. This is slightly lower than expected from a previous determination of $\rho = 0.9$ g/cm^3 at the freezing point of butanol (~184 K) (obtained with the conventional method [1]) assuming a linear temperature dependence.

To summarize, the method described allows a precise determination of the hydrogen density in hydrocarbon samples at $\leq$4 K. The method can be used for any kind of material with a temperature independent molecular composition and may even be applicable to the measurement of the density of 'beads' targets.

References

[1] S. Mango, Ö. Runólfsson, M. Borghini, NIM 72 (1969) 45-50

[2] P. Haffter, PSI-experiment Z-89-06

List of Participants

Ball, J.	Saclay /France
Baum, G.	Bielefeld /Germany
Brandt, B. van den	Wisconsin /U S A
Bueltmann, S.	Bielefeld /Germany
Bunyatova, E. I.	Dubna /USSR
Crabb, D. G.	Michigan /U S A
Court, G.	Liverpool /England
Delheij, P. P. J.	TRIUMF /Canada
Deregel, J.	Saclay /France
Dhawan, S.	Yale /U S A
Durand, G.	Saclay /France
Dutz, H.	Bonn /Germany
Finger, M.	Prague /Czechoslovakia
Gelder, H. van	Delft /Netherlands
Goertz, S.	Bonn /Germany
Gruebler, W.	Zuerich /Switzerland
Hautle, P.	Zuerich /Switzerland
Hayashi, N.	Nagoya /Japan

Henstza,A.	Leiden /Netherlands
Horikawa,N.	Nagoya /Japan
Hughes,V.W.	Yale /U S A
Iwata,T.	Nagoya /Japan
Jarmer,J.	Los Alamos /U S A
Johannson,H.	CERN /Switzerland
Karnaukhov,I.M.	Kharkov /USSR
Knop,W.	Geesthacht /Germany
Koijk,J.van den	Delft /Netherlands
Konter,J.A.	PSI /Switzerland
Kovaljov,A.I.	Leningrad /USSR
Kraemer,D.	Bonn /Germany
Krisch,A.D.	Michigan /U S A
Krumpolc,M.	Illinois /U S A
Liburg,M.	Dubna /USSR
Mango,S.	PSI /Switzerland
Mathie,E.L.	TRIUMF /Canada
Meyer,W.	Bonn /Germany
Nakanishi,T.	Nagoya /Japan
Nakasuka,N.	Nagoya /Japan
Niinikoski,T.O.	CERN /Switzerland
Penttila,S.	Los Alamos /U S A
Postma,H.	Delft /Netherlands
Prokofiev,A.	Leningrad /USSR

Raymond,R.S.	Michigan /U S A
Reichertz,L.	Bonn /Germany
Reicherz,G.	Bonn /Germany
Rijllart,A.	CERN /Switzerland
Ritt,S.	Karlsruhe /Germany
Sahling,S.	Dresden /Germany
Sorokin,P.V.	Kharkov /USSR
Takala,S.	CERN /Switzerland
Thiel,W.	Bonn /Germany
Thomas,A.	Bonn /Germany
Trentalange,S.	UCLA /U S A
Vovchenko,V.	Leningrad /USSR
Wenkebach,W.Th.	Leiden /Netherlands
Wessler,M.	PSI /Switzerland

Index of Contributors

MIX
Papier aus verantwortungsvollen Quellen
Paper from responsible sources
FSC® C105338

If you have any concerns about our products,
you can contact us on
ProductSafety@springernature.com

In case Publisher is established outside the EU,
the EU authorized representative is:
Springer Nature Customer Service Center GmbH
Europaplatz 3, 69115 Heidelberg, Germany

Printed by Libri Plureos GmbH
in Hamburg, Germany